200
Advances in Polymer Science

Editorial Board:
A. Abe · A.-C. Albertsson · R. Duncan · K. Dušek · W. H. de Jeu
J.-F. Joanny · H.-H. Kausch · S. Kobayashi · K.-S. Lee · L. Leibler
T. E. Long · I. Manners · M. Möller · O. Nuyken · E. M. Terentjev
B. Voit · G. Wegner · U. Wiesner

Advances in Polymer Science

Recently Published and Forthcoming Volumes

Polymers for Regenerative Medicine
Volume Editor: Werner, C.
Vol. 203, 2006

Peptide Hybrid Polymers
Volume Editors: Klok, H.-A., Schlaad, H.
Vol. 202, 2006

**Supramolecular Polymers ·
Polymeric Betains**
Vol. 201, 2006

Ordered Polymeric Nanostructures at Surfaces
Volume Editor: Vancso, G. J., Reiter, G.
Vol. 200, 2006

Emissive Materials · Nanomaterials
Vol. 199, 2006

Surface-Initiated Polymerization II
Volume Editor: Jordan, R.
Vol. 198, 2006

Surface-Initiated Polymerization I
Volume Editor: Jordan, R.
Vol. 197, 2006

Conformation-Dependent Design of Sequences in Copolymers II
Volume Editor: Khokhlov, A. R.
Vol. 196, 2006

Conformation-Dependent Design of Sequences in Copolymers I
Volume Editor: Khokhlov, A. R.
Vol. 195, 2006

Enzyme-Catalyzed Synthesis of Polymers
Volume Editors: Kobayashi, S., Ritter, H., Kaplan, D.
Vol. 194, 2006

Polymer Therapeutics II
Polymers as Drugs, Conjugates and Gene Delivery Systems
Volume Editors: Satchi-Fainaro, R., Duncan, R.
Vol. 193, 2006

Polymer Therapeutics I
Polymers as Drugs, Conjugates and Gene Delivery Systems
Volume Editors: Satchi-Fainaro, R., Duncan, R.
Vol. 192, 2006

Interphases and Mesophases in Polymer Crystallization III
Volume Editor: Allegra, G.
Vol. 191, 2005

Block Copolymers II
Volume Editor: Abetz, V.
Vol. 190, 2005

Block Copolymers I
Volume Editor: Abetz, V.
Vol. 189, 2005

Intrinsic Molecular Mobility and Toughness of Polymers II
Volume Editor: Kausch, H.-H.
Vol. 188, 2005

Intrinsic Molecular Mobility and Toughness of Polymers I
Volume Editor: Kausch, H.-H.
Vol. 187, 2005

Polysaccharides I
Structure, Characterization and Use
Volume Editor: Heinze, T.
Vol. 186, 2005

Advanced Computer Simulation Approaches for Soft Matter Sciences II
Volume Editors: Holm, C., Kremer, K.
Vol. 185, 2005

Ordered Polymeric Nanostructures at Surfaces

Volume Editors: G. Julius Vancso, Günter Reiter

With contributions by
K. Albrecht · M. Al-Hussein · M. Alonso · C. Aparicio · F. J. Arias
M. Charles-Harris · G. H. Degenhart · B. Dordi · E. Engel
C. L. Feng · M. P. Ginebra · A. Girotti · S. Golze · N. Hadjichristidis
M. A. Hempenius · W. H. de Jeu · I. Korczagin · R. G. H. Lammertink
M. Moeller · A. Mourran · M. Navarro · S. Pispas · J. A. Planell
J. Reguera · G. Reiter · J. C. Rodríguez-Cabello · D. I. Rozkiewicz
H. Schönherr · Y. Séréro · A. Shovsky · J.-U. Sommer · G. J. Vancso

The series *Advances in Polymer Science* presents critical reviews of the present and future trends in polymer and biopolymer science including chemistry, physical chemistry, physics and material science. It is adressed to all scientists at universities and in industry who wish to keep abreast of advances in the topics covered.
As a rule, contributions are specially commissioned. The editors and publishers will, however, always be pleased to receive suggestions and supplementary information. Papers are accepted for *Advances in Polymer Science* in English.
In references *Advances in Polymer Science* is abbreviated *Adv Polym Sci* and is cited as a journal.

Springer WWW home page: springer.com
Visit the APS content at springerlink.com

Library of Congress Control Number: 2006921178

ISSN 0065-3195
ISBN-10 3-540-31921-2 Springer Berlin Heidelberg New York
ISBN-13 978-3-540-31921-4 Springer Berlin Heidelberg New York
DOI 10.1007/11605294

This work is subject to copyright. All rights are reserved, whether the whole or part of the material is concerned, specifically the rights of translation, reprinting, reuse of illustrations, recitation, broadcasting, reproduction on microfilm or in any other way, and storage in data banks. Duplication of this publication or parts thereof is permitted only under the provisions of the German Copyright Law of September 9, 1965, in its current version, and permission for use must always be obtained from Springer. Violations are liable for prosecution under the German Copyright Law.

Springer is a part of Springer Science+Business Media

springer.com

© Springer-Verlag Berlin Heidelberg 2006
Printed in Germany

The use of registered names, trademarks, etc. in this publication does not imply, even in the absence of a specific statement, that such names are exempt from the relevant protective laws and regulations and therefore free for general use.

Cover design: *Design & Production* GmbH, Heidelberg
Typesetting and Production: LE-TEX Jelonek, Schmidt & Vöckler GbR, Leipzig

Printed on acid-free paper   02/3100 YL – 5 4 3 2 1 0

Volume Editors

Dr. G. Julius Vancso
Materials Science and Technology
of Polymers
MESA⁺ Institute for Nanotechnology
University of Twente
P.O. Box 217
7500 AE Enschede
The Netherlands
g.j.vancso@tnw.utwente.nl

Dr. Günter Reiter
Directeur de recherche at the CNRS
Institut de Chimie des Surfaces et Interfaces
(ICSI) CNRS
15, rue Jean Starcky, BP 2488
68057 Mulhouse Cedex, France
G.Reiter@uha.fr

Editorial Board

Prof. Akihiro Abe
Department of Industrial Chemistry
Tokyo Institute of Polytechnics
1583 Iiyama, Atsugi-shi 243-02, Japan
aabe@chem.t-kougei.ac.jp

Prof. A.-C. Albertsson
Department of Polymer Technology
The Royal Institute of Technology
10044 Stockholm, Sweden
aila@polymer.kth.se

Prof. Ruth Duncan
Welsh School of Pharmacy
Cardiff University
Redwood Building
King Edward VII Avenue
Cardiff CF 10 3XF, UK
DuncanR@cf.ac.uk

Prof. Karel Dušek
Institute of Macromolecular Chemistry,
Czech
Academy of Sciences of the Czech Republic
Heyrovský Sq. 2
16206 Prague 6, Czech Republic
dusek@imc.cas.cz

Prof. W. H. de Jeu
FOM-Institute AMOLF
Kruislaan 407
1098 SJ Amsterdam, The Netherlands
dejeu@amolf.nl
and Dutch Polymer Institute
Eindhoven University of Technology
PO Box 513
5600 MB Eindhoven, The Netherlands

Prof. Jean-François Joanny
Physicochimie Curie
Institut Curie section recherche
26 rue d'Ulm
75248 Paris cedex 05, France
jean-francois.joanny@curie.fr

Prof. Hans-Henning Kausch
Ecole Polytechnique Fédérale de Lausanne
Science de Base
Station 6
1015 Lausanne, Switzerland
kausch.cully@bluewin.ch

Prof. Shiro Kobayashi
R & D Center for Bio-based Materials
Kyoto Institute of Technology
Matsugasaki, Sakyo-ku
Kyoto 606-8585, Japan
kobayash@kit.ac.jp

Prof. Kwang-Sup Lee
Department of Polymer Science &
Engineering
Hannam University
133 Ojung-Dong
Daejeon 306-791, Korea
kslee@hannam.ac.kr

Prof. L. Leibler
Matière Molle et Chimie
Ecole Supérieure de Physique
et Chimie Industrielles (ESPCI)
10 rue Vauquelin
75231 Paris Cedex 05, France
ludwik.leibler@espci.fr

Prof. Timothy E. Long
Department of Chemistry
and Research Institute
Virginia Tech
2110 Hahn Hall (0344)
Blacksburg, VA 24061, USA
telong@vt.edu

Prof. Ian Manners
School of Chemistry
University of Bristol
Cantock's Close
BS8 1TS Bristol, UK
ian.manners@bristol.ac.uk

Prof. Martin Möller
Deutsches Wollforschungsinstitut
an der RWTH Aachen e.V.
Pauwelsstraße 8
52056 Aachen, Germany
moeller@dwi.rwth-aachen.de

Prof. Oskar Nuyken
Lehrstuhl für Makromolekulare Stoffe
TU München
Lichtenbergstr. 4
85747 Garching, Germany
oskar.nuyken@ch.tum.de

Prof. E. M. Terentjev
Cavendish Laboratory
Madingley Road
Cambridge CB 3 OHE, UK
emt1000@cam.ac.uk

Prof. Brigitte Voit
Institut für Polymerforschung Dresden
Hohe Straße 6
01069 Dresden, Germany
voit@ipfdd.de

Prof. Gerhard Wegner
Max-Planck-Institut
für Polymerforschung
Ackermannweg 10
Postfach 3148
55128 Mainz, Germany
wegner@mpip-mainz.mpg.de

Prof. Ulrich Wiesner
Materials Science & Engineering
Cornell University
329 Bard Hall
Ithaca, NY 14853, USA
ubw1@cornell.edu

Advances in Polymer Science
Also Available Electronically

For all customers who have a standing order to Advances in Polymer Science, we offer the electronic version via SpringerLink free of charge. Please contact your librarian who can receive a password or free access to the full articles by registering at:

springerlink.com

If you do not have a subscription, you can still view the tables of contents of the volumes and the abstract of each article by going to the SpringerLink Homepage, clicking on "Browse by Online Libraries", then "Chemical Sciences", and finally choose Advances in Polymer Science.

You will find information about the

– Editorial Board
– Aims and Scope
– Instructions for Authors
– Sample Contribution

at springer.com using the search function.

Foreword to the 200th Volume

Advances in Polymer Science celebrates the 200th Volume of the Series. This asks for a moment of recollection and consideration, since *Advances in Polymer Science* is a true mirror of scientific progress, intellectual penetration and scholarly apprehension of one of the most successful fields of current research. Polymer Science is not only the fundament of the plastics industry but it has also gained reputation as an enabling factor of advanced technologies from microelectronics to bioengineering. Moreover, it has shown to be a unique challenge to theoretical and experimental physics. The bridging function to molecular biology is obvious. It is this context which is properly and meticulously mirrored by the last 100 Volumes of this Series. While the focus rests on concise descriptions of special topics, specialized and selected methods, careful discussion of synthetic and analytical approaches to new challenges in materials preparation and performance written by established experts in their respective sub-discipline, a complete picture of the state of the art emerges nevertheless when one draws the sum over the nearly 400 individual review papers published since 1991 in Volumes 101 to 200 of this Series. All of the contributions are critical reflections on the respective topic and give pertinent information not only to the expert but also to the newcomer to the field to whom guidance to the large and diverse body of available literature is provided. Emerging subfields or novel directions of research are discussed for the first time in a comprehensive manner. *Advances in Polymer Science* contributes with this style and dedication to the scientific fundament in an international context. While the first 100 Volumes were dominated by contributions from mainly Europe, and the classical languages of science of the last century, namely English, French and German were all represented, this has changed completely. Authors from all continents are well represented. English has become the sole medium of expression.

The strongest impact to the contributions for the last 100 Volumes had the disappearance of the east-west conflict. A large body of research experience and literature from Eastern Europe and Russia became suddenly available and has found its way into *Advances in Polymer Science* in the form of a considerable number of expert reviews. These reviews reflect the immense work performed over many years in research centres and institutes which were not accessible to the rest of the scientific world. The freshness and originality of the approach to define and tackle research problems comes as a surprise to many of us who have experienced the times when such information was not available at all or only in a rudimentary form. These reviews are to be considered as an important widening of the horizon of polymer science.

Similarly we see a considerable increase of contributions from the US and Canada in the last 100 Volumes indicating both the cosmopolitan nature of polymer science and the appreciation which this Series receives internationally as a depository of pertinent information and expertise.

A few remarks concerning the subdisciplines covered by the last 100 Volumes may be of interest in so far as directions and currents in Polymer Science are pinpointed.

Contrary to the believe of many critical bystanders **Polymer Synthesis** is still strongly going ahead. This is demonstrated by the ca. 100 individual reviews published in this Series since 1991 devoted to progress in synthetic methods, catalysis of polymerisations, biogenic macromolecules and their synthetic modification. Many of these contributions cover novel developments in synthesis allowing precision synthesis of novel macromolecular structures not available before. The next important field concerns **analytical methods** including rheology and solution properties of polymers. Nearly 60 reviews are devoted to this context, and – as is typical for this sub-discipline – **polymer physics** aspects are integral components of the reviews.

While both of these areas are the classical realm of *Advances in Polymer Science* we see the strong emergence of another area in the last 100 Volumes: **biomedically relevant polymers**. This includes synthesis of biocompatible polymers, polymeric drugs, application and evaluation of polymers in the medical field and in bioengineering which are summarized in well over 40 reviews.

Application of polymers and polymer based **hybrid materials** including aspects of processing and processing related properties constitute the central theme of another group of reviews. This includes reviews on inorganic/organic and polymer-polymer blends as well as reviews on degradation of practically relevant polymer compositions. This group of ca. 40 reviews illuminates the crossover between fundamental science and application driven research that is so typical for many sub-disciplines of our field.

The same could be said for the group of review papers dealing with **complex polymer systems** such as networks, gels and similar complex macromolecular topologies characterized by an interplay of regularity and randomness. Physical properties and phenomena related to the complexibility of structure are highlighted as well, fracture behaviour, relaxation phenomena and elasticity are key words to describe the features which are discussed among others in this group of ca. 40 reviews.

The very strong evolution of **theory and simulation** is reflected in another group of review papers. Theory of polymer systems has evolved as a segment of theoretical physics and has had a very strong impact on how experiments are conducted and data sets are explained and evaluated in many areas of polymer science. The development of simulation methods based on sound fundaments of theory have equally contributed to significant progress which is reviewed in 35 articles found among the nearly 400 reviews that appeared since 1991.

While this is certainly a more recent development properly reflected in this series a somewhat traditional field which is normally associated with polymer physics is still relevant and finds attention in form of review articles. It is the field of **crystallization of polymers** including phenomena of nucleation and crystal growth as well as studies into the crystallography of polymers and liquid crystal phases of polymers which finds attention by well over 20 reviewers. Nearly the same attention is given to polymers of particular chain architecture which in consequence create special supramolecular or hierarchical order phenomena. **Blockcopolymers**, micelle forming polymers, polymer based membranes and similar topologies are in focus. These areas are strongly associated with the current trend in materials science to emphasize the **nanoscale** characteristics of such structures which is also of central interest to the ca. 15 contributions dealing with polymers covalently or by physisorption attached to solid planar surfaces, the physical properties of the **brush-like structures** formed, and pattern production found in the interaction of polymers with surfaces.

It is somewhat sad to see that a classical field covered very well by the first 100 Volumes of this series has almost disappeared: the **kinetics** of polymerisation reactions is only covered by less than 10 review papers which sheds light onto the research situation where such studies have moved out of polymer science and into the field of chemical engineering.

In conclusion, *Advances in Polymer Science* has emerged as the prime source of relevant data and expertise presented in a critical form to all researchers in the field of polymer science. It has reflected the changes in research targets and research style in a precise and unambiguous manner and thus gives testimony of a very lively and quite progressive field in constant change and motion. We are anxious to see how the field will evolve as reflected in the next 100 Volumes of this series.

Mainz, March 2006   Gerhard Wegner

Preface

Modern technologies often demand access to functional structures, mostly at surfaces, at length-scales below 1 µm (1–100 nm). In various cases such structures must be ordered in a regular and tuneable fashion, while corresponding systems must also exhibit specific physical properties. The fabrication of hierarchically ordered structures at multiple length-scales, which serve as functional platforms, remains an experimental challenge. Such platforms, often based on polymers, are important both for fundamental studies investigating relationships between the nanostructures and the resulting physical properties, as well as for promising (potential) applications. Examples where such nanostructures are needed include faster and denser microelectronic systems, X-ray optics, nanolithography, and bioactive surfaces.

Structural order across many length-scales can be created by using self-organising materials, whose assembly is controlled and directed by various molecular interactions (and their combination) as well as by external constraints. Self-organisation occurs from nanoscopic to macroscopic scales, controlled by phenomena like microphase separation, adsorption or crystallisation. Ordered, nanoporous materials may be obtained in sol-gel processes or by aggregation (crystallisation) of colloidal particles. Competing factors controlling structure formation include chemical differences, conformational entropy, spatial constraints (molecular shape effects) and external (electric, magnetic, hydrodynamic flow) fields. Supramolecular materials built from highly regular, molecular nanostructures characterized by specific chemistries, like functional end-groups or molecular shape (flat objects, rods, spheres, tubules) may possess interesting viscoelastic, electrical, or optical properties and may sensitively respond to various external stimuli (sensors). Nanostructured surfaces may exhibit highly selective interactions with molecules in the environment and may act as biosensors.

Fundamental processes of nature work on the nanoscale rather than on the microscale. In order to study these processes (self-assembly, aggregation, ordering, etc.), as well as to mimic them, we must also perform our investigations on the nanometer scale, using well-defined, regular, tuneable and controllable systems. In corresponding mesoscopic devices we expect to encounter new physical phenomena (increasing importance of intermolecular forces, quantum effects, etc.), which may modify—or even govern—thermodynamic, mechanical, electrical and optical properties. Employing such *molecular* phenomena may provide the following advantages for the fabrication of modern devices based on ordered polymeric nanostructures:

Simplicity, as the system has the tendency to create such patterns in a self-regulating way, based on intrinsic phenomena like self-assembly, microphase separation, or morphological instabilities induced by intermolecular interactions.

Accuracy, as the size of the molecules determines the characteristics of the nanoscopic features.

Speed, as on molecular length-scales intrinsic intermolecular interactions may be much more significant than external forces. Consequently, we expect faster motion towards the equilibrium states.

What do we mean when we speak of *nanostructures*? Besides random (non-ordered) structures (e.g., nanoporous surfaces) that are built up from nanometer objects (crystallites, elements of microphase-separated block copolymer morphologies, etc.) we mainly consider ordered (symmetric, periodic, regular) structures exhibiting structural hierarchy on multiple length-scales containing sub-micrometer (molecular) features and order on a longer (e.g., micrometer) length-scale. Accurate control of lateral position and orientation of these structures with nanometer precision is of crucial importance. A combination of several processes (microphase separation, crystallisation, surface induced effects, spontaneous formation of organised surface structures at nanometer scales, hydrodynamic flow patterns, dewetting, etc.) guided by additional parameters (patterned substrates, temperature and field gradients) presents a highly promising approach to obtain nanostructures. As we demonstrate in this volume, interdisciplinary investigations carried out by physicists and chemists, in close collaboration with applied research, represent feasible pathways to reach this goal.

An intensive search is underway for new, simple, fast and versatile routes to create nanoscopically ordered patterns on surfaces. Several approaches have been explored and many of them are based on polymers. The molecular structure and functionality, hence also intermolecular (interchain) interactions of polymers, can be "tailored", which are essential factors guiding the self-assembly process at the molecular level. The size of a polymer (its molecular weight) can be varied, which provides a tuneable length-scale. Block copolymers allow one to combine different properties (stiff and flexible blocks, amorphous and crystalline (crystallisable) blocks, electrically conducting and insulating blocks, etc.) in a single molecule, thus introducing diversity at a molecular level. Accordingly, polymeric nanostructures are precise at the nanometer-level (size of the molecules), are flexible (different shapes such as spherical or lamellar morphologies, as well as more complex architectures are possible), allowing one to obtain various geometries (two-dimensional versus three-dimensional structures, circular versus linear patterns, assembly on flat or curved substrates, etc.). Corresponding platforms can be fabricated in a highly reproducible fashion, encompassing fast processes. Other classes of materials can be incorporated as well, e.g. a build up of metallic structures (decoration) on patterned polymer surfaces by external means like evaporation can be achieved.

Diblock copolymers composed of two chemically different molecules, chemically linked together via one end, exhibit various morphologies (spheres, cylinders, lamellae) at a molecular level, controlled by the size of the two blocks and by their chemical nature. At surfaces, these morphologies are usually "masked" by the preferential segregation of one component to the surface resulting in uniform surface properties and microdomains aligned parallel to the surface. In order to nonetheless achieve patterning of the surface by block copolymers, the substrate should be prepatterned on the length-scale of the polymer, or coated with a thin layer of anchored random copolymers to provide a neutral surface, equally attractive for both blocks. Application of, e.g. high electric fields (or other orienting effects) allows one to also overcome the surface effect and to align block copolymer morphologies perpendicular to the surface. The use of more sophisticated architectures (such as triblock copolymers, miktoarm star copolymers, etc.) can yield fascinating, novel and complex morphologies. Mixtures of block copolymers and homopolymers may also provide interesting micro- and nanostructures, and thus offer additional routes for patterning. In addition, block copolymers allowing for selective decomposition (either via oxidation, sputtering, ion beam etching or appropriate temperature treatment) open possibilities to modify or to enhance the intrinsic nanostructures. Deposition of other materials (e.g. evaporating metals or semiconducting materials) onto such modified structures open new pathways towards the fabrication of organic–inorganic nanostructures.

Making use of nanoscale phenomena such as self-assembly, we can build designer systems with tailored, purpose-oriented properties, effectively combining mesoscopic sciences with supramolecular chemistry. Employing these phenomena in an "intelligent" way presents a highly promising approach for putting together useful, modern and "smart" devices in inexpensive and accurate ways. In addition to structuring, we must also consider the kinetics of the corresponding fabrication processes. Depending on how fast the molecules order, they may not be able to reach the thermodynamically predicted state and instead get "trapped" in metastable states. It is not obvious if kinetic aspects favour order or rather produce disorder. As an additional step, polymeric structures can be modified or treated using approaches like selective degradation, as mentioned. This way the polymer patterns can be sensitively changed in the vertical direction without modifying its lateral distribution.

Obviously, besides the attempt to answer fundamental questions, we are also interested in taking advantage of the acquired knowledge in various applications. Thus, the use of engineered, nanostructured polymeric platform surfaces in the context of application-motivated problems in biomedicine, biology, materials science and nanotechnology is treated by various contributions in this collection of articles.

Mulhouse, France　　　　　　　　　　　　　　　　　　　　Günter Reiter
Enschede, The Netherlands　　　　　　　　　　　　　　　　Julius Vancso

Contents

The Formation of Ordered Polymer Structures at Interfaces:
A Few Intriguing Aspects
J.-U. Sommer · G. Reiter 1

Designed Block Copolymers for Ordered Polymeric Nanostructures
N. Hadjichristidis · S. Pispas 37

Surface Micelles and Surface-Induced Nanopatterns Formed
by Block Copolymers
K. Albrecht · A. Mourran · M. Moeller 57

Liquid Crystallinity in Block Copolymer Films
for Controlling Polymeric Nanopatterns
W. H. de Jeu · Y. Séréro · M. Al-Hussein 71

Surface Nano- and Microstructuring with Organometallic Polymers
I. Korczagin · R. G. H. Lammertink · M. A. Hempenius · S. Golze ·
G. J. Vancso 91

Genetic Engineering of Protein-Based Polymers:
The Example of Elastinlike Polymers
J. C. Rodríguez-Cabello · J. Reguera · A. Girotti · F. J. Arias ·
M. Alonso 119

Organic and Macromolecular Films and Assemblies as (Bio)reactive
Platforms: From Model Studies on Structure–Reactivity Relationships
to Submicrometer Patterning
H. Schönherr · G. H. Degenhart · B. Dordi · C. L. Feng ·
D. I. Rozkiewicz · A. Shovsky · G. J. Vancso 169

Development of a Biodegradable Composite Scaffold for Bone Tissue Engineering: Physicochemical, Topographical, Mechanical, Degradation, and Biological Properties
M. Navarro · C. Aparicio · M. Charles-Harris · M. P. Ginebra ·
E. Engel · J. A. Planell 209

Author Index Volumes 101–200 233

Subject Index 261

Adv Polym Sci (2006) 200: 1–36
DOI 10.1007/12_015

Published online: 25 October 2005

The Formation of Ordered Polymer Structures at Interfaces: A Few Intriguing Aspects

Jens-Uwe Sommer (✉) · Günter Reiter

Institut de Chemie des Surfaces et Interfaces, CNRS-UHA, 15, rue Jean Starcky, 68057 Mulhouse Cedex, France
ju.sommer@uha.fr

1	**Introduction**	2
2	**Copolymer Ordering at Surfaces**	3
2.1	Overview of the Problem of Nano-Structure Ordering in Copolymer Films	3
2.2	Ordering of Symmetric Copolymers at Neutral Surfaces: Monte Carlo Simulations	6
2.3	Ordering of Surface Patterns in Cylindrical Copolymers	10
3	**Crystallization Effects in Copolymers**	13
3.1	Varying Lamellar Spacings by Crystallizing Block Copolymers at Different Temperatures	14
3.2	Individual Crystallization and Melting of Polymer Nano-Crystals in Spherical Domains of a Block Copolymer Mesostructure	16
4	**Polymer Crystallization in Ultra-Thin Films: Diffusion Controlled Pattern Formation and Morphogenesis**	18
5	**Search for Ordering by Crystallization onto Patterned Substrates**	26
6	**Conclusions**	30
7	**Appendix: Experimental Procedures**	31
References		33

Abstract Nano-meter sized ordered structures emerge naturally in thin copolymer films. Control of this order and in particular the orientation of copolymer mesophases is possible, for example by modifying substrate interactions. Near neutral surfaces lamellae forming diblock copolymers are forced into a perpendicular order. Using the results of Monte Carlo simulations, it can be shown that this effect is guided by the kinetic pathway the system takes during the microphase ordering process. Microphase separation is prone to defect structures, which inhibit long-range order required for potential applications in nano-technology. A better understanding of the dynamics of the healing-out of defects is necessary to allow us to develop procedures which can lead to long-range ordered copolymer films. As an example, aspects of this dynamic in cylinder-forming copolymers are discussed. Besides microphase separation, polymer crystallization presents another possibility for the creation of soft-ordered systems. It is shown how crystallization in diblock copolymers influences the microphase morphology in thin films and eventually leads to highly ordered surface patterns, which may be guided by external constraints such as three-phase contact lines. Crystallization of polymer monolayers results in growth patterns, which are controlled by temperature and polymer architecture. We present a generic lattice model where the crystallization behavior of thin polymer films is linked to the interplay between internal chain order and growth. Using this model we show that anisotropy of the substrate leads to oriented dendritic growth patterns with order on multiple length scales.

Keywords AFM · Block copolymers · Computer simulations · Long range order · Polymer crystallization · Striped substrates

1
Introduction

The formation of long-range ordered nano-structures at surfaces represents one of the major challenges in nano-technology. Polymers are suitable to reach this goal because of the nanometer range of their molecular extensions and the diversity in which different and complex polymer molecules can be provided by modern synthetic chemistry. In order to obtain ordered surface patterns, the specific effects of confined geometries and surface fields come into play. Moreover, ordered surface patterns are not necessarily equilibrium structures but can emerge from non-equilibrium processes like meta-stable states controlled by kinetic pathways. In fact, equilibrium states in thin polymer films are frequently inhibited or difficult to obtain because of large kinetic barriers between alternative non-equilibrium states and the optimal thermodynamic state. Therefore, in the search for ordered structures in polymer films one has to consider the role of non-equilibrium states and kinetic pathways.

Surface patterns formed by block copolymers are considered as a promising means to create functional nanoscopic structures needed for the fabrication of miniaturized devices. However, a crucial step in order to approach this goal and to make applications feasible is the formation of structures exhibiting long-range order. Unfortunately, the mesoscopic patterns are prone to defects as an already present very small perturbation can destroy soft-order. To obtain long-range order several potential pathways have been proposed in the literature, such as graphoepitaxy [18, 43], in-plane electric fields [29, 49], temperature gradients [3], or directional solidification [40].

One of the main difficulties in obtaining long-range order of surface patterns is due to the smallness of the interaction energies per chain arising from entropic effects. Thus, significantly larger energies would provide much more stable patterns. The energies involved in crystallization of polymers are much stronger. However, here the formation of metastable structures, which can relax and thus change in time, represents the main crux of this approach for the formation of ordered surface patterns. Therefore, a profound understanding of the basic processes involved in polymer crystallization and means to direct the growth of polymer crystallization are needed to provide the necessary control for obtaining long-range order.

Here, we will discuss physical aspects of structure formation in thin polymer films induced by microphase ordering in diblock copolymers, pattern formation in crystallizable polymer films, as well as the interplay between both.

2
Copolymer Ordering at Surfaces

2.1
Overview of the Problem of Nano-Structure Ordering in Copolymer Films

Beyond a characteristic temperature T_{ODT} [14, 20], diblock copolymers self-assemble into microphase-separated morphologies. The theoretically predicted equilibrium states of such ordered phases correspond to perfect super-lattices with various symmetries depending on the volume ratio of both blocks. In particular, lamellar (symmetric diblocks), cylindrical (asymmetric diblocks), and spherical (strongly asymmetric diblocks) morphologies are expected and have been observed [2]. For typical copolymers, the characteristic length scale of these microphases is in the range from about 10 nm up to some 100 nm. It is justified to say that in the bulk, microphase-separated block-copolymers represent ordered nano-structures.

However, quite soon experimental studies showed that perfect long-range order of these nanostructures is difficult to achieve. This is in particular true for thin films, where methods such as shear alignment are not applicable [11, 29, 57]. The actual morphology of microphase-separated block-copolymers is dominated by defects of various kinds [10]. Corresponding computer simulations have proved that only relatively low energetic efforts are needed to form such defects, see also Fig. 1. In contrast, if one wants to heal out such defects, a very high energetic effort has to be invested to allow for the necessary cooperative rearrangements of large parts of the sample. Typically, disordered, poly-domain structures are dominating which have to be classified as meta-stable. The results of the energetic analysis of defect structures from computer simulations of block copolymer mesostructures are illustrated in Fig. 1. Here, using the bond fluctuation model (for more details see next section), the interaction energy for defect structures is compared with perfectly ordered morphologies.

The situation becomes even more complex when these copolymers are interacting with interfaces like in thin films cast onto substrates [16, 17, 24, 26, 30, 41, 53, 54]. Generally, at both interfaces (substrate, vacuum), preferential wetting by one block occurs. For symmetric diblock copolymer chains, this

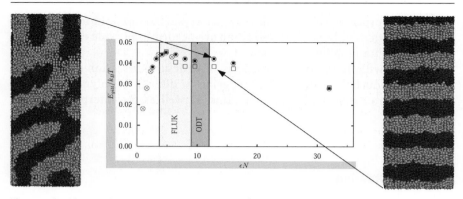

Fig. 1 The figure shows results from large-scale computer simulations of symmetric block-copolymers of chains made of 24 (12 + 12) monomer units using the Bond-Fluctuation-Model (BFM) [4]. The *left part* displays a typical morphology of the ordered phase as obtained from the disordered phase by step-wise increasing the interaction strength (purely repulsive interactions) up to the indicated value. Structures containing large defects can be observed. The periodic boundaries in the *xy*-direction (*left and right in the figure*) allow for any orientation of the lamellae. Both, the *top* and the *bottom* interface are reflecting for both types of monomers (ideally neutral interfaces). One can observe a perpendicular order of the lamellae at both reflecting boundaries which coexists with a different orientation in the center. Even after long relaxation times the defects in the morphology are not healed out completely. The *middle part* of the figure shows the plot of the interaction energy (E_{gitt}) per lattice unit (measured in units of kT) for different values of the local interaction constant ε. The symbol ⊗ corresponds to the defect containing structure (*left picture*), while the symbol □ corresponds to a sample of completely aligned lamellae as displayed on the right. The full alignment has been achieved by varying boundary conditions and subsequent relaxations. One observes that the gain in surface tension (determining the interaction energy in the ordered state) is a tiny fraction of kT which is completely irrelevant for the strongly segregated state. The additional symbol ● corresponds to lamellae oriented towards one interface, obtained by using a strong selective attraction potential at this interface (picture not shown here)

preferential interfacial interaction will favor a morphology where lamellae are oriented flat-on with respect to the substrate. Such order parallel to the substrate is also expected for the case of cylindric morphologies. Typically, such orientation of lamellae or cylinders parallel to the substrate also leads to rough surface topographies (island-hole-patterns) as they are frequently observed in thin films [6, 44]. For example, optical micrographs indicate that perfectly smooth surfaces are difficult to obtain [1]. In the case of symmetric block copolymers this effect is frequently attributed to the incommensurability of the lamellar period with the film thickness. In some cases, however, varying the temperature (and thereby changing the lamellar period as a consequence of a temperature dependent interaction parameter) may lead to comparatively smooth surfaces without island-hole patterns as can be seen in Fig. 2.

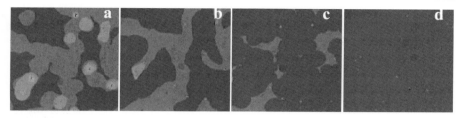

Fig. 2 Evolution of the surface topography, visualized by optical microscopy. Images are from a thin film (approx. 140 nm) of PB$_h$ - *PEO*(3.7–3.6) for different temperatures. **a** 120 °C, **b** 105 °C, **c** 88 °C, and **d** 75 °C. Lighter gray levels correspond to thicker regions. Note the discrete changes of the gray levels. The images are 100×100 μm^2 in size

Nonetheless, there are also strong indications that even smooth films do not necessarily represent equilibrium situations as the actual surface topography depends on the thermal treatment the sample has experienced. In fact, the existence of corrugated block copolymer surfaces is difficult to explain based on equilibrium arguments only. As for the case of defect structures, island-hole patterns might reflect meta-stable states which energetically differ only slightly from the ideally smooth surface states.

A route to achieve regular surface patterns in copolymer films is to *reorient* lamellar or cylinder morphologies in the direction perpendicular to the surface. Since one of the ultimate goals is to replace one or both polymer phases by other materials such as metals, such ordered structures have to be accessible from the free surface. In addition, in many cases a pronounced aspect ratio (depth-to-lateral extension) is needed. Promising candidates for achieving such a goal would be lamellae or cylinders oriented strictly perpendicular to the substrate. It was first shown by Russell and coworkers that strong electric fields can orient the interfaces of block-copolymers in the direction of the external field. Applied to thin films it has been demonstrated that cylindrical phases can be oriented almost perfectly in the direction perpendicular to the surface [42]. Unfortunately, this method is very much restricted to cylindrical morphologies since in the case of lamellar morphologies a second directing process in the plane of the film is needed as we will discuss below.

It has been indicated by experiments that there is another way to achieve perpendicular order even for films containing lamellar mesophases. The use of substrate coatings made of random copolymers allows us to neutralize the interactions of the two blocks with the substrate [16, 23, 25]. This leads to perpendicular orientation of the lamellae in the vicinity of the so-prepared substrate. It is worth noting that this effect is not simply related to some incommensurability between the lamellar period and the film thickness as has been discussed by several authors, see [52]. Remarkably, perpendicular orientation at neutral surfaces can persist against other orientations (e.g. induced by the opposite surface) as has been suggested also by experiments [25]. Furthermore, there are indications from computer simulations (presented below) that ordering at neutral surfaces can start before bulk ordering occurs. To explain the preference of perpendicular lamellae at neutral surfaces, Pickett, Witten, and Nagel [33] have discussed effects of nematic alignment of monomers and stretching of chains at neutral walls. Beside equilibrium arguments, however, it will be necessary to consider the pathway of the formation of ordered microphases to gain deeper insights into the nature of the ordering phenomena of copolymers at near neutral surfaces. In particular computer simulations can be used to reveal the molecular origin of ordering of copolymers at surfaces and in thin films.

2.2
Ordering of Symmetric Copolymers at Neutral Surfaces: Monte Carlo Simulations

The microphase separation of block copolymers has motivated many computer simulational studies, see [8, 9, 15, 31, 58]. As a key result of these studies deviations from mean field theory [20] have been observed early [9].

As pointed out already, at surfaces usually one of the species is preferred and it is rather difficult to provide non-selective surface conditions. By contrast, neutral surfaces are a natural choice in computer simulations (using reflecting boundary conditions) which are therefore most suitable to investigate the effects discussed in the previous section.

To investigate and to understand the molecular origin of orientation effects near neutral walls, we have applied a large-scale Monte Carlo simulation using the bond fluctuation model (BFM) [15, 45]. The simulation method is illustrated in Fig. 3. The size of the simulated lattice is $L_x \times L_y \times L_z = 100 \times 100 \times 50$ in lattice units. Periodic boundary conditions are applied in the xy-direction and repulsive walls in the z-direction. The copolymers are symmetric ($N_A = N_B$) and have a total length of $N = 16$–48 repeat units. Repeat units correspond roughly to Kuhn segments in real systems and are denoted as monomers in the following. The lattice density is 0.6224 which corresponds to a dense melt in the framework of the BFM. In total 37 344 monomers (corresponding to 1556 chains for $N = 24$) are simulated. The simulation starts with an athermal equilibrated system which is cooled successively below the ODT. Thus, the chains have to self-assemble into the ordered state. For further details, see [45].

In marked contrast to the case of spinodal decompositions in polymer blends (and also in contrast to the mean field theory of microphase separation), the microphase ordering is preceded by a stretching process of the chains. These findings concur with simulation results obtained earlier [9]. In Fig. 4 we have displayed various measures of the extension of the individual copolymer chains, namely the mean distance between the center of mass (COM) of both blocks, R_{AB}, the radius of gyration of the total chain, R_g, and the radius of gyration of a single block (which, by symmetry is the same for A and B), R_A. Our simulations indicate that the microphase separation pro-

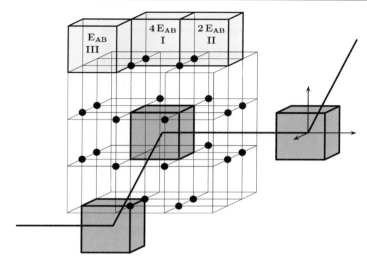

Fig. 3 In the BFM model each monomer occupies a unit cube in a 3D simple cubic lattice. Bonds between monomers are taken in a range between 2 and $\sqrt{10}$ lattice units which ensure cut-avoiding during local moves. In a Monte Carlo step a monomer is chosen randomly and tries to move into one of the six possible nearest neighbor positions. The move is rejected if the new lattice places are occupied by other monomers (excluded volume). If the energy difference between the new and the old monomer position is positive, the move is only accepted with the corresponding Metropolis rate. The interaction energy between different monomer species (A and B) is calculated in three shells around the given monomer as indicated in the figure. We implement only repulsive interaction between different monomer species. All other interactions are athermal

cess can be subdivided into two steps: First, stretching of individual chains where both blocks are locally separated ("polarization" in terms of the spatial separation of the two species within each chain). The chains can be thought to be in a dumbbell-like form. Second, formation of long-range order formed by "polarized" and stretched chains. Note that in the strongly segregated state additional stretching of chains is much less dramatic, see Fig. 4. As a consequence of this multi-stage process, the morphology of microphase-separated copolymers should be affected by environment effects which influence the stretching process of the chains *during* microphase separation. Along this line, we argue that neutral walls provide an *entropic surface field* which facilitates stretching in the direction parallel to the wall and, therefore, the formation of perpendicular lamellae near the walls. Our data even indicate a surface transition, where pre-ordering of perpendicular morphologies at neutral surfaces takes place before bulk ordering.

The above conclusions can be supported by analyzing separately the behavior of chains in the surface region and the bulk region, as displayed in Fig. 5. Here, we consider a chain to belong to the surface region if at least one monomer touches the surface. According to this definition a particular

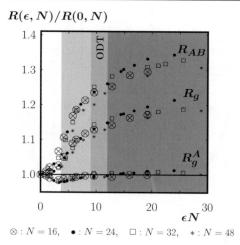

Fig. 4 The figure displays the change in the extensions of single symmetric block copolymer chains during the process of microphase separation, i.e. upon increasing the strength of the local interaction which is expressed by the scaling variable εN. Here, N denotes the number of monomers per chain. The different chain lengths which have been simulated are indicated by the symbols. The upper set of curves correspond to the mean distance between the COM of both blocks, the middle set of curves correspond to the radius of gyration of the whole chain, while the lower set of curves corresponds to the radius of gyration of individual blocks. A good scaling behavior with εN can be observed for all chain lengths. The ODT is located at about $\varepsilon N \simeq 10.5$, see [15]. Strong stretching of the whole copolymer chains already takes place far below the ODT (*upper curves*). This is accompanied by a weak contraction of each individual block (*lower curves*)

chain can change between surface and bulk region during the course of its motion. We note that the simulation box is wide enough to allow for about 2/3 of $N = 24$ chains to belong to the bulk part. A snapshot of the surface region is displayed on the right hand side in Fig. 5. In the left part of Fig. 5, the end-to-end distance distribution G is plotted for surface and bulk chains, respectively. Already in the disordered state, (a) and (b), the distribution of the z-component (a) of G is much narrower for surface chains compared to the distributions in the directions parallel to the surface (b). The logarithm of this distribution is proportional to the free energy of a single chain with a given end-to-end distance. Therefore, the peaked and narrow distribution of the end-to-end distance, Fig. 5a, implies a higher force to stretch chains in the surface layer in the direction perpendicular to the walls. This fact has been verified explicitly in our simulations by applying an external force to the chains [45].

Part (c) and (d) of Fig. 5 show the distributions for an interaction strength slightly below the ODT. In the surface layer (Fig. 5c) only the directions parallel to the surface are influenced. A fact which indicates the preferred stretching of chains in the direction parallel to the surface. Moreover, for the

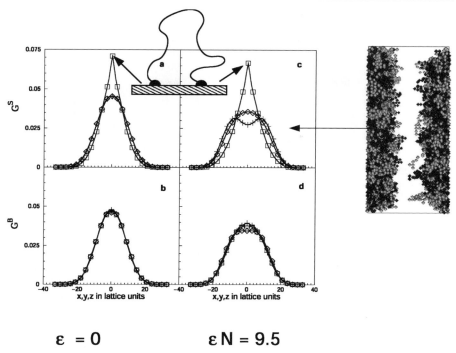

Fig. 5 The end-to-end distance distribution G is plotted for chains in the surface layer (upper two plots a and c) and for chains in the bulk (lower part b, d) for $N = 24$. Chains belong to the surface layers if at least one monomer is in contact to a surface. In our simulation box, about 2/3 of the chains have no surface contacts and are counted as bulk chains. Since the distributions in the surface layer and also in the ordered state are anisotropic we distinguish between the spatial components: $\square = z$-direction, $+ = x$-direction, and $\diamond = y$-direction. The distribution in the surface layer is clearly anisotropic even in the disordered state $\varepsilon = 0$. The sharp (non-Gaussian) peak indicates a preferential location of end-points close to the surface which is expected in dense polymer systems. For an interaction constant close but still below the ODT (c, d) the surface chains already display long-range order in the x-direction while the distribution of the z-component is practically not influenced. By contrast, the bulk chains, (part d), are still isotropic but stretching (without inducing any long-range order) is displayed by a broadening of the distribution functions

parallel components we also observe the first signs of long-range order (double peak behavior of the x-distribution) before bulk order is reached.

The morphology for a thin copolymer film is shown in Fig. 6 before and after microphase separation, viewed through one of the repulsive surfaces. Perpendicular orientation of lamellae is clearly obtained near the reflecting and neutral surfaces. On the other hand, orientation is not perfect and the long-range order of the two-dimensional lamellar morphology at the surface is highly disturbed by defects, see Fig. 6.

Fig. 6 The surface of the simulated films is shown before (*left*) and after (*right*) microphase separation for $N = 24$. The morphology in the ordered phase is dominated by meta-stable defect structures

As we have shown, computer simulations give strong indications for an entropic surface field which is the consequence of the anisotropy of the chain fluctuations in the surface layer. Even without the presence of an additional external field, this favors the formation of lamellae perpendicular to the surface and explains the observation of experiments. It is worth noting that perpendicular order is selected during the kinetic pathway of microphase ordering, in particular during the fluctuation-phase below the ODT, where the chains are stretched but no macroscopic order is created. This does not necessarily require that perpendicular orientation is clearly preferred in thermodynamic equilibrium. However, once formed, there is an enormous free energy barrier between perpendicular and parallel orientation of lamellae, so that switching to parallel orientation is not observed.

As can be seen from the above discussion, the great challenge for the formation of perfect, self-organized surface patterns is therefore to find ways for creating in-plane long-range order of copolymer morphologies in thin films. From the viewpoint of the physics of phase transitions, such long-range order could be induced by nucleation and growth processes or, more generally speaking, by applying other non-equilibrium preparation methods such as temperature and concentration gradients or zone-casting techniques [51] in combination with neutralized substrates.

2.3
Ordering of Surface Patterns in Cylindrical Copolymers

In the last section we showed that the ordering of symmetric copolymers faces two major problems: First, it is difficult to obtain perpendicular orientation of lamellae. Second, in almost all cases long-range lateral order of the periodic surface pattern is missing. The first problem can be solved by using neutralized substrates. However, obtaining long-range lateral order remains a difficult task. Here, we will turn to asymmetric copolymers, in particular to cylinder forming systems. When employing selective boundary conditions, cylinder-phases ordered parallel to the surface will emerge. Although the usefulness of parallel ordered cylinder phases in nano-technology is limited (for instance to preparing templates to form metallic nano-structures, here the perpendicular order of cylinders is more interesting) in comparison to lamellae forming systems the cylindrical geometry allows more easily the study of the dynamics of long-range ordering [10, 13]. The aim of this section is to present some results concerning the dynamics of the formation of long-range order at the surface of thin asymmetric copolymer films starting from non-equilibrium short-range ordered films.

For our experimental realization of a cylindrical geometry we used a short asymmetric diblock copolymer of hydrogenated polybutadiene-block-poly(ethylene oxide), PB_h-PEO(3.7-1.1). We employed the phase contrast obtained by tapping mode AFM to visualize the mesophase pattern. The advantage of short molecules is the relatively low viscosity which allows for comparably fast rearrangements in the mesophase patterns. As in any such block copolymer system, preparing thin films by spincoating results in poorly ordered patterns containing many defects. However, annealing these films at elevated temperatures provides a way of improving the order. We note that our system consists of two highly incompatible blocks with a rather high ODT. Thus, when annealing at temperatures as high as 200 °C, the system remains in the two phase region. The increased mobility at high temperatures allows the system to remove the defects and to establish increasingly larger domains of perfectly aligned cylinders. In Fig. 7 we present typical images demonstrating the evolution of order with annealing time. A large number of defects which exist at the beginning (see Fig. 7A) is removed relatively quickly and domains of well-aligned cylinders appear. Continuing to eliminate defects at the domain boundaries leads to a growth in domain size. In Fig. 7D the size of the domain is already larger than the size of the image. At

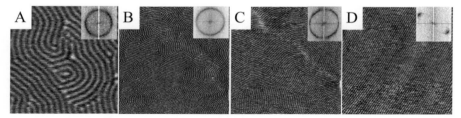

Fig. 7 Typical tapping mode atomic force microscopy *phase* images from a thin diblock copolymer film PB_h-PEO(3.7-1.1) showing the evolution of the pattern during annealing at 180 °C. Annealing time for **A** and **B** 5 min, **C** 30 min, and **D** 120 min. The insets gives the corresponding Fourier transforms. The size of the images B, C and D is $1 \times 1 \,\mu m^2$. Image (**A**) is a zoom-in on (**B**) with the size of $300 \times 300 \, nm^2$

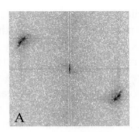 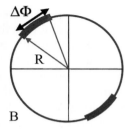

Fig. 8 The Fourier transform shown in part **A** is taken from Fig. 7D, showing two rather narrow reflexes. The scheme of part **B** indicates how the width $\Delta\Phi$ of these reflexes has been determined for a fixed radius R

the longest annealing times at the highest temperatures used (i.e. 200 °C) we observed domain sizes of up to 5 μm.

We characterized the temporal evolution of the patterns by determining the angular width $\Delta\Phi$ of the reflexes of the Fourier transforms of the AFM phase images. In Fig. 8 we give a schematic representation of the definition of $\Delta\Phi$ and in Fig. 9 we give an example for how $\Delta\Phi$ was measured. Along this definition we would obtain for a completely random orientation of the cylinders $\Delta\Phi = 180$ degrees. In Fig. 10 we show that in the course of annealing $\Delta\Phi$ changes initially rather quickly but with the increase in domain size the evolution slows. From our experiments we conclude that the alignment of cylinders can certainly be improved by simply annealing the sample. However, perfect order over large distances is extremely difficult to obtain as the healing out of defects involves rearrangements of progressively larger do-

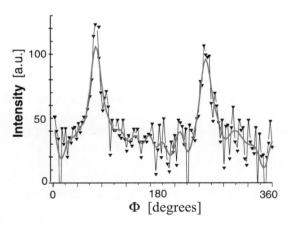

Fig. 9 Intensity (in arbitrary units) of the Fourier transform shown in Fig. 8 for a fixed radius R as a function of the angle Φ. Data points are the results directly obtained from the Fourier transform, the full line represents a smoothed curve obtained by adjacent averaging. The width $\Delta\Phi$ of the distribution is taken at half-height of the peaks

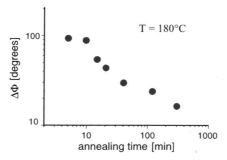

Fig. 10 Width of distribution, see Fig. 8, plotted as a function of annealing time

mains which accordingly need more time. Order on areas of some μm^2 may be obtained quite easily by annealing. However, to achieve such order over reasonable time scales is only possible if the incompatibility of the two blocks is high and the viscosity of the polymer is low.

3
Crystallization Effects in Copolymers

Crystallization presents a promising way to bring high and long-ranged order to surface patterns of block copolymer mesophases. One has to note that the typical energies involved in "soft-ordered" polymers such as microphase-separated copolymers are of the order of several tens of kT per chain. For chains made of 10000 monomers this yields a typical interaction energy of only a few 10^{-3} kT per monomer. Compared to the energy of thermal fluctuations (kT) this value is extremely small. Consequently, polymer morphology, even in strongly segregated block-copolymers, is dominated by thermally controlled conformational statistics. On the other hand, these conformational degrees of freedom will be strongly reduced by crystallization processes which lead to well-ordered crystal lattices. Thus, crystallization forces are incomparably stronger than the typical forces in microphase-separated amorphous polymers. As a consequence, the morphology of copolymer systems can be completely changed by crystallizing one or more components [12]. In addition, one has to take into account that crystallization of polymers involves a hierarchy of ordered structures crossing a multitude of length scales, starting from the crystal unit cell, over nanometer-sized crystallites up to hundreds of microns for spherulites. In block copolymer systems, these length-scales are superposed to, and sometimes in competition with, the mesoscopic structures due to the interaction (incompatibility) of the blocks. Consequently, crystallizable block copolymers in thin films or at surfaces present interesting systems for the creation of patterned substrates exhibiting multiple length scales.

3.1
Varying Lamellar Spacings by Crystallizing Block Copolymers at Different Temperatures

In Fig. 11 we show an AFM image of the surface of a weakly asymmetric block copolymer (PB_h-PEO(3.7-2.9)), where the PEO component is in the crystalline state. There, the most spectacular result is represented by the parallel line pattern visible in the phase image. These lines are supposed to represent vertically oriented lamellae which are separated by a well-defined distance of 22 ± 1 nm, a value of the order of the size of the molecules. We have to conclude that the molecules formed lamellae which are oriented vertically with respect to the substrate. The difference in viscoelastic properties (the crystalline PEO domains are much harder than the liquid PB_h domains) is responsible for the contrast in the AFM phase image. Interestingly, the vertical lamellae resulting from the crystallization process were preferentially aligned along the borders of dewetted regions. The perfectness of this alignment could be even further improved by reducing the crystal growth rate, i.e. by crystallizing at higher temperatures. We note that the spacing between these lamellae is highly constant and not dependent on film thickness.

The morphologies of polymer crystals (and the corresponding lamellar spacings) generally represent meta-stable structures which are always significantly affected by the kinetics of the crystallization process. We investigated this influence of the growth kinetics by varying the crystallization temperature and thus the crystal growth rate. In Fig. 12 we compared typical results for four crystallization temperatures T_c differing only by a few

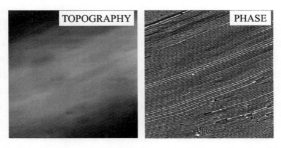

Fig. 11 Typical tapping mode atomic force microscopy images from a thin diblock copolymer film PB_h-PEO(3.7-2.9) crystallized at 35 °C, taken close to a three-phase contact line (visible in the *upper left corner*) created by a dewetting process which occurred in the molten state. Topography and phase contrast are shown. The size of the images is $2 \times 2\,\mu m^2$. The maximum height is 150 nm. Phase contrast reflects the different viscoelastic properties between the hard crystalline PEO and the soft amorphous PB_h regions, indicated by "lines" with light and dark shades of gray, respectively. A characteristic distance of about 22 ± 1 nm, caused by the sequence of "hard" and "soft" regions, can be seen in the phase image. This suggests that the lamellae were oriented vertically

degrees. A small change in T_c, from 38 °C (part A of Fig. 12) to 45 °C (part C of Fig. 12), resulted in a doubling of the characteristic spacing. For T_c between 20 °C and about 40 °C we observed a spacing of about 22 ± 1 nm. Increasing T_c to 48 °C led to the complete loss of a characteristic separation distance between lines of low and high stiffness, as demonstrated in part D of Fig. 12. It should be noted that at 48 °C the topography image (not shown here, see Figure 9 of [34]) and the phase image were clearly correlated. In particular, depressions between stripe-like features appeared as dark lines in the phase image while the stripes (independent of their width!) are represented by uniformly light parts.

On the basis of simple geometric arguments (we know the number of the monomers and their size in the crystalline state as well as the lamellar spacing observed by AFM) we conclude that for temperatures up to about 40 °C we have once folded interdigitated PEO blocks (sandwiched between PB_h layers) resulting in a repeat period of about 22 nm. This value has to be re-

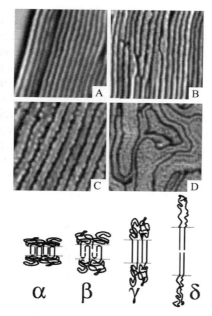

Fig. 12 Comparison of AFM phase-images for thin films of PB_h-PEO(3.7-2.9), crystallized at different temperatures: **A** 38 °C, **B** 41 °C, **C** 45 °C, and **D** 48 °C. The size of each image is 400×400 nm². The characteristic spacing increases from **A** 22 ± 1 nm to **B** 34 ± 1 nm and **C** 44 ± 1 nm and is lost for **D**. These spacings can be related to different levels of chain folding as indicated by the sketches $\alpha, \beta, \gamma, \delta$. While α, β and γ can explain the spacing in **A**, **B** and **C**, respectively, δ can probably not be realized due to unfavorable, or even impossible, strong stretching of the amorphous block. Thus, **D** is probably due to the re-orientation of the vertical lamellae of **C** into a horizontal alignment. Such alignment allows for lateral spacings of any width which are not related to the size of the molecules

lated to the maximum length of the fully extended crystalline block alone which is 18.4 nm. For one fold we thus obtain a length of 9.2 nm for the PEO block. Taking into account that the lamellar period L consists of amorphous and crystalline layers (ABA sequence), and further assuming that the PEO blocks are interdigitating we obtain a value between 20–25 nm for L, depending on the conformation of the PB_h block. At higher temperatures the system starts to remove the fold and eventually, at around 45 °C (growth proceeds at a speed of less than 1 nm/sec as this temperature is only about 9 degrees below the melting point), and we end up in a state of fully extended but interdigitated PEO blocks with a length of 18.4 nm. In Fig. 12 we have sketched these ideas.

In contrast to molten (amorphous) systems, the resulting lamellar spacing in the crystalline state cannot simply be described by thermodynamic arguments alone. It also depends strongly on the kinetics of the crystallization process. As long as the equilibrium between the loss of entropy due to the stretching of the PB_h blocks and the gain in crystallization energy of the PEO blocks is not obtained, the amorphous PB_h blocks have no possibility to control the number of folds of the PEO blocks. In such a case, the PEO blocks crystallize more or less unaffected by the PB_h blocks. Thus, crystallization of the PEO blocks sets the lateral separation of PB_h-PEO(3.7-2.9) junction points and thus controls the degree of stretching of the PB_h blocks. The fact that the patterns evolve with temperature (Fig. 12) proves that the once folded state at low temperatures is NOT an equilibrium state but is the result of the kinetics of the crystallization process. The equilibrium situation is most likely obtained if crystallization takes place very slowly at a temperature only a few degrees below the equilibrium melting temperature.

3.2
Individual Crystallization and Melting of Polymer Nano-Crystals in Spherical Domains of a Block Copolymer Mesostructure

In another set of experiments we used the asymmetric block copolymer PB_h-PEO(21.1-4.3), where PEO presents the minority phase of about 17 volume percent and forms spherical cells with a diameter of about 12 nm. A typical AFM image of the surface of a thin film prepared from this polymer is shown in Fig. 13. The first point to realize is the ability of the AFM phase mode to distinguish even between two liquid polymers, the liquid PEO cells embedded within the liquid PB_h matrix.

The key question we aim to answer is how crystallization proceeds under such extreme confinement. One has to realize that within such small spherical cells with a diameter of about 12 nm only about 145 PEO blocks each having about 100 monomers are contained. In addition, each block is attached to a much longer amorphous block forming the confining matrix around the PEO cells. Previous experiments [21, 22] already showed that the overall

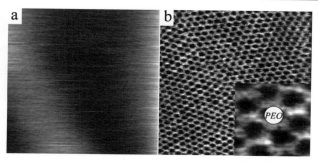

Fig. 13 AFM topography **a** and phase **b** images showing the surface of an about 100 nm thick PB_h-PEO(21.1-4.3) film. While the film is rather smooth (the maximum height variation is 8 nm), the different viscoelastic properties of PB_h and PEO allow us to identify the distribution of the two blocks in the hexagonally packed mesostructure of spherical PEO cells in an amorphous PB_h matrix. The size of the images is 500×500 nm^2. The inset in **b** shows a magnification of a 65×65 nm^2 section of **b** with one PEO cell indicated schematically

growth kinetics of highly asymmetric block copolymers forming spherical or cylindrical mesophases differs qualitatively from the kinetics in unconfined geometries. As crystallization did not destroy the mesophases, it was concluded [21, 22] that crystallization was initiated by homogeneous nucleation, separately in each compartment. However, until now it was not possible to verify by direct visualization of the crystallization process if nucleation was really independently occurring in each compartment.

Employing tapping mode AFM phase contrast to distinguish between crystalline and amorphous cells, we observed in direct space that crystallization occurred in a random manner, cell-by-cell (see Fig. 14). Systematic AFM experiments [35] using samples held at -23 °C for increasing times showed that the fraction of crystalline cells increased with crystallization time, following a relation which can be approximated by $n = n_\infty(1 - e^{-t/\tau})$. Here, n_∞ and

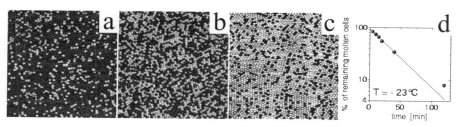

Fig. 14 AFM phase images (1×1 μm^2) showing the variation in the number and the distribution of crystalline cells in a thin PB_h-PEO(21.1-4.3) film after crystallization at -23 °C for **a** 5 min, **b** 15 min, and **c** 120 min, respectively. **d** Shows the percentage of still remaining molten cells as a function of time of crystallization

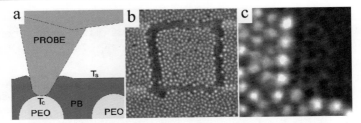

Fig. 15 Controlled melting of crystalline PEO cells with an AFM tip. **a** – Schematic drawing of tip-crystalline cell contact at high tapping force. **b** – AFM phase image (1×1 μm^2) showing a rectangular pattern sketched on the sample surface by local melting of the crystalline cells under the tip. The phase contrasts have the following meaning: *light dots* = crystalline cells, *dark dots* = molten cells, intermediate contrast = amorphous PB$_h$ matrix. **c** – AFM phase image (200×200 nm^2) showing brighter PEO cells at the boundary between crystalline and molten region which may correspond to an increased size or better-ordered crystals inside these cells

τ are the maximum fraction of crystallizable cells and a characteristic time of the process, respectively. Fitting this equation to the data at early times yielded $\tau = 35$ min. No discernible indications of any correlation or coupling between cells could be found. Crystallization of one cell did not induce crystallization of neighboring cells. Our results demonstrate that the probability for nucleating a crystal in a cell is not visibly affected by the state (either molten or crystalline) of the neighboring cells. Thus, at all densities of crystalline cells, the PEO spheres crystallized independently, with a random spatial distribution of the cells.

As we show in Fig. 15, the independence of the individual crystalline cells can also be tested by local melting via a hot AFM tip. Under appropriate conditions (see [56] for more details), only crystalline cells in direct contact with this hot tip melt while neighboring cells stay crystalline.

4
Polymer Crystallization in Ultra-Thin Films: Diffusion Controlled Pattern Formation and Morphogenesis

While the above results demonstrated that crystallization can establish long-range order in diblock copolymer films, a complete understanding of the underlying ordering processes is still lacking. In particular, the pathways polymers follow during crystallization represent a longstanding but still intensively debated phenomenon. In order to shed more light onto these ordering processes, fundamental studies on simplified systems—e.g. monolayers of homopolymers—have been performed.

Here, we first present a brief description of the theoretical model and corresponding simulations which provide us with a solid basis for the understanding of polymer crystallization in quasi-two dimensions. Simultaneously performed experiments on ultra-thin films (monolayers) of rather short polymers, obtained by autophobic dewetting in thin PEO or PS-PEO(3-3) films [36–38], will be presented subsequently.

Crystallization of homogeneous polymer films leads to structuring of the film. When the film becomes thinner than the characteristic length-scale of the polymer chains, various diffusion-controlled crystallization patterns can be observed. This is due to the fact that crystalline lamellae in such ultra-thin films grow within the surface plane (i.e. flat-on), with the crystalline stems oriented upright. Short chains, which are strongly adsorbed before crystallization, have to desorb partially and subsequently stretch out considerably during crystallization. As a result, locally a significant jump in area density is observed. This desorption-crystallization process of short chains in ultra-thin films is sketched in Fig. 16.

In order to obtain a deeper understanding of the formation and the resulting properties of the patterns obtained, we have developed a generic model which reflects the basic features of the complex growth and reorganization processes of polymer crystallization. Here, the qualitative model proposed in

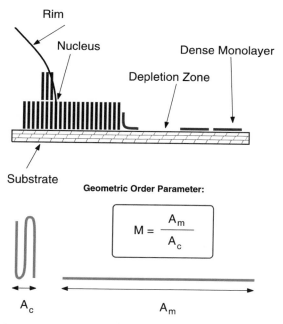

Fig. 16 Sketch of the crystallization process of short chains which are flatly adsorbed onto a substrate. Crystallization is started by heterogeneous nucleation which is usually realized by thicker parts of the film such as the rim of the region surrounding the polymer monolayer. An order parameter can be defined as the ratio of the area per chain occupied before and after crystallization, see *lower part* of the figure

Fig. 16 is casted into a simple but quantitative lattice algorithm. The basic idea is that individual chains successively increase their internal order (characterized by the degree of chain folding) during the crystallization process. The more the chain is ordered (the fewer folds it has) the lower is the surface area needed for this chain. The ultimate degree of order is represented by the completely stretched chain which only occupies a surface area proportional to the cross-section of one stem a_0 (area of a crystalline unit cell), see Fig. 16. Let A_0 be the area of the corresponding liquid chain, flatly adsorbed onto the surface. Then, $M = A_0/a_0 \sim N \gg 1$ chains can occupy the same area A_0 in the crystalline state. By contrast, in a simple growth model [27] the area per particle remains constant and it is the original dilution of particles which is responsible for the various diffusion-controlled patterns [28, 55].

The key difference between polymer crystallization in quasi-two dimensions and simple growth processes is the additional possibility to change the internal degree of order of polymer chains in the crystalline phase which is expressed by the local jump in the area density. Even when the starting point is a dense polymer monolayer, this process leads to diffusion-controlled growth because the gain of free surface area by each crystallization event will effectively dilute the system after a small period of growth. The simplest generalization of the original lattice models for diffusion controlled growth [27] is therefore a model which allows for *multiple occupation* of lattice cells in the crystalline state. Such multiple occupation exactly reflects the increase of internal chain order during the polymer crystallization process. In our algorithm we *map the internal chain order onto the multiple occupation number of lattice cells in the crystalline (aggregated) state*. Two parameters control the behavior of our model in its simplest version: First, the energy of crystal bonds which has to be overcome if a chain detaches from the crystal or changes its internal state by reorganization. We represent this effect by a temperature variable $T = -E/\ln p_0$, where E is the binding energy per ordered unit and p_0 is the probability to jump against such a barrier in a Monte Carlo simulation. Second, the (mainly entropic) resistance of a chain to increase its internal degree of order, i.e. to remove chain folds. Since chains do not spontaneously orient and stretch, such a transition must compete with the higher entropy the chains have in the disordered, molten state. The microscopic (molecular) origin of this barrier is rather complex and involves various effects such as a possible pre-ordering of chains adsorbed at the crystal growth front, precursor states or some other mesomorphic states which are locally established before the chains "lock-in" to the crystal phase [32, 48]. Since our model is established on a length scale larger than the individual segments we don't have to account for all these details. We rather assume that all these aspects can be represented by a single barrier parameter which reflects the major properties of the disorder-order transition within a single chain. Such simplification implies a certain *universality* of the polymer crystallization process, representing a non-equilibrium growth and reorganization process. In the simplest version of our model we assume a constant barrier parameter p_S for increasing the cell occupation number by one, i.e. for increasing the order of the chains by a quantity proportional to the size of one lattice unit occupied by an adsorbed chain. The model is sketched in Fig. 17. For more details, see [46].

Both model parameters T and p_S influence the morphology of the crystal. In Fig. 18 we show a few realizations of the growth model obtained by varying the crystallization temperature and keeping the entropy barrier constant. The most impressive feature resulting from the increase in crystallization temperature is the widening of the finger structures. This effect is much more dramatic than the thickening of the lamellae caused by the higher internal order (stretching) of the chains.

Another interesting feature of the growth processes, including the possibility of internal reorganization (improvement of order), is the appearance of inhomogeneous surface profiles of the crystals. In Fig. 19 we display a three-dimensional view on a part of the growth pattern shown in Fig. 18c. A striking observation is that high rims at the edges surround the more folded

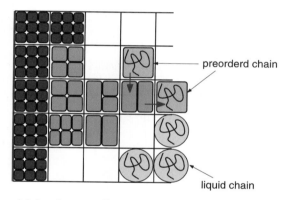

Fig. 17 Lattice model for the crystallization of polymers in quasi-two dimensions. The scale of coarse graining corresponds to the size of a whole polymer chain in the adsorbed state and is represented by one lattice site. In the disordered, liquid state the chains can diffuse freely subjected only to the excluded volume conditions. In order to enter the crystal phase, the chain has to be stretched which is modeled as a multiple occupation of lattice cells. Here, more than one chain (all assumed to have the same degree of order) have to share the same lattice site. Internal ordering of chains can take place only against a barrier probability of p_S. The binding energy of chains in the crystalline state depends on the degree of order of the neighboring chains on the lattice. If a chain reaches a high internal order within a surrounding having at least the same degree of order, it is unlikely for it to melt again. The average degree of order (which corresponds to the crystal thickness) is a meta-stable, self-organized parameter. The meta-stability of this state is not only reflected by the average degree of order (average thickness) but also by the morphology of the growth structure. We note that our model is able to reflect thermal reorganization processes during and after growth

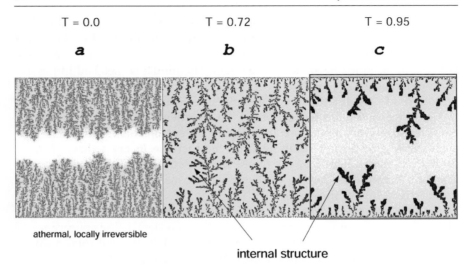

Fig. 18 A series of growth morphologies for different temperatures is displayed. The entropic barrier is set to $p_S = 0.6$, a value which allows for considerable reorganization processes, as they are observed also in experiments. Increasing the temperature leads not only to thicker lamellae but also to much wider finger structures. The interplay between the internal ordering (chain stretching) and morphology changes represents the key for the understanding of structure formation in polymer crystals

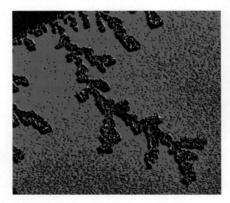

Fig. 19 The surface topography of the simulated crystal is displayed. The elevations at the edges of the finger structures are in a state of higher order compared to the interior of the crystal. This effect is caused by the higher mobility (fewer nearest neighbors) of chains at the boundary of the crystal which makes them prone to reorganize towards a more stable state

chains in the interior of the crystal. The dynamic origin of this effect is easy to understand in terms of our model: towards the end of the crystallization process the density of free chains decreases. Thus, the growth rate decreases and reorganization processes become more important. This concerns in particular the chains directly at the crystal edge which have fewer crystalline neighbors than within the crystal lattice and therefore can change their state more easily. Consequently, the boundaries "self-anneal" and confine the interior of the crystal. Such morphologies are very stable in the simulation model and appear for various choices of parameters. As one particular result of this process, hole-rim patterns can be obtained when the crystal has grown under strong under-cooling, leading to comparatively thin and unstable lamellae.

After the observation of such morphologies in our simulations, close inspection of the experimental results allowed us to confirm that the corresponding states are also found in experiments. In Fig. 20 we show a few results for PEO-2k and PS-PEO(3-3) systems which show the typical surface morphology as expected from the simulations. It has to be noted that single crystals surrounded by elevated structures and hole formation within crystalline structures have been observed also for other systems, see for instance [50, 59].

These reorganization processes represent a form of morphogenesis which we can observe directly after growth or after long relaxation times. They are only a first indication of the potential of reorganization of polymer crystals. If we increase the temperature close to (but still below) the melting point, morphological transformations are accelerated and additional morphologies can be observed. In Fig. 21 the result of such reorganizations induced by a temperature jump is presented. Here, we started from the growth pattern already displayed in some detail in Fig. 19. The applied temperature jump

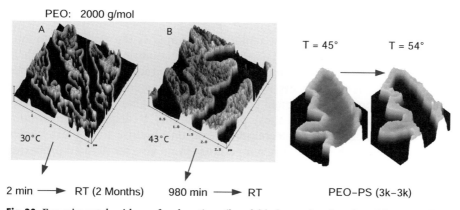

Fig. 20 Experimental evidence for elevations (less folded states) at the edge of the crystalline domains and for the "rim-hole" patterns related to the predictions of the simulation model. The *left part* shows the formation of high elevations at edges and rim-hole patterns in short chain PEO-2k crystals after long relaxation times at room temperature (RT). In the *right part*, analogous morphologies are obtained by annealing two-dimensional PS-PEO(3-3) crystals. Basically, the copolymer shows the same properties as the homopolymer. However, the non-crystallizable polystyrene block affects the ordering process of PEO chains in the crystal and therefore influences both the entropic barrier and the melting enthalpy (expressed by the effective temperature T in the simulations)

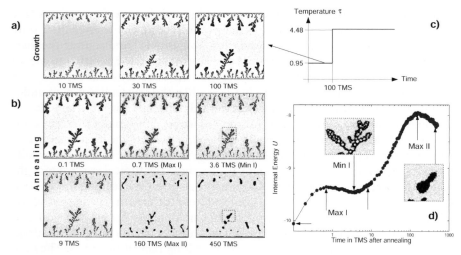

Fig. 21 a Three consecutive stages during growth. All times are given in units of 1000 Monte Carlo Steps (TMS). The last picture in the first row corresponds to the pattern shown in Fig. 19. It serves as the starting point for the annealing process. Different values of the order parameter are coded in the gray scale. Completely stretched chains correspond to black dots on the lattice. **b** Morphogenesis for six different times after the temperature jump shown in **c**. The evolution of the internal energy (negative number of interacting units per lattice place) is displayed in **d**. The *arrows* indicate the times where the morphology snap-shots of **b** have been taken. Two characteristic stages of morphogenesis are magnified

is shown in Fig. 21c. The most remarkable finding is the non-monotonous change of the internal energy, U, of the non-equilibrium system which is shown in Fig. 21d. At short times after the temperature jump, U increases due to detachment of chains at the edge of the crystal as a result of the changed attachment-detachment probability of the higher temperature. However, already after a rather short period of time, U displays a maximum (Max I) and then turns to decrease. The second picture in Fig. 21b shows the reason for this behavior: the temperature jump also accelerated processes in the interior of the finger structures, which locally improved the internal order (local increase of the height of the lamella) leading, in turn, to the formation of holes within the crystal. This "hole-rim" phase is a first metastable morphology caused by annealing. After an additional period of time the gain in U by improving chain order can no longer compensate for the still continuing loss in U by detaching chains at the periphery of the crystal. The internal energy starts to increase again. The third picture in Fig. 21b represents the time range of the morphogenesis where the internal energy shows a minimum (Min I). At later times, the fingers are broken up into fragments. The structure starts to decompose as shown by the fourth picture of the annealing series. At even longer times, a third process sets in which stops the structural decay. Parts of the remaining crystal structure take advantage of the locally higher concentration of chains (a consequence of the transformation of the original morphology) and start to form compact droplet-like patches which are more stable in the high temperature environment. The transition to more compact structures is reflected by a second maximum (Max II) in the internal energy. This droplet phase is the second meta-stable morphology caused by annealing. The droplets show various dynamic features such as surface diffusion, shape fluctuations and, coalescence processes which drive the evolution towards even larger and more stable structures (and lower internal energy). Note that the decrease of internal energy corresponds to an entropy increase of the heat bath, while the decreasing energy must be accompanied by an increasing disorder of the system. The droplet phase shows a liquid-like behavior at large length scales (shape fluctuation, coalescence) while at small length scales it corresponds to the state of highest ordered (fully stretched) chains. For more details, see [47].

Such liquid-like behavior at large length scales of crystallized polymers can be observed directly in experiments on PEO monolayers as we show in Figs. 22, 23 and 24.

The morphogenesis observed in simulations, and confirmed in experiments, proposes also novel interpretations of old experimental facts such as the "double peak" behavior of melting curves obtained in DSC experiments at constant heating rates [47]. Our studies on crystallization of quasi-two dimensional crystallization processes therefore provide new insights into the process of polymer crystallization in general.

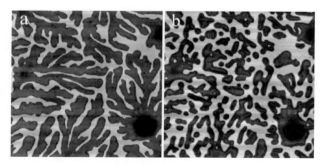

Fig. 22 Morphological transformations visible in a monolamellar crystal of PEO-7.6k grown at 35 °C for 10 min **a** before and **b** after the sample was annealed for 5 min at 45 °C. The size of the images is $5 \times 5\,\mu m^2$. Note the tendency for transformation of the tree-like shape of the fingers into an arrangement of circular (droplet-like) structures with a slightly higher region at the periphery. The defect zone in the right lower corner served as a marker for an unambiguous identification of the investigated area even after annealing. This allows us to follow the pathways within the "fingers" the system has chosen for the relaxation of the folded chains

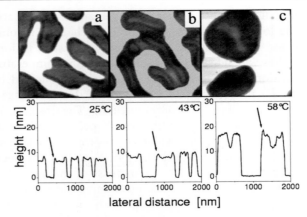

Fig. 23 Typical results obtained by tapping mode AFM for the relaxation of crystalline PEO-7.6k mono-lamellar crystalline domains. The size of the images is $1 \times 1\ \mu m^2$. **a** After crystallization at room temperature RT = 25 °C, **b** 5 min/43 °C, and **c** 5 min/54 °C. Increasingly darker shades of gray cover a range from 0 to 50 nm in height. Please note that the width of the slightly higher region at the periphery is increasing with annealing temperature. At the same time the tree-like shape transformed into an arrangement of circular (droplet-like) structures. Typical cross-sections corresponding to images **a**, **b**, and **c** are shown below. Note the elevations at the edges (*marked with arrows*) and the different heights of the crystalline domains

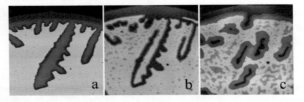

Fig. 24 Annealing close to the melting temperature: Tapping mode AFM images, measured at room temperature, showing the morphological changes induced by annealing finger-like patterns resulting from crystallization of monolayers of PS-PEO(3-3) in a pseudo-dewetted hole (*substrate is light, fingers are dark*). **a** Directly after crystallization at 45 °C, **b** after annealing for 1 min at 55 °C and **c** after annealing for 1 min at 56 °C. The size of the images is $10 \times 10\ \mu m^2$

5
Search for Ordering by Crystallization onto Patterned Substrates

The "finger-like" structures which we have discussed in the last section do not show any orientation. Here, we consider the possibility to obtain ordered structures from growth processes using anisotropic substrates. Such substrate properties can be achieved by patterning the surface with periodic or aperiodic arrays of stripes (lanes) with an alternating affinity with respect to the adsorbed chains as shown in Fig. 25. Similar pre-patterned substrates have

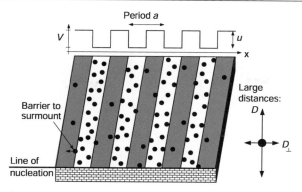

Fig. 25 A striped surface is modeled by a periodic surface potential $V(x)$ which defines the interaction between a polymer chain and the surface. Polymers deposited onto the substrate will concentrate in the lanes of higher affinity. If a chain changes from a lane of higher affinity to a lane of lower affinity, a potential barrier u has to be surmounted. On scales larger than the periodicity a of the stripe pattern the diffusivity becomes anisotropic. In our simulations we use 400 lanes with a period of 12 lattice units and a width of 6 lattice units per lane

been recently used to obtain highly ordered copolymer soft-order [39]. Our simulation model for polymer crystallization is perfectly suitable to test the possibility to align polymer crystals, to investigate the resulting morphologies and to optimize the parameters. A deeper understanding of the underlying physical processes can help further to develop new strategies for creating ordered nano-structures using polymer crystallization.

In the subsequently shown simulation results we considered a setup as sketched in Fig. 25. The preferred growth of the crystals is along the lanes of higher affinity but the period of the stripe pattern is so small that exchange of chains between different lanes is indispensable. As a consequence, the transport properties differ according to the direction of the displacements: fast diffusion along the lanes, denoted by D, and slow diffusion crossing the lane boundaries, denoted by $D_\perp$. Crystallization now proceeds in the direction along the lanes which is induced by a nucleation line oriented perpendicular to the lanes. The difference in property between low and high affinity lanes is introduced by a potential barrier u, see Fig. 25. If a particle tries to enter into a region of low affinity, the necessary move is accepted only with a probability

$$p_\perp = e^{-u/kT}. \qquad (1)$$

In the following we restrict our discussion to the case of $p_\perp = 0.2$. For more details, see [7]. The initial coverage of the lattice has to be smaller than unity which corresponds to an incomplete occupation of the low-affinity lanes.

In Fig. 26 we compare the growth patterns for our striped substrate with the corresponding isotropic (non patterned) substrate. It can be seen clearly that both morphologies are quite different. While on the non-patterned sub-

Fig. 26 A snapshot is shown of the crystal morphology obtained after 64 000 MCS (64 TMS) for the pre-patterned substrate (*lower part*) compared to an isotropic substrate (*upper part*). The model parameters are $p_S = 0.6$, $p_0 = 0.25$ with an initial melt coverage of 2/3. With our choice of $p_\perp$ this gives an initial occupation of 0.5 and 0.83 for the low-affinity lanes and the high-affinity lanes, respectively

strate the growth results in thick fingerlike structures, the patterned surface exhibits highly directed anisotropic needle-crystal growth for which the main branch of each crystal is located on a high affinity lane. Although side branching occurs frequently, these branches are always much less developed in comparison to the main branch. The dominance of the main branch results in forward oriented growth of each crystal. Note that in contrast to the isotropic substrate where a significant amount of free polymers are still present, the free particle reservoir on the pre-patterned substrate is almost completely exhausted. This demonstrates the faster and more efficient growth under anisotropic diffusion conditions.

The possibility of crossing lanes is responsible for the instability (sidebranching) of the growth front. The faster growth along the lanes where no potential barriers have to be surmounted leads to preferential orientation of the fingers in the direction of higher mobility. In addition, the competition between neighboring growing fingers for available free molecules is responsible for a *hierarchy of characteristic length scales* in the direction perpendicular to the growth direction. In this competition some crystals survive. The distance between these surviving crystals displays a characteristic length scale as indicated by the structure function in Fig. 27. The structure function is calculated in the direction perpendicular to the fingers for different distances from the seed (starting point). As can be already seen in the lower part

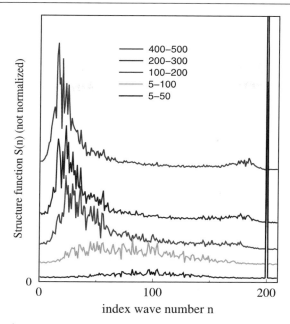

Fig. 27 Structure function $S(n)$ in the direction perpendicular to the fingers as a function of the index of the wave number $q = 2\pi n/L$ (L – lattice size). The different curves correspond to measurements at different distances from the seed (starting point) as indicated in the figure in lattice units. With increasing distance to the seed the maximum gets sharper and is shifter to the left, i.e. to larger distances

of Fig. 26, a gradient pattern is obtained as a result of the competition for free chains. Close to the seed almost all high affinity lanes are crystallized. Later on, during growth fewer and fewer of these crystals survive. Exactly this effect leads to the shift of the maximum of the structure function in Fig. 27 when taken at larger distances from the seed. Moreover, a closer analysis reveals that the distribution of surviving fingers is not random [7], that is the distance between growing crystals becomes larger and more periodic (regular) as growth proceeds. This is also reflected in Fig. 27 by the sharpness of the peaks at distances of more than 100 lattice units from the seed.

To conclude, we have shown that on a pre-patterned substrate the competition between growing crystals for the surrounding melt chains leads to characteristic, well defined, and to some extent even regular and ordered crystal morphologies. For the growth parameters used in this work, these morphologies are characterized by periodic features which are an order of magnitude larger than the initial periodicity which was imposed onto the patterned substrate. To us, the use of anisotropic substrates seems to represent a promising way to create regular surface structures based on fast, diffusion controlled pattern formation processes.

6
Conclusions

It was the aim of this work to highlight a few aspects of spontaneous pattern formation in thin polymer films with the focus on the formation of long-range ordered nano-structured surfaces. Here, we have focused on two aspects: microphase separation in diblock copolymers and crystallization. The major challenge in using microphase separation in block copolymers to form ordered surface patterns is to understand and control the interplay between surface effects such as preferential adsorption of one species and geometric constraints with the process of microphase ordering. Perpendicular ordering of lamellar patterns for instance is induced by neutral surfaces. This can be explained with the particular pathway copolymers have to take into the ordered state where chains are stretched and oriented before long-range order is established. Neutral surfaces facilitate stretching of chains parallel to the surface, thus leading to lamellar ordering perpendicular to the surface. Yet, the resulting patterns contain many defects and further ordering mechanisms are necessary to obtain long-range order. Candidates for such secondary ordering processes are crystallization or the implication of temperature gradients. Computer simulations also show that defects in the soft-order of block copolymers are difficult to heal since the energy effort to create defects is comparatively low but the free energy barrier between defect states and perfectly ordered states is very high. Thus, generally speaking, meta-stable states can be very persistent.

One strategy to control long-range order is to use crystallization effects. Crystallization of one component in block copolymers can lead to transformations between different copolymer morphologies and reorientations of existing microphase separation patterns. In thin films, three phase contact lines can be used to align patterns resulting from crystallization of one component.

Generally, polymer crystallization yields meta-stable patterns which are particularly pronounced in thin films where diffusion-controlled structures can be formed. Moreover, polymer crystallization in thin films opens new perspectives to study the crystallization process experimentally as well as theoretically. Since crystallization is the strongest driving force for pattern formation in polymers we have investigated some general aspects using simplified models and computer simulations. We have demonstrated that polymer crystallization in thin films can be mapped into a combined growth and reorganization process using a simple lattice algorithm. When the substrate is not isotropic (anisotropic diffusion of adsorbed molecules), needle crystals emerge instead of randomly branched morphologies. These needle crystals grow much faster compared with the isotropic system. Diffusion control leads to selection of needle crystals and results in a characteristic "survival" distance between the crystals. Thus, ordered structures can be obtained from growth processes on anisotropic substrates.

Growth processes as they emerge from crystallization have an enormous potential for pattern formation at multiple length scales. Again, the example of polymer crystallization can be applied to demonstrate this feature: in the bulk, banded, spherulitic patterns are formed which display order at scales up to the cm-scale. A deeper understanding of non-equilibrium growth mechanisms will be an essential key for creating ordered structures in polymer systems.

7
Appendix: Experimental Procedures

For our experimental studies on polymer crystallization we used low molecular weight (M_w) poly(ethylene oxide) (PEO), either as a homopolymer or attached to an amorphous polystyrene or hydrogenated polybutadiene block. These block copolymers are abbreviated by PS-PEO and PB_h-PEO, respectively. PEO is a well-investigated polymer [5, 19, 36, 37]. Molecular details of all investigated polymers are given in Table 1.

For many of the crystallization experiments presented here we used adsorbed polymer monolayers. To obtain such monolayers the following procedure was applied. First, we spincoated a dilute toluene solution of dried polymer onto a clean silicon wafer, resulting in thin films of a thickness varying between about 50 and 100 nm. These films were annealed in the molten state which led to autophobic pseudo-dewetting [36, 37]. Thereby, cylindrical holes were formed in the film. These holes contained an adsorbed monolayer at their bottom [37]. The autophobic behavior is caused by rather strong adsorption due to the hydroxyl groups present on the surface of the Si-wafer. These hydroxyl groups were created by a UV-ozone treatment and allowed for strong interaction with PEO, thus favoring strong adsorption. The resulting thin films were first molten for about 10 min at temperatures above 70 °C to allow for the formation of a monolayer. Then, the monolayer was crystallized

Table 1 Characteristics of the polymers used in our studies

Sample	M_w (PS or PB_h)	M_w (PEO)	M_w/M_n
PEO-2k	–	2000	< 1.1
PEO-7.6k	–	7600	< 1.1
PS-PEO(3–3)	3000	3000	1.1
PB_h-PEO(3.7-1.1)	3700	1100	< 1.1
PB_h-PEO(3.7-2.9)	3700	2900	< 1.1
PB_h-PEO(3.7-3.6)	3700	3600	< 1.1
PB_h-PEO(21-4)	21 100	4300	1.15

at temperatures T_c between room temperature and the melting temperatures of the various polymers used. All samples were crystallized at constant temperature. In order to check for relaxation processes within the crystalline state, samples were sequentially annealed for some minutes at temperatures higher than T_c, but always below the equilibrium melting point (i.e. 64 °C for PEO-7.6k and 57 °C for PS-PEO(3-3)). After each annealing step, the sample was quickly cooled to room temperature and analyzed by Atomic Force Microscopy (AFM).

In order to obtain rather well-ordered amorphous mesostructures, thin PB_h-PEO(21-4), PS-PEO(3-3), PB_h-PEO(3.7-2.9), or PB_h-PEO(3.7-3.6) films were annealed in the molten state at temperatures up to 150 °C before crystallization experiments were performed. It is important to mention that in contrast to PS-PEO(3-3), PB_h-PEO(3.7-2.9), or PB_h-PEO(3.7-3.6) samples *annealed* PB_h-PEO(21-4) samples did *not* crystallize at ambient conditions, even after storage at room temperature for many months. Only after cooling the samples to temperatures below about −20 °C, was crystallization observed. Melting, on the other hand, occurred at temperatures above about +40 °C. Consequently, at room temperature the samples neither crystallized nor melted. This fact enabled AFM measurements of partially crystallized samples at room temperature, preserving the structure obtained by crystallization at low temperatures.

All samples were first analyzed by optical microscopy. Thermal treatment was also performed under the optical microscope. The samples were placed onto an enclosed hotplate, purged with nitrogen, under a Leitz-Metallux 3 optical microscope. The crystallization temperature at the hot stage was controlled to within 0.1 degrees. No polarization or phase contrast was used. Contrast is due to the interference of the reflected white light at the substrate/film and film/air interface, resulting in well-defined interference colors which can be calibrated with a resolution of about 10 nm. This is sufficient to allow us to visualize the growth of crystalline structures in the monolayer region. For the thickness range studied in this work, the interference patterns became the darker the thicker the film (or crystals) were. We have followed the progression of the crystal growth front in real time by capturing images with a CCD camera.

Atomic Force Microscopy (AFM) measurements were performed with a Nanoscope IIIa/Dimension 3000 (Digital Instruments) in the tapping mode at ambient conditions, using the electronic extender module simultaneously allowing phase detection and height imaging. We used Si-tips (model TESP) with a resonant frequency of about 300 kHz. Scan-rates were between 0.2 and 4 Hz. The free oscillation amplitude of the oscillating cantilever was typically around 50 nm, the setpoint amplitude (damped amplitude, when the tip was in intermittent contact) was slightly lower. In all measurements, topographic (height mode) and viscoelastic (phase-mode) data were recorded simultaneously. It should be noted that semicrystalline polymers are well suited for the use of the "phase-mode" as the differences in viscoelastic properties between crystalline and amorphous regions are large [35].

Most of the AFM measurements were done at ambient conditions. However, when trying to melt nanometer sized polymer crystals we raised the temperature of the oscillating cantilever of the AFM (up to nominally 80 °C) as well as the sample itself (up to nominally 40 °C).

Acknowledgements We kindly acknowledge the POLYNANO collaboration (Research and Training Network, HPRN-CT1999-00151). During the last five years we have enjoyed many fruitful discussions with all members of this collaboration and the results of this work are closely related to our common work within POLYNANO. We wish to thank Drs. C. Vasilev and G. Dorenbos for their contributions to this work.

References

1. Bassereau P, Brodbreck D, Russell T, Brown H, Shull K (1993) Topological coarsening of symmetric diblock copolymer films: Model 2d systems. Phys Rev Lett 1(11):1716–1719
2. Bates FS, Fredrickson GH (1990) Block copolymer thermodynamics—theory and experiments. Annu Rev Phys Chem 41:525–557
3. Bodycomb J, Funaki Y, Kimishima K, Hashimoto T (1999) Single-grain lamellar microdomain from a diblock copolymer. Macromolecules 32(6):2075–2077
4. Carmesin I, Kremer K (1988) The bond fluctuation method: A new effective algorithm for the dynamics of polymers in all spatial dimensions. Macromolecules 21(9):2819–2823
5. Cheng S, Wu S, Hen JH, Zhuo Q, Quirk R, von Meerwall E, Hsiao B, Habenschuss A, Zschak P (1993) Isothermal thickening and thinning processes in low molecular weight poly(ethylene oxide) fractions crystallized from the melt. 4. End-group dependence. Macromolecules 26(19):5105–5117
6. Coulon G, Collin B, Ausserre D, Chatenay D, Russel T (1990) Islands and holes on the free-surface of thin diblock copolymer films: 1. Characteristics of formation and growth. Journal de Physique 51(24):2801–2811
7. Dorenbos G, Sommer J-U, Reiter G (2003) Polymer crystallization on pre-patterned substrates. J Chem Phys 118(2):784–791
8. Fried H, Binder K (1991) The microphase separation transition in symmetric diblock copolymer melts: A Monte Carlo study. J Chem Phys 94(15):8349–8366
9. Fried H, Binder K (1991) Non-gaussian conformational behavior in diblock copolymer melts: Is the rpa valid? Eur Phys Lett 16(3):237–242
10. Hahm J, Lopes W, Jaeger H, Silbener S (1998) Defect evolution in ultrathin films of polystyrene-block-polymethylmethacrylate diblock copolymers observed by atomic force microscopy. J Chem Phys 109(23):10111–10114
11. Hahn H, Lee JH, Balsara NP, Garetz BA, Watanabe H (2001) Viscoelastic properties of aligned block copolymer lamellae. Macromolecules block copolymers, shear alignment. 34(25):8701–8709
12. Hamley IW, Fairclough JPA, Terrill NJ, Ryan AJ, Lipic PM, Bates FS, Towns-Andrews E (1996) Crystallization in oriented semicrystalline diblock copolymers. Macromolecules 29(27):8835–8843

13. Harrison C, Adamson D, Cheng Z, Sebastian J, Sethuraman S, Huse D, Register R, Chaikin P (2000) Mechanisms of ordering in striped patterns. Copolymer, ordering, patterns. Science 290(5496):1558–1560
14. Helfand E, Wassermann ZR (1976) Block copolymer theory. 4. Narrow interphase approximation. Macromolecules 9(6):879–888
15. Hoffmann A, Sommer JU, Blumen A (1997) Statics and dynamics of dense copolymer melts: A Monte Carlo simulation study. J Chem Phys 106(16):6709–6721
16. Kellogg GJ, Walton DG, Mayes DG, Lambooy P, Russell TP, Gallagher PD, Satija SK (1996) Observed surface energy effects in confined diblock copolymers. Phys Rev Lett 76(14):2503–2506
17. Kikuchi M, Binder K (1994) Microphase separation in thin films of the symmetric diblock-copolymer melt. J Chem Phys 101(4):3367–3377
18. Kim S, Solak H, Stoykovich MP, Ferrier N, de Pablo J, Nealey P (2003) Epitaxial self-assembly of block copolymers on lithographically defined nanopatterned substrates. Nature 424(6947):411–414
19. Kovacs A, Straupe C (1980) Isothermal growth, thickening, and melting of poly(ethylene oxide) single-crystals in bulk. 3. Bilayer crystals and the effect of chain ends. J Cryst Growth 48(2):210–226
20. Leibler L (1980) Theory of microphase separation in block copolymers. Macromolecules 13:1602–1617
21. Loo YL, Register R, Ryan T (2000) Polymer crystallization in 25-nm spheres. Phys Rev Lett 84(18):4120–4123
22. Lotz B, Kovacs A (1969) Phase transitions in block-copolymers of polystyrene and polyethylene oxide. ACS Polym Prepr 10:820–825
23. Mansky P, Liu Y, Huang E, Russell TP, Hawker C (1997) Controlling the polymer-surface interactions with random copolymer brushes. Science 275:1458–1460
24. Mansky P, Russell TP, Hawker CJ, Mays J, Cook DC, Satija SK (1997) Interfacial segregation in disordered block copolymers: Effect of tuneable surface potential. Phys Rev Lett 79(2):237–240
25. Mansky P, Russell TP, Hawker CJ, Pitsikalis M, Mays J (1997) Ordered diblock copolymer films on random copolymer brushes. Macromolecules 30:6810–6813
26. Matsen MW (1997) Thin films of block copolymer. J Chem Phys 106(18):7781–7789
27. Meakin P (1998) Fractals, scaling and growth far from equilibrium. Volume 5 of Cambridge Nonlinear Science Series. Cambridge University Press, Cambridge
28. Miyashita S, Saito Y, Uwaha M (1997) Experimental evidence of dynamical scaling in a two-dimensional fractal growth. J Phys Soc Japan 66(4):929–932
29. Morkved T, Lu M, Urbas A, Ehrichs E, Jaeger H, Mansky P, Russell T (1996) Local control of microdomain orientation in diblock copolymer thin films with electric fields. Science 273:931–933
30. Morkved TL, Jaeger HM (1997) Thickness-induced morphology changes in lamellar diblock copolymer ultrathin films. Europhys Lett 40(6):643–648
31. Murat M, Grest GS, Kremer K (1999) Statics and dynamics of symmetric diblock copolymers: A molecular dynamics study. Macromolecules 32(3):595–609
32. Olmsted PD, Poon WCK, McLeish TCB, Terrill NJ, Ryan AJ (1998) Spinodal-assisted crystallization in polymer melts. Phys Rev Lett 81(2):373–376
33. Picket GT, Witten TA, Nagel SR (1993) Equalibrium surface orientation of lamellae. Macromolecules 26:3194–3199
34. Reiter G, Castelein G, Hoerner P, Riess G, Sommer JU, Floudas G (2000) Morphologies of diblock copolymer thin films before and after crystallization. Eur Phys JE 2:319–334

35. Reiter G, Castelein G, Sommer HU, Röttele A, Thurn-Albrecht T (2001) Direct visualization of random crystallization and melting in arrays of nanometer-size polymer crystals. Phys Rev Lett 87(22):226101
36. Reiter G, Sommer JU (1998) Crystallization of adsorbed polymer monolayers. Phys Rev Lett 80(17):3771–3774
37. Reiter G, Sommer JU (2000) Polymer crystallization in quasi-2 dimensions: Part i: Experimenal results. J Chem Phys 112(9):4376–4383
38. Reiter G, Vidal L (2003) Crystal growth rates of diblock copolymers in thin films: Influence of film thickness. Eur Phys JE 12:497–505
39. Rockford L, Mochrie S, Russell T (2001) Propagation of nanopatterned substrate templated ordering of block copolymers in thick films. Macromolecules 34(5):1487–1492
40. Rosa CD, Park C, Thomas E, Lotz B (2000) Microdomain patterns from directional eutectic solidification and epitaxy. Nature 405(6785):433–437
41. Russell TP (1996) On the reflectivity of polymers: Neutrons and X-rays. Physica B 221:267–283
42. Schaffer E, Thurn-Albrecht T, Russell TP, Steiner U (2000) Electrically induced structure formation and pattern transfer. Nature 403(6772):874–877
43. Segalman R, Yokoyama H, Kramer E (2001) Graphoepitaxy of spherical domain block copolymer films. Adv Mater 13(15):1152–1155
44. Singh N, Kudrle A, Sikka M, Bates F (1995) Surface-topography of symmetrical and asymmetric polyolefin block-copolymer films. Journal de Physique II 5(3):377–396
45. Sommer JU, Hoffmann A, Blumen A (1999) Block copolymer films between neutral walls: A monte carlo study. J Chem Phys 111(8):3728–3732
46. Sommer JU, Reiter G (2000) Polymer crystallization in quasi-2 dimensions: Part ii: Kinetic models. J Chem Phys 112(9):4384–4393
47. Sommer JU, Reiter G (2001) Morphogegesis of lamellar polymer crystals. Europhys Lett 56(5):755–761
48. Strobl G (2000) From the melt via mesomorphic and granular crystalline layers to lamellar crystallites: A major route followed in polymer crystsallization? Eur Phys JE 3:165–183
49. Thurn-Albrecht T, Schotter J, Kastle CA, Emley N, Shibauchi T, Krusin-Elbaum L, Guarini K, Black CT, Tuominen M, Russell TP (2000) Ultrahigh-density nanowire arrays grown in self-assembled diblock copolymer templates. Science 290:2126–2129
50. Tian MW, Loos J (2001) Investigations of morphological changes during annealing of polyethylene single crystals. J Polym Sci B 39(7):763–770
51. Tracz A, Jeszka JK, Watson MD, Pisula W, Mullen K, Pakula T (2003) Uniaxial alignment of the columnar super-structure of a hexa (alkyl) hexa-peri-hexabenzocoronene on untreated glass by simple solution. J Am Chem Soc 125:1682
52. Tsori Y, Andelman D (2001) Diblock copolymer thin lms: Parallel and perpendicular lamellar phases in the weak segregation limit. Eur Phys JE 5:605–614
53. Turner MS (1992) Equilibrium properties of a diblock copolymer lamellar phase confined between flat plates. Phys Rev Lett 69(12):1788–1791
54. Turner MS, Johner A, Joanny JF (1995) Wetting behaviour of thin diblock films. J Physique I 5:917–932
55. Uwaha M, Saito Y (1989) Aggregation growth in a gas of finite density: Velocity selection via fractal dimension of diffusion limited aggregation. Phys Rev A 40(8):4716–4723
56. Vasilev C, Heinzelmann H, Reiter G (2004) Controlled melting of individual, nanometer-sized, polymer crystals confined in a block copolymer mesostructure. J Polym Sci B Polym Phys 42(7):1312–1320

57. Wang H, Newstein MC, Krishnan A, Balsara NP, Garetz BA, Krishnamoorti BHR (1999) Ordering kinetics and alignment of block copolymer lamellae under shear flow. Macromolecules 32(11):3695–3711
58. Weyersberg A, Vilgis TA (1993) Phase transitions in diblock copolymers: Theory and Monte Carlo simulations. Phys Rev E 48(1):377–390
59. Winkel A, Hobbs J, Miles M (2000) Annealing and melting of long-chain alkane single crystals observed by atomic force microscopy. Polymer 41(25):8791–8800

Designed Block Copolymers for Ordered Polymeric Nanostructures

Nikos Hadjichristidis[1] (✉) · Stergios Pispas[1,2]

[1]Department of Chemistry, University of Athens, Panepistimiopolis Zografou, 15771 Athens, Greece
hadjichristidis@chem.uoa.gr, pispas@eie.gr

[2]Theoretical and Physical Chemistry Institute, National Hellenic Research Foundation, Vas. Constantinou Ave., 11635 Athens, Greece

1	Introduction	38
2	Synthesis of Model Block Copolymers and Self-Organization in Solution and in Bulk	41
2.1	Block Copolymers with Functional Groups at Specific Sites	41
2.1.1	End-Functionalized Diblock Copolymers	41
2.1.2	Junction-Point-Functionalized Diblock Copolymers	43
2.1.3	Postpolymerization Functionalization	43
2.2	ABC Miktoarm Stars	46
2.3	Linear ABC Terpolymers	48
2.4	Amphiphilic Block Copolymers	50
2.5	Amorphous-Side-Chain Liquid Crystalline Block Copolymers	52
3	Conclusions	53
	References	54

Abstract Design concepts and strategies concerning the preparation of a variety of block copolymer architectures are presented. These aspects are discussed in relation to the self-organization of novel block copolymers on surfaces, utilizing a series of self-assembling processes, including microphase separation, aggregation, amphiphilicity, crystallization, and liquid-crystal formation.

Keywords Anionic polymerization · Block copolymers · Microphase separation · Self-assembly · Surfaces

Abbreviations
ABA   Linear triblock copolymer
ABC   Linear triblock terpolymer
AFM   Atomic force microscopy
DMADPE 1-(4-Dimethylaminophenyl)-1-phenylethylene
DMAPLi 3-(Dimethylamino)propyl lithium
DTMAB Dodecyltrimethylammonium bromide
MMA   Methyl methacrylate
MPPHM 6-[4-(4-Methoxyphenyl)phenoxy]hexyl methacrylate
n-BuLi *n*-Butyl lithium

ODT	Order-disorder transition
P2VP	Poly(2-vinylpyridine)
PB	Polybutadiene
PBO	Poly(butylene oxide)
PCL	Poly(ε-caprolactone)
PEO	Poly(ethylene oxide)
PHEMA	Poly(2-hydroxyethyl methacrylate)
PI	Polyisoprene
PILi	Polyisoprenyl lithium
PMMA	Poly(methyl methacrylate)
PS	Polystyrene
PSLi	Polystyryl lithium
PtBMA	Poly(*tert*-butyl methacrylate)
RI	Refractive index
SAXS	Small-angle X-ray scattering
s-BuLi	*Sec*-Butyl lithium
SDS	Sodium dodecylsulfate
SEC	Size-exclusion chromatography
TEM	Transmission electron microscopy
T_g	Glass transition
THF	Tetrahydrofuran
UV	Ultraviolet

1
Introduction

Block copolymers can be microphase separated in bulk, on surfaces and thin films to give fascinating nanostructured morphologies or can self-associate in selective solvents to form supramolecular structures (micelles) with interesting properties that can serve as models for molecular assembly and molecular recognition in biological systems [1-5]. Manipulating morphology and self-organization of copolymers can lead to tailor-made materials for applications. Our understanding of natural laws will benefit from the study of model systems, and the results of such studies will aid the design of useful products.

The current interest in the self-assembly of polymeric systems on surfaces stems from the many opportunities that these assemblies present for the preparation of novel functional nanomaterials, i.e., for drug delivery, in catalysis and nanoreactor technology, and for molecular templating. The interesting aspect of these systems is that their properties and structure can be manipulated by a number of parameters such as: a) chemical structure, composition, and architecture, b) preparation methods and microengineering techniques used, and c) nature and properties of the underlying substrate and its interactions with the polymer chains [1-5].

So far a large number of studies has been published dealing with the behavior of block copolymers in thin films and surfaces [4-13]. Most of them concern the properties of simple diblock copolymers. Studies in this area were devoted to block copolymer microphase separation in reduced dimensionality, i.e., in the presence of one or two boundary surfaces or in ultrathin films (thickness lower than the characteristic structural dimensions in bulk) and to establishing the relationships between block copolymer assembly under these conditions and the molecular characteristics of the copolymers. Directional microdomain growth due to the presence of a solid substrate, the evolution of the microstructure as a function of time, and the air/material surface structure were also investigated.

Fewer studies have been devoted to the behavior of more complex macromolecular structures. The architectures that have attracted the interest of investigators lately are those of linear triblock copolymers (ABA) and terpolymers (ABC).

Investigations on the self-assembly in thin films of an amorphous P2VP-PS-P2VP triblock copolymer, forming cylindrical microdomains in bulk, showed that the orientation of microdomains due to the P2VP/substrate interactions persisted in the entire film in contrast to the diblock case [14]. This was viewed as a result of the formation of an interconnected structure in the triblock coming from the formation of loops within the microdomains. More recently, AFM and SAXS measurements on a PEO-PBO-PEO amorphous-semicrystalline triblock thin film revealed the presence of a semicrystalline PEO monolayer at the substrate, comprised of unfolded chains, and PBO blocks at the air/polymer surface in a looped conformation [15].

A surface reconstruction was observed in thin films of a PS-PB-PMMA triblock terpolymer with a low fraction of PB, the first study devoted to ABC triblocks [16]. In this sample the surface had a number of defects including curved lamellae and disclinations. The lamellae period on the free surface was double in comparison to the bulk value. Surface reconstructions were also observed during more recent investigations on PS-PB-PMMA, PB-PS-PMMA, and PS-PB-P*t*BMA triblock terpolymers. The variation in block copolymer architecture and chemical nature of the blocks produced changes in the interactions between blocks and polymer chains and confining surfaces (due to different surface energies of the components) [17, 18]. In this way, it was possible to show that the observed reconstructions are caused by a complex balance of enthalpic and entropic contributions to the free energy of the system. An analogy with classical crystals was made by the authors.

The microdomain morphology of thin films from a PS-P2VP-P*t*BMA triblock terpolymer was investigated through swelling with solvents having different affinities for the three blocks [19–22]. In this case the polar P2VP middle block is adsorbed on the polar Si/SiO_2 substrate. An ABCBA lamellae structure could be identified as the equilibrium structure. The treatment with different solvents resulted in different nonequilibrium surface structures as well as in different lamellae spacings. Microdomain evolution starts from the surface and proceeds throughout the film. When THF and $CHCl_3$ were used with a more asymmetric PS-P2VP-P*t*BMA copolymer, different morphologies could be identified depending on the solvent nature and treatment. Additionally, the solvent drying rate was found to affect the orientation of the domains. Slow solvent drying resulted in a lamellae alignment parallel to the substrate, whereas fast drying created a lamellae orientation perpendicular to the substrate. Mechanical stresses applied to block copolymer phases during the evaporation process were suggested to be responsible for the observed behavior. The effect of the nature of the substrate on the self-assembly of the same triblock terpolymers was also investigated [23]. The different interaction strengths between the P2VP middle block and the Si or polyimide surface resulted in different configurations of the P2VP block and hence a different structure of the layer in contact with the substrate. This particular structure of the first layer influenced the lamellae structure within the polymer film.

In a systematic study involving a series of PS-P2VP-PMMA and PS-PHEMA-PMMA terpolymers, where the substrate-absorbing middle block was varied in length, it was shown that the lateral spacing and the morphology of the microphase-separated structures could be controlled by the film thickness and the lengths of the end blocks [24].

In another study thin patterned films of PS-P2VP-PMMA block terpolymers were used as substrates for PS/PMMA homopolymer blends [25]. It was found that for thin enough films of the blend phase, separation of the homopolymers was suppressed.

The nanopatterned surfaces obtained by PS-PB-PCL block terpolymer films on Si and mica were investigated by AFM [26]. The polymer surface of the films was found to be covered with amorphous and crystalline PCL. The PCL/PB interfaces were parallel to the substrate.

The large variety of surface structures observed so far in the aforementioned block copolymer systems indicates the large potential of tailoring polymer surfaces by a combination of well-defined block copolymer, architecturally complex materials, as well as, by a number of physicochemical parameters of the systems. Studies along these lines are definitely needed since the full potential of the existing and new block copolymer systems has not been evaluated.

Synthetic efforts along these lines are focused on the synthesis of block copolymers having various architectures, thereby providing promising materials for achieving the desired goals, in terms of properties of polymeric surfaces [27, 28]. In the case where nanopatterned surface structures are desired, the chemical structure of the polymers can be designed in a way that will facilitate the self-assembly of the molecules on substrates. In most cases these macromolecules contain one part that is able to interact with model surfaces and one or two that interact with the environment, crystallize, or form mesostructures due to incompatibility reasons. The concept of altering macromolecular-substrate interactions and structure formation due to environmental changes on the free surface of the films is also taken into account in the design of the chemical systems. A number of principles, directions, and examples can be drawn from the self-organization of macromolecules in solution and in bulk. The synthetic goals are accomplished by the choice and variation of the chemical nature of each part of the copolymers and architecture of the macromolecule using appropriate synthetic methodologies. Among them, living anionic polymerization methods have been proven to be the most powerful and successful [29, 30].

2
Synthesis of Model Block Copolymers and Self-Organization in Solution and in Bulk

2.1
Block Copolymers with Functional Groups at Specific Sites

Nonpolar block copolymers with functional polar groups combine the properties of a block copolymer with those of functionalized homopolymers. End functionalized polymers can self-associate in solution in a manner analogous to low-molecular-weight surfactants, whereas the solid-state properties of the precursor polymer are greatly influenced by the presence of small amounts of polar groups since their aggregation persists in bulk [31]. The interaction of a polymer chain with a surface in solution can be altered and sometimes controlled by the nature and placement of specific functional groups.

Linear block copolymers of styrene and isoprene with polar dimethylamino, sulfobetaine, or phosphorous zwitterionic-liquid crystalline groups in predetermined positions in the polymeric chain can be synthesized by anionic polymerization high-vacuum techniques. It is well known that anionic polymerization produces polymers with predetermined molecular weights, narrow molecular weight/compositional distributions and well-defined architectures. These polar groups can be introduced at one chain end (at the polystyrene or the polyisoprene side), at both ends, or at the junction point between the blocks. Due to their polar nature, among other things, they can interact with polar groups on surfaces and influence polymer chain behavior near a surface. The synthetic strategies followed in each case are given below.

2.1.1
End-Functionalized Diblock Copolymers

For the synthesis of end-functionalized block copolymers with one dimethylamino end group 3-(Dimethylamino)propyl lithium (DMAPLi) was used as initiator, in a nonpolar solvent (i.e., benzene). The diblocks were prepared by sequential addition of monomers starting with the block to which the end group was going to be linked [32, 33]. In the case where the dimethylamino group is at the PI chain end isoprene is polymerized first. After completion of the polymerization of the first block, styrene is added to the polymerization mixture together with a small quantity of tetrahydrofuran in order to promote the crossover reaction. The polydienes prepared under these conditions have a large 1,4 content and subsequently a low glass transition temperature. The reaction sequence is given in Scheme 1.

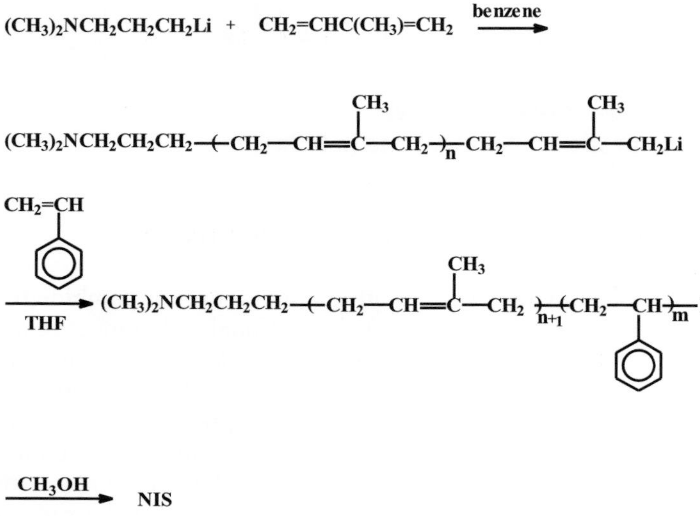

Scheme 1

The synthetic scheme outlined above can be used for the preparation of triblock copolymers with two dimethylamino end groups if the living functional diblock chains are coupled with Me_2SiCl_2.

The synthesis of the diblock copolymers with two dimethylamino end groups was accomplished by following the synthetic route illustrated in Scheme 2 [34]. Isoprene was polymerized first using DMAPLi as initiator followed by the addition of styrene and a small amount of THF in order to accelerate the crossover reaction. After completion of the styrene polymerization, an excess of 1-(4-dimethylaminophenyl)-1-phenylethylene (DMADPE), vacuum-distilled from n-BuLi, was introduced to the polymerization mixture as a solution in benzene (DMADPE/Li = 1.2/1). The capping reaction between PSLi living ends and DMADPE was allowed to proceed for 3 d at room temperature. The living chains were deactivated with degassed methanol.

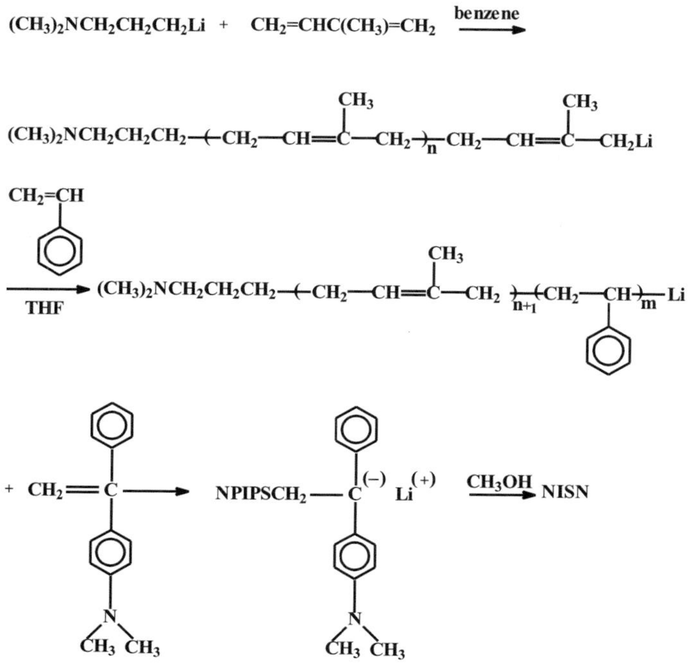

Scheme 2

2.1.2
Junction-Point-Functionalized Diblock Copolymers

Junction-point-functionalized block copolymers were synthesized according to Scheme 3 [34]. Styrene was polymerized first using s-BuLi as initiator. After completion of polymerization a small excess of DMADPE was introduced to the reaction mixture (DMADPE/Li = 1.2/1). The reaction was left for completion for 3 d at room temperature. Isoprene was then added, polymerized, and terminated by addition of degassed methanol. In the case of the triblock copolymer synthesis, after completion of isoprene polymerization, a predetermined amount of a Me_2SiCl_2 solution in benzene was introduced to the reaction mixture (Cl/Li = 1/2.2). The coupling reaction was essentially complete in 3 d. Solvent/nonsolvent fractionation was employed in order to separate the triblock from excess diblock.

2.1.3
Postpolymerization Functionalization

(i) Conversion of the tertiary amine groups to sulfobetaine zwitterions: The dimethylamino groups, located at the ends or the junction point of the diblock

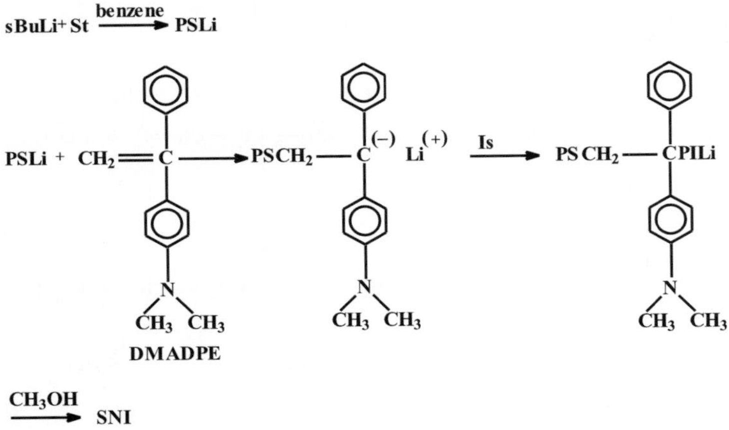

Scheme 3

copolymer chain, were converted to sulfobetaine zwitterions by reaction with excess cyclopropanesultone [32–34]. The reaction took place in dilute THF solutions at 70 °C for several days to ensure complete conversion (Scheme 4).
(ii) Conversion of the tertiary amine groups to zwitterionic-mesogenic end groups: The conversion of the terminal amine group to a zwitterionic-mesogenic one was achieved by reaction with excess 2-{6-[4-(4-methoxyphenylazo)phenoxy]hexyloxy}-2-oxo-1,3,2λ^5-phospholane (amine/phospholane = 1/10) [35]. The latter compound was synthesized by following procedures found in the literature. A 2% w/v solution of the amine-capped polymer in THF was allowed to react for several days at 70 °C under a nitrogen atmosphere according to the following Scheme 5.

In all cases ^{1}H-NMR spectroscopy was used in order to check the yield of the transformation reaction.

The association behavior of the functionalized block copolymers was investigated in detail in nonpolar solvents selective for the macromolecular chains. In this case starlike aggregates were formed as determined by the hydrodymanic properties of the aggregates [31–35]. Their aggregation number decreased as the molecular weight of the polymer increased (i.e., the weight fraction of the insoluble polar chains decreased) due to excluded volume repulsions. In the case of nonpolar solvents selective for the block that does not carry the functional polar group, aggregation numbers and the size of the mi-

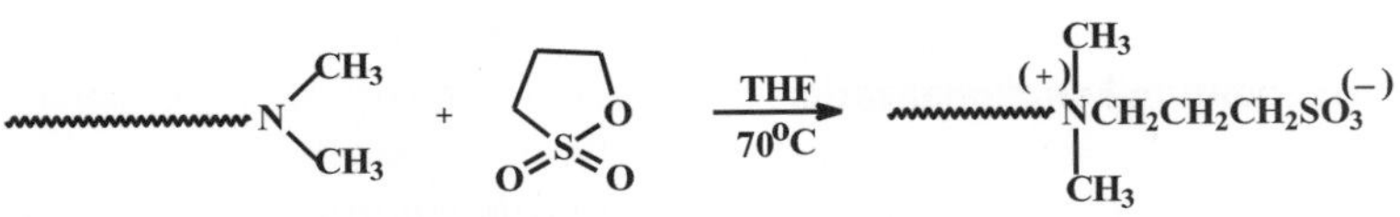

Scheme 4

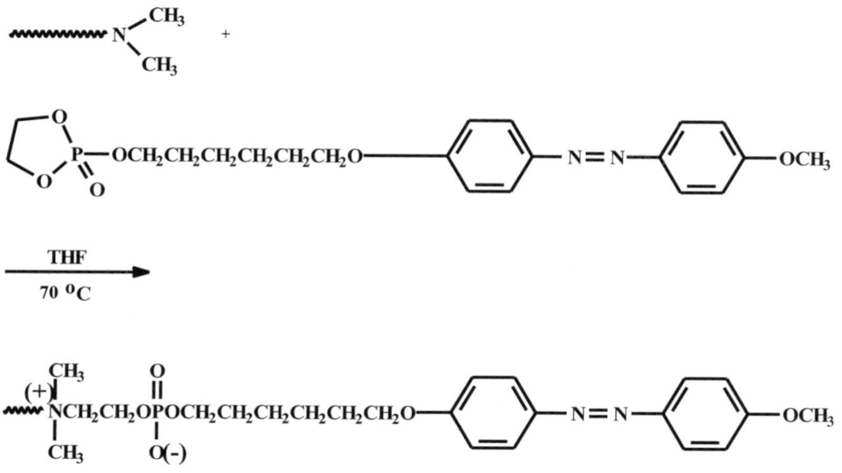

Scheme 5

celles formed are increased due to the presence of the polar groups in their core. Due to the extra association of the polar groups inside the cores the thermal stability of the micelles is increased compared to the case of nonfunctional block copolymers [36].

The presence of the polar groups also influences the self-organization of the materials in the bulk [37, 38]. An increase in the order-disorder transition (ODT) and in the stability of the cylindrical PS domains was observed in the case where the PS block was carrying a zwitterionic end group in comparison to the dimethylamino functionalized homologs. The viscoelastic behavior of the systems were also affected showing an increased plateau modulus when the zwitterionic groups were placed in the PI chain end, due to the aggregation of the polar groups. This aggregation also slows down the end-to-end chain-relaxation process of the PI blocks due to their confinement between the polar aggregates and the glassy PS domains. A similar increase in the ODT was observed in the case of junction-point-functionalized zwitterionic block copolymers in accordance with existing theoretical predictions.

Recently, dimethylamino and zwitterion end-functionalized block copolymers were used in the construction of "smart" polymer surfaces [39]. These surfaces were found to respond almost reversibly to external stimuli, being able to alter the wetting characteristics of a polymer surface when it is exposed to a humid environment. In this process the ability of a diblock copolymer, having the hydrophilic group attached to the mostly surface active block, to accumulate at the polymer/air interface was utilized. The end group is hidden below the polymer surface when in contact with dry air while it is present on the surface when the material is exposed to water vapor. Due to the increase of the concentration of the polar groups on the polymer surface, the water contact angle is reduced. This process could be reversed and repeated again by successive exposures of the surface to dry and wet environment. Interestingly, the response was found to be more rapid after the first cycle, showing evidence of an "educated" response of the polymeric material to the external stimulus.

2.2
ABC Miktoarm Stars

ABC miktoarm-star terpolymers are star-shaped macromolecules consisting of three chemically different chains connected together at a common junction point [40]. This particular molecular topology induces some constraints in the ways that the different arms can be arranged in a periodic structure and gives the molecule the ability to "choose" which arms interact directly in the segregated state and/or with the substrate and the free surface environment in thin films. Therefore, interesting morphologies in bulk and in thin films are anticipated. Furthermore, the incompatibility between the three arms of the macromolecule, a parameter that also plays a major role in microdomain formation, can be tuned by judicius choice of the monomers to be used in the polymer synthesis. The more diverse the chemical nature and, therefore, the properties of the three arms, the more interesting the interplay in the course of microphase separation is expected to be.

In this respect, miktoarm-star terpolymers, comprised of PS, PI, and PMMA arms, show great scientific interest since all three blocks have different properties (in terms of polarity, solubility, glass transition, etc.). Their synthesis was accomplished using controlled chlorosilane chemistry [41]. The controlled substitution of two chlorine atoms of trichloromethylsilane in order to synthesize the macromolecular linking agent (PI)(PS)MeSiCl has been achieved by first reacting excess of CH_3SiCl_3 with living PILi in order to replace one Cl atom. Excess CH_3SiCl_3 was removed in the vacuum line, over a period of several days, and then PSLi was added slowly (titration) until the second Cl atom of the macromolecular linking agent was replaced by PS as evidenced by size-exclusion chromatography (SEC). The resulting monofunctional macromolecular chlorosilane solution was slowly added to a dilute solution containing a stoichiometric amount of α,ω-dilithium initiator (I). The procedure was followed with SEC in order to ensure complete reaction of the monofunctional macromolecular clorosilane with only one active center of the initiator. The macromolecular initiator, so created, was used to polymerize MMA at $-78\,°C$ in THF giving the desired ABC miktoarm-star terpolymer. A schematic representation of the sequence of reactions used to synthesize the (PS)(PI)(PMMA) μ-star terpolymer is given in Scheme 6.

The bulk morphology of these materials was investigated by TEM and SAXS [42, 43]. The miktoarm stars showed new and interesting morphologies whose symmetry depends on the detailed characteristics of the samples, i.e., composition and molecular weight of the arms. For the samples with the

Is + sec-BuLi ⟶ PILi

PILi + excess CH$_3$SiCl$_3$ ⟶ PISi(CH$_3$)Cl$_2$ + LiCl + CH$_3$SiCl$_3$

St + sec-BuLi ⟶ PSLi

PISi(CH$_3$)Cl$_2$ + PSLi $\xrightarrow{\text{titration}}$ (PI)(PS)MeSiCl + LiCl

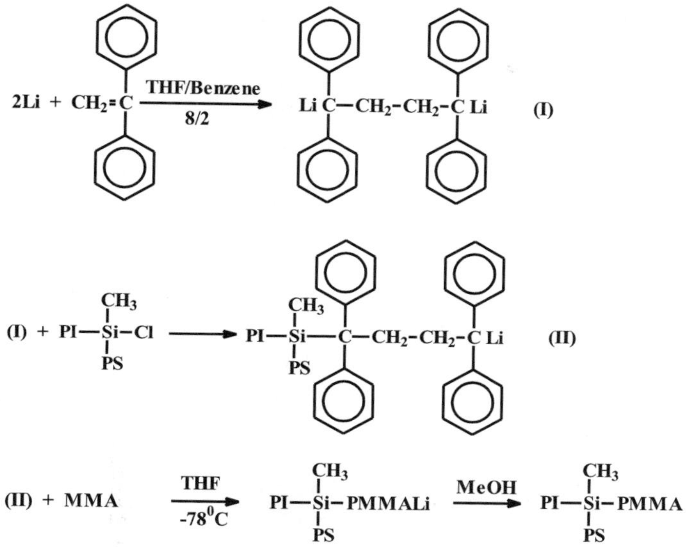

2Li + CH$_2$=C(Ph)$_2$ $\xrightarrow[8/2]{\text{THF/Benzene}}$ LiC(Ph)$_2$-CH$_2$-CH$_2$-C(Ph)$_2$Li   (I)

(I) + PI-Si(CH$_3$)(PS)-Cl ⟶ PI-Si(CH$_3$)(PS)-C(Ph)$_2$-CH$_2$-CH$_2$-C(Ph)$_2$Li   (II)

(II) + MMA $\xrightarrow[-78^\circ C]{\text{THF}}$ PI-Si(CH$_3$)(PS)-PMMALi $\xrightarrow{\text{MeOH}}$ PI-Si(CH$_3$)(PS)-PMMA

Scheme 6

higher asymmetry, three-microphase two-dimensionally periodic structures were observed with an inner PI column surrounded by a PS annulus in a matrix of PMMA. For two of these samples the PI/PS and PS/PMMA interfaces were cylindrical. For the other two, the PI/PS and PS/PMMA interfaces exhibited a unique nonconstant mean curvature having a diamond-prism-like shape (Fig. 1). Presumably, these microstructures where chosen in order to eliminate the unfavorable contacts between PI and PMMA phases. In these morphologies the star junction points should lie on the PI/PS interface.

The two more symmetric miktoarm stars showed an even more complex morphology. In this microdomain structure the PI phase forms convex triangular prism-shaped columns centered on sites with threefold symmetry. The PI prism columns are separated by PS columns forming a hexagonal lattice. The PS microdomains are centered on sites with twofold symmetry. The PI and PS microdomains surround the PMMA cylindrical columns ordered in a hexagonal array with sixfold symmetry. The structure is two-dimensionally periodic (Fig. 2). In this morphology the junction points are allowed to be situated in lines where the three different types of microdomains coincide.

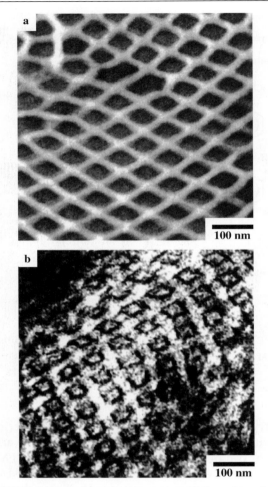

Fig. 1 Bright-field TEM images of styrene-isoprene-methyl methacrylate (SIM) 3-miktoarm star SIM-57/85/49 stained **a** with OsO_4 and **b** with RuO_4. In the OsO_4-stained micrographs, the *dark regions* correspond to the PI phase that forms rhomboid columns, whereas in the RuO_4-stained micrographs, the *gray regions* correspond to the PI phase and the *dark* to the PS phase

2.3
Linear ABC Terpolymers

In contrast to miktoarm stars, linear triblock terpolymers are comprised of three blocks with different chemical natures arranged in a sequential manner. In this case two junction points exist between blocks within the same chain inducing additional constraints in the arrangement of the blocks in an ordered mesostructure. In analogy to ABC miktoarm stars, the ordered

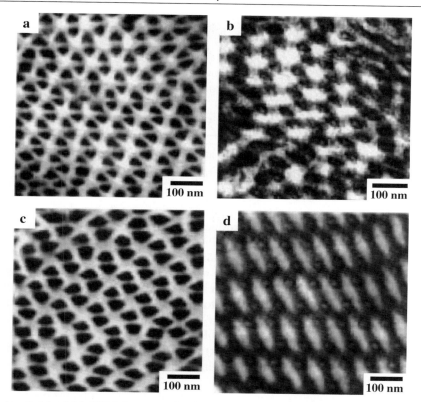

Fig. 2 Bright-field TEM images of selected regions of different styrene-isoprene-methyl methacrylate (SIM) 3-miktoarm stars, viewing approximately along the one-dimensionally continuous axis of the structures: **a** SIM-92/60/94 stained with OsO$_4$; **b** SIM-92/60/94 stained with RuO$_4$; **c** SIM-72/77/109 stained with OsO$_4$; **d** SIM-72/77/109 stained with RuO$_4$. In the two OsO$_4$-stained micrographs, the *dark regions* correspond to the PI phase, which forms triangular prism-shaped columns, whereas in the two RuO$_4$-stained micrographs, the *dark regions* correspond to the PI/PS phase boundary, the *gray regions* to the PS and PI phases, and the *lightest regions* to the PMMA phase

mesostructure can be controlled by the composition and relative block lengths as well as the strength of interaction between blocks, a parameter that is directly related to the chemical nature of the blocks. These kinds of materials have been investigated in terms of their bulk morphologies, as was mentioned in the previous section.

In a continuation of the studies reported so far triblock terpolymers of polyisoprene (PI)—poly(2-vinyl pyridine) (P2VP)—polyethylene oxide (PEO) was synthesized using benzyl potassium as the initiator and sequential addition of monomers [44]. The polymerization of isoprene and 2-vinylpyridine was carried out in THF at -78 °C. Ethylene was distilled into the solution and polymerized at 50 °C for 5 d. Termination was effected by

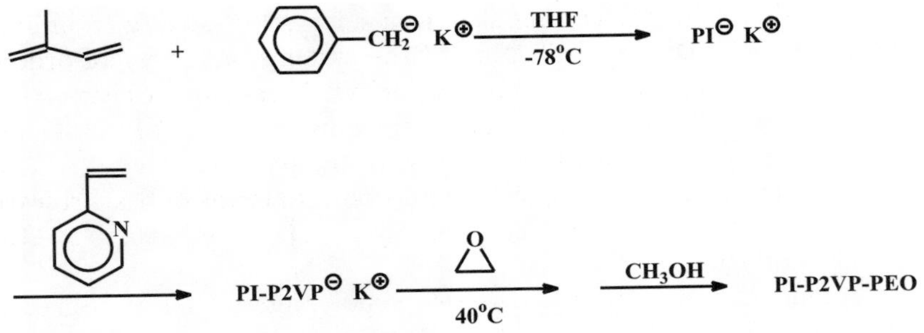

Scheme 7

addition of methanol. SEC analysis of each final block copolymer showed a single narrow peak, free of byproducts. The reaction sequence is summarized in Scheme 7.

The choice of the particular monomers is based on the fact that PI is a nonpolar low T_g polymer, P2VP is a relatively more polar material with a coordinating ability, due to the presence of nitrogen in the side phenyl ring, with high T_g and the potential to be transformed into a water-soluble polyelectrolyte by postpolymerization reaction. On the other hand, PEO is a semicrystalline polymer, water-soluble (polar), and biocompatible. The incorporation of all these diverse properties in the same molecule makes these materials very interesting in terms of their self-organization in solution, bulk, and thin films.

Multiblock copolymers comprised of more than three different monomers have been recently synthesized, and their chemical topology and new and interesting morphologies observed already in bulk promise a variety of ordered microstructures on surfaces and thin films [45–47].

2.4
Amphiphilic Block Copolymers

Amphiphilic block copolymers, i.e., block copolymers consisting of a hydrophobic and a hydrophilic block, have attracted the interest of scientists lately due to their solubility in water and their potential applications [48, 49]. Block copolymers containing a hydrocarbon or a heteroatom containing hydrocarbon block as the water-insoluble block and an anionic, cationic, or neutral water-soluble block have been investigated so far with respect to their capability for micelle formation in water [48–56]. Among these, block copolymers having a poly(ethylene oxide) (PEO) water-soluble block comprise a special category due to the particular properties of PEO, i.e., water solubility, neutral character, and biocompatibility. Studies have also been devoted to the interactions of aggregates of amphiphilic block copolymers with low-molecular-weight surfactants. The interaction of amphiphilic block copolymers with solid surfaces has also been a subject of investigation in recent years [57–62]. Due to the nature of the systems, a variety of ordered structures on surfaces can be created in principle. These surfaces have a high potential of being biocompatible or at least showing some similarities with biological surfaces and interfaces, depending on the block copolymer system, and thus creating a link to biological systems both in terms of basic research and applications. In this concept, block copolymer of polydienes and PEO are promising candidates.

Poly(butadiene-*b*-ethylene oxide) block copolymers were synthesized by anionic polymerization techniques [63]. Polymerizations were conducted in THF using *n*-BuLi as the initiator. Butadiene was polymerized first at −78 °C. The resulting polybutadiene (PB), under this experimental conditions, has a high percentage of 1,2 microstructure (∼ 80% as determined by NMR spectroscopy), but still its T_g is lower than room temperature enabling direct dissolution of the block copolymers in water. After completion of the polymerization of the first monomer, ethylene oxide was added at 40 °C in the presence of phosphazine base (Scheme 8).

The aggregation behavior of a series of well-defined PB1,2-PEO block copolymers was investigated in dilute aqueous solutions by static and dynamic-light scattering as well as by viscometry [63]. The aggregation number, size, and polydispersity of the micelles in pure water were found to depend on the molecular weight and the composition of the parent block copolymers. Apart from spherical micelles micellar clusters and cylindrical micelles were found in some cases. Measurements at different temperatures indicated a shrinkage of the spherical micelles by increasing temperature, probably due to worsening of the thermodynamic quality of water toward the poly(ethylene oxide) micellar corona. The presence of a low-molecular-weight salt, NaCl, in solution had minimal effects in the structural characteristics of the aggregates. The chemical nature and concentration of low-molecular-weight surfactants, namely sodium dodecylsulfate (SDS) and

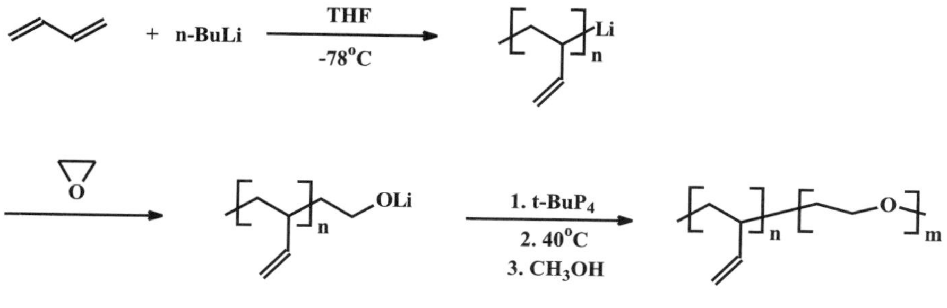

Scheme 8

dodecyltrimethylammonium bromide (DTMAB), in the micellar solutions was found to have a profound effect on the type, size, shape, and polydispersity of the aggregates formed in each case. These effects, coupled with the control of the micellar characteristics through the molecular characteristics of the individual copolymers, can lead to the control of the aggregate structure through chemical and physicochemical means. The deposition of the different aggregates formed in these solutions on surfaces may lead to the formation of patterned, polymer-containing surfaces whose shapes and dimensions would be determined by the aggregate structure in solution.

Another interesting feature of these block copolymers is the fact that the double bonds present in the PB block can be used for crosslinking the PB part [64, 65]. This can lead to the creation of nanoobjects, i.e., through crosslinking of the PB of aggregates in aqueous solution. Alternatively, crosslinking of the PB microdomains in the bulk or thin film state can create a network, producing variable confinement of the crystallizable PEO chains and thus affect their crystallization characteristics.

2.5
Amorphous-Side-Chain Liquid Crystalline Block Copolymers

Block copolymers containing liquid crystalline and amorphous blocks are interesting materials due to the resulting interplay of self-organization at two different levels, i.e., microphase separation between the dissimilar blocks and microstructure organization of the liquid crystalline segments [66–70]. Taking advantage of this interplay one can tailor the nanostructure of these systems at two length scales through chemical design. Although several studies concerning the bulk morphology and thermotropic phase behavior of coil-liquid crystalline block copolymers have been published, the behavior of such systems in thin films and surfaces have received little attention.

Along these lines of thinking block copolymers of styrene and 6-[4-(4-methoxyphenyl)phenoxy]hexylmethacrylate (MPPHM) were synthesized by anionic polymerization. Styrene was polymerized first in benzene using s-BuLi as the initiator. After completion of polymerization the living PSLi chains were end-capped with diphenylethylene and the solvent was changed to THF. Anhydrous LiCl was introduced to the reactor, and the temperature was lowered to – 40 °C. At that temperature a solution of purified MPPHM in THF was introduced slowly. After complete reaction of the liquid crystalline monomer, polymerization was terminated with methanol[70]. The procedure is outlined schematically in Scheme 9.

It must be emphasized that all samples involved in structure-property relationship studies should be thoroughly characterized in terms of molecular weights, average composition, molecular weight, and compositional and architectural uniformity [1]. A variety of characterization methods can and should be employed, as was done in the cases outlined above. Intermediate

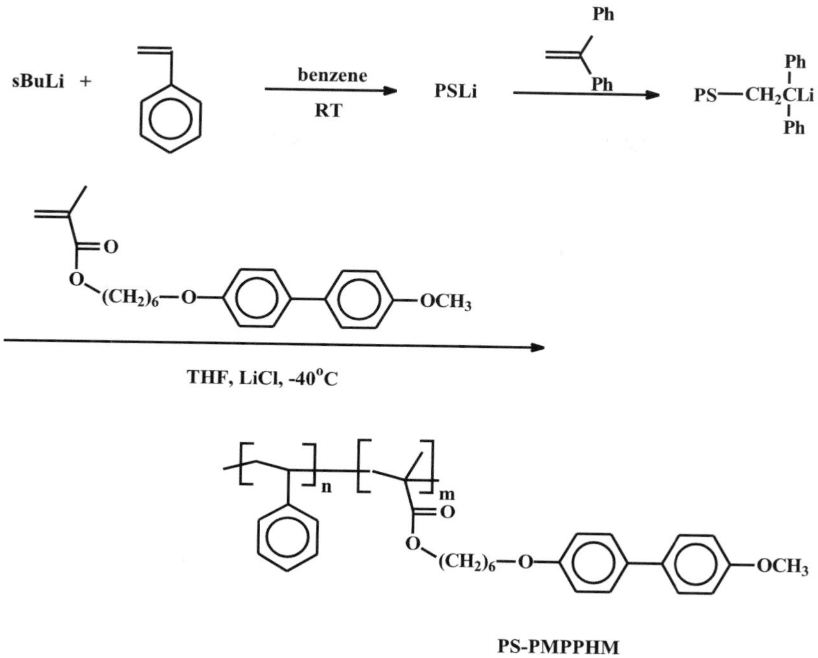

Scheme 9

and final products of the synthetic reaction schemes were analyzed by SEC, using RI and UV detection, to determine sample homogeneity and molecular-weight distributions, in tetrahydrofuran. Number average molecular weights of each part and of the whole molecules were obtained by membrane osmometry in toluene. Weight average molecular weights were obtained by low-angle laser-light scattering in a common good solvent (usually tetrahydrofuran). The average composition of the samples was determined by ^{1}H-NMR spectroscopy or UV spectroscopy. Good agreement between the expected and obtained molecular characteristics as well as between the different methods employed verifies the successful synthesis of the desired block copolymers and the low molecular weight and compositional and architectural heterogeneity of the macromolecules.

3
Conclusions

Polymer chemistry is able to produce a large variety of different block copolymer architectures. With the right choice of synthetic routes and chemical structure of the blocks, different self-assembly mechanisms can be encrypted in the same molecule. The properties of these novel materials, and especially their self-organization on surfaces, remain to be explored.

References

1. Hadjichristidis N, Pispas S, Floudas G (2002) Block copolymers: synthetic strategies, physical properties and applications. Wiley, New York
2. Hamley IW (1998) The Physics of Block Copolymers. Oxford University Press, Oxford
3. Lohse DJ, Hadjichristidis N (1997) Curr Opin Colloid Interf Sci 2:171
4. Krausch G (1995) Mater Sci Eng Rep 14:1
5. Cox JK, Eisenberg A, Lennox RB (1999) Curr Opin Colloid Interf Sci 4:52
6. Anastasiadis SH, Russell TP, Satija SK, Majkrzak CF (1989) Phys Rev Lett 62:1852
7. Russell TP, Coulon G, Deline VR, Miller DC (1989) Macromolecules 22:4600
8. Mayes AM, Russell TP, Bassereau P, Baker SM, Smith GS (1994) Macromolecules 27:749
9. Carvalho BL, Thomas EL (1994) Phys Rev Lett 73:3321
10. Mansky P, Russell TP, Hawker CJ, Pitsikalis M, Mays JM (1997) Macromolecules 30:6810
11. Huang E, Pruzinsky S, Russell TP, Mays JM, Hawker CJ (1999) Macromolecules 32:5299
12. Liu Y, Zhao W, Zheng X, King A, Singh A, Rafailovich MH, Sokolov J, Dai KH, Huang E, Mansky P, Russell TP, Harrison C, Chaikin PM, Register RA, Hawker CJ, Mays JW (2000) Macromolecules 33:80
13. Rockford L, Mochrie SGJ, Russell TP (2001) Macromolecules 34:1487
14. de Jeu WH, Lambooy P, Hamley IW, Vakin D, Pedersen JS, Kjaer K, Seyger R, van Hutten P, Hadziioannou G (1993) J Phys II (France) 3:139
15. Hamley IW, Wallwork ML, Smith DA, Fairclough JPA, Ryan AJ, Mai SM, Yang YW, Booth C (1998) Polymer 39:3321
16. Stocker W, Beckmann J, Stadler R, Rabe JP (1996) Macromolecules 29:7502
17. Rehse N, Knoll A, Magerle R, Krausch G (2003) Macromolecules 36:3261
18. Rehse N, Knoll A, Konrad M, Magerle R, Krausch G (2001) Phys Rev Lett 87:035505
19. Elbs H, Drummer C, Abetz V, Krausch G (2002) Macromolecules 35:5570
20. Fukunaga K, Hashimoto T, Elbs H, Krausch G (2002) Macromolecules 35:4406
21. Elbs H, Fukunaga K, Stadler R, Sauer G, Magerle R, Krausch G (1999) Macromolecules 32:1204
22. Fukunaga K, Elbs H, Magerle R, Krausch G (2000) Macromolecules 33:947
23. Fukunaga K, Hashimoto T, Elbs H, Krausch G (2003) Macromolecules 36:2852
24. Boker A, Muller AHE, Krausch G (2001) Macromolecules 34:7477
25. Fukunaga K, Elbs H, Krausch G (2000) Langmuir 16:3474
26. Balsamo V, Collins S, Hamley IW (2002) Polymer 43:4207
27. Pitsikalis M, Pispas S, Hadjichristidis N, Mays JW (1998) Adv Polym Sci 135:1
28. Hadjichristidis N, Pispas S, Pitsikalis M, Iatrou H, Vlahos C (1999) Adv Polym Sci 142:71
29. Hadjichristidis N, Pitsikalis M, Pispas S, Iatrou H (2001) Chem Rev 101:3747
30. Hadjichristidis N, Iatrou H, Pispas S, Pitsikalis M (2000) J Polym Sci Part A Polym Chem 38:3211
31. Hadjichristidis N, Pispas S, Pitsikalis M (1999) Prog Polym Sci 24:875
32. Pispas S, Hadjichristidis N (1994) Macromolecules 27:1891
33. Pispas S, Hadjichristidis N, Mays JW (1994) Macromolecules 27:6307

34. Pispas S, Hadjichristidis N (2000) J Polym Sci Part A Polym Chem 38:3791
35. Pispas S, Hadjichristidis N (2000) Macromolecules 33:1741
36. Pispas S, Allorio S, Hadjichristidis N, Mays JW (1996) Macromolecules 29:2903
37. Floudas G, Fytas G, Pispas S, Hadjichristidis N, Pakula T, Khokhlov AR (1995) Macromolecules 28:5109
38. Pispas S, Floudas G, Hadjichristidis N (1999) Macromolecules 32:9074
39. Anastasiadis SH, Retsos H, Pispas S, Hadjichristidis N, Neophytidis S (2003) Macromolecules 36:1994
40. Hadjichristidis N (1999) J Polym Sci Part A Polym Chem 37:857
41. Sioula S, Tselikas Y, Hadjichristidis N (1997) Macromolecules 30:1518
42. Sioula S, Hadjichristidis N, Thomas EL (1998) Macromolecules 31:5272
43. Sioula S, Hadjichristidis N, Thomas EL (1998) Macromolecules 31:8429
44. Ekizoglou N, Hadjichristidis N (2001) J Polym Sci Part A Polym Chem 39:1198
45. Ekizoglou N, Hadjichristidis N (2002) J Polym Sci Part A Polym Chem 40:2166
46. Bellas V, Iatrou H, Pitsinos G, Hadjichristidis N (2001) Macromolecules 34:5376
47. Takahashi K, Hasegawa H, Hashimoto T, Bellas V, Iatrou H, Hadjichristidis N (2002) Macromolecules 35:4859
48. Alexandridis P, Lindman B (eds) (2000) Amphiphilic block copolymers. Self assembly and applications. Elsevier, Amsterdam
49. Jonsson B, Lindman B, Holmberg K, Kronberg B (1998) Surfactants and polymers in aqueous solutions. Wiley, Chichester
50. Moffitt M, Khougaz K, Eisenberg A (1996) Acc Chem Res 29:95
51. Zhang L, Eisenberg A (1996) J Am Chem Soc 118:3168
52. Zhang L, Eisenberg A (1995) Science 268:1728
53. Forster S, Hermsdorf N, Bottcher C, Lindner P (2002) Macromolecules 35:4096
54. Qin A, Tian M, Ramireddy C, Webber SE, Munk P (1994) Macromolecules 27:120
55. Guenoun P, Davis HT, Tirrell M, Mays JW (1996) Macromolecules 29:3965
56. Xu R, Winnik MA, Hallett FR, Riess G, Croucher MD (1991) Macromolecules 24:87
57. Spatz JP, Sheiko S, Moller M (1996) Macromolecules 29:3220
58. Webber GB, Wanless EJ, Armes SP, Baines FL, Biggs S (2001) Langmuir 17:5551
59. Walter C, Harrats C, Muller-Buschbaum P, Jerome R, Stamm M (1999) Langmuir 15:1260
60. Meiners JC, Quintelritzi A, Mlynek J, Elbs H, Krausch G (1997) Macromolecules 30:4945
61. Regenbrecht M, Akari S, Forster S, Mohwald H (1999) J Phys Chem B 103:6669
62. Webber GB, Wanless EJ, Butun V, Armes SP, Biggs S (2002) Nanoletters 2:1307
63. Pispas S, Hadjichristidis N (2003) Langmuir 19:48
64. Won YY, Davis HT, Bates FS (1999) Science 283:960
65. Maskos M, Robin Harris J (2001) Macromol Rapid Commun 22:271
66. Mao G, Ober CK (1997) Acta Polymerica 48:405
67. Chiellini E, Galli G, Angeloni AS, Laus M (1994) Trends Polym Sci 2:244
68. Sanger J, Gronski W, Maas S, Stuhn B, Heck B (1997) Macromolecules 30:6783
69. Osuji CO, Chen JT, Mao G, Ober CK, Thomas EL (2000) Polymer 41:8897
70. Yamada M, Iguchi T, Hirao A, Nakahama S, Watanabe J (1995) Macromolecules 28:50

Surface Micelles and Surface-Induced Nanopatterns Formed by Block Copolymers

Krystyna Albrecht · Ahmed Mourran · Martin Moeller (✉)

Deutsches Wollforschungsinstitut an der RWTH Aachen e.V., Pauwelsstrasse 8, 52074 Aachen, Germany
moeller@dwi.rwth-aachen.de

1	Introduction	58
2	Deposition of Block Copolymer Micelles from Selective Solvents	59
3	Two-Dimensional Surface Micelles Prepared from Slightly Selective Solvents at the Water/Air Interface	60
4	Surface-Induced Nanopatterns	61
5	Conclusions	68
	References	69

Abstract Lateral phase segregation of block copolymers in ultra-thin films results in highly ordered arrays at interfaces. In this brief review we focus on PS-b-P2VP and PS-b-P4VP block copolymers. Selective solvents and polar solid substrates lead to a brush surface formed by adsorption of free polymer chains in solution followed by adsorption of whole micelles upon withdrawal of the sample from the solution. These laterally highly ordered patterns are obtained by rapid solvent evaporation and are thermodynamically not stable. In the second case presented two-dimensional surface micelles of poly(styrene)-$block$-poly(N-alkyl-4-vinylpyridinum iodide) are formed by spreading a solution in a slightly selective solvent mixture onto the water/air interface. The resulting pattern morphology strongly depends on the surface pressure and the length of the alkyl chain used for quaternization. These patterns can be transferred to solid substrates without structural changes. In contrast, polymer deposition onto mica from nonselective solvents is controlled by the polymer-surface interaction and the resulting patterns are induced by the surface (Surface-Induced NanoPATterns, SINPATs). The PVP block adsorbs strongly on mica in a two-dimensional coil whereas the PS block dewets the adsorbed PVP layer forming isolated clusters. These patterns are thermodynamically stable. As one possible application, Ti/SINPAT composites have been successfully used as etch masks allowing the transfer of regular structures of nanometer size into substrates.

Keywords Adsorption · Block copolymers · Nanopattern · Self assembly · SFM

Abbreviations
LB   Langmuir-Blodget
PnBA   poly(n-butyl acrylate)
PnBMA   poly(n-butyl methacrylate)

PtBMA	poly(*t*-butyl methacrylate)
PHEMA	poly(2-hydroxyethyl methacrylate)
PMMA	poly(methyl methacrylate)
PEO	poly(ethylene oxide)
PODCMA	poly(octadecyl methacrylate)
PS	poly(styrene)
PVP	poly(vinylpyridine)
P2VP	poly(2-vinylpyridine)
P4VP	poly(4-vinylpyridine)
R_g	radius of gyration
SFM	scanning force microscopy
SINPAT	surface-induced nanopattern
TEM	transmission electron microscopy
XPS	X-ray photoelectron spectroscopy

1
Introduction

Confining polymeric chains in a layer thinner than their natural length scale, the radius of gyration, will considerably alter their conformation and the resulting physical properties compared to the bulk properties [1–3]. The effect of the confinement on the interfacial properties is determined to a large extent by the detailed chemical structure of the polymeric chains. One of the most convincing examples is given by end-functionalized polymer chains that can adopt different conformations when they are covalently linked to a substrate. Depending on the number of chains per area and the macromolecular length, an extended chain "brush" or a coiled chain "mushroom" may occur representing two extreme configurations [4]. This conformational change has an effect on the static as well as the dynamic properties of the surface. Such decorated surfaces can be created by diblock copolymers, in which the shorter block adsorbs strongly onto the surface from the solution and the other longer block is repelled away from the surface and extended in the solution [5]. However, the homogeneous "hairy" layer decorating the surface becomes questionable when the surface is exposed to an external stimulus, for example a solvent with a different polarity or temperature. In particular, it has been predicted [6] and observed [7] that under certain conditions a surface of end-grafted chains in a poor solvent undergoes a structural transition from isolated clusters via densely packed clusters ("dimples") to a laterally homogeneous layer. It has been argued that in a poor solvent the end-grafted layer is unstable towards lateral inhomogeneities in the monomer density parallel to the grafting plane. Therefore, the end-grafted polymer chains clump together and form a dimpled surface with a characteristic length-scale proportional to the chain extension [8]. Several approaches have been proposed to control the microphase structure in a grafted layer. The most elegant combines conformational changes with phase separation in grafted layers of at least two immiscible species of chains [9]. In contrast to covalently linked chains, selective adsorption of a diblock copolymer engenders surface layers with rich structures. In particular, a pronounced affinity of one block to the substrate confines the anchoring chain to adopt a quasi two-dimensional conformation, while the second block is repelled from the surface [10].

In this article we focus on the formation of long-range ordered laterally-segregated structures of PS-b-PVP block copolymers at interfaces. When a selective solvent is used, micelles form in solution and are subsequently deposited onto the substrate. In contrast, polymer deposition from nonselective solvents is controlled by the polymer-surface interaction and the resulting patterns are surface induced (Surface Induced NanoPATterns, SINPATs). In this brief review, these two borderline cases as well as intermediate processes are discussed and compared.

2
Deposition of Block Copolymer Micelles from Selective Solvents

In selective solvents diblock copolymers undergo a phase separation and form micelles in which the core is composed of the insoluble block and the corona by the soluble block. These micelles are usually spherical and have a narrow size distribution. However, their shape and size may change under certain conditions. Several reviews have dealt with the micellization of block copolymers [11–13]. Since the core of such micelles can be loaded with inorganic precursor salts, the adsorption of these micelles onto solid substrates provides a way to control the deposition of inorganic nanoparticles [14]. Meiners et al. have shown that the adsorption of PS-b-P2VP micelles onto mica from solutions in toluene results in the spontaneous formation of spherical micelles with a high long-range order [15]. Due to the selectivity of the solvent these micelles consist of insoluble P2VP cores surrounded by a stretched corona of soluble PS blocks. In addition scanning force microscopy (SFM) images showed that a laterally homogeneous polymer film of 4 nm thickness is adsorbed underneath the micelle layer. From this it was concluded that the obtained film morphology originates from a two-step adsorption process. First a homogeneous copolymer brush is formed in solution on the mica surface with P2VP as the anchoring block while the PS chains are swollen in the solvent. Subsequently, upon sample removal the brush collapses and entire micelles adsorb. The latter step was proved by comparison of the aggregation number of micelles in the solution determined by dynamic light scattering with the aggregation number estimated from SFM measurements after solvent evaporation. Surface plasmon spectroscopy experiments confirmed these conclusions [16]. This two-step adsorption model applies only to polar substrates. For less polar substrates micelles adsorb directly onto the surface since the driving force for brush formation is reduced. However, the resulting film morphology, a hexagonally ordered array of micelles, is not in thermodynamical equilibrium since it is obtained by rapid solvent evaporation. Annealing of the samples leads to a transformation of the ordered micelles to a lamellae structure oriented parallel to the substrate and which is chemically homogeneous [17].

Sohn et al. prepared free-standing micellar monolayer films of PS-b-P4VP by the spin coating of toluene solutions [18]. Due to strong interactions between the coronas of the micelles the films could be removed from the mica by immersion into water and transferred onto various substrates. In addition, core-corona inversion could be induced without changing the micellar packing by dipping free standing films into core-selective solvents.

Graphoepitaxial deposition of protonated PS-b-P2VP micelles from acidic aqueous solution onto lithographically patterned Si/SiO_2 surfaces was reported by Hahn et al. [19]. No ordering of the micelles was observed on the elevated sections. In contrast to that, in the depressions the micelle density was significantly higher and the ordering was improved. Since the films were prepared by dip coating this graphoepitaxial effect was attributed to capillary forces that pull the polymer solution into the depressions where the micelles are trapped by thinning the meniscus [20].

3
Two-Dimensional Surface Micelles
Prepared from Slightly Selective Solvents at the Water/Air Interface

In a series of publications Eisenberg et al. [21–27] demonstrated the preparation of two-dimensional surface micelles on the water/air interface using quarternized PS-b-P4VP block copolymers. Poly(styrene)-$block$-poly(N-alkyl-4-vinylpyridinum iodide) dissolved in slightly selective chloroform/isopropanol mixtures was spread onto water using Langmuir–Blodget (LB) techniques. Evaluation of surface pressure graphs in combination with transmission electron microscopy (TEM) and SFM analysis was used to characterize the LB films. $(PS)_{260}$-b-$(4$-decyl-4-$PVP^+I^-)_{240}$ spontaneously self-assembled on the water-air interface to form highly regular aggregates. These aggregates consist of a core composed of PS blocks and a corona of highly extended dec-4-PVP^+I^-. At low surface pressure the average core-to-core edge distance is comparable to the length of two extended PVP chains. Depending on the surface pressure the ionized P4VP chains can either lay flat on the water surface around the PS aggregate ("starfish" state) or submerge in the aqueous phase ("jellyfish" state). A transition from starfish to jellyfish conformation occurs at high surface pressure [21]. These surface aggregates were found to be stable and could be deposited as monomolecular film on both hydrophilic or moderately hydrophobic substrates [22].

Variation of the alkyliodides used for quarternization revealed that the starfish-jellyfish transition strongly depends on the length of the alkyl chain. If the alkyl chain is short (methyl to n-butyl) the transition takes place at very small surface pressures since the PVP chains are not surface adsorbed. A plateau formation within the surface pressure vs. surface area graph does not occur until the side chains were n-hexyl groups [23]. Since the transition does not depend on the length of the PVP block [24] it is merely determined by the solubility of the polyelectrolyte block. Besides the starfish-jellyfish transition a morphological change within the low pressure regime was observed. This transition depends on the block length ratio of PS and quarternized PVP blocks expressed as mol percentage of styrene units. Up to a content of 86% styrene starfish micelles are found as described above. In the range of 86–94 mol % styrene content rodlike PS structures arise and above 94 mol % closed planes of PS are formed [25]. As rod and plane morphologies require PVP chains that are short compared to PS they only show little or no plateau formation in the higher pressure region [24].

Similar 2D surface micelles were observed at the water-air interface with nonionic block copolymers like PS-b-PnBMA, PS-b-PtBMA, PS-b-PtBA [28, 29], PS-b-PMMA [30], PODCMA-b-PMMA [31] and PS-b-PEO [32].

4
Surface-Induced Nanopatterns

A different way of preparing laterally segregated block copolymer patterns on solid substrates by means of self assembly was presented by Spatz et al. [10]. Heterogeneous, highly ordered patterns of PS-b-P2VP on mica were prepared by casting from an ultra-diluted ($c = 0.05$ g/L) solution in chloroform as the nonselective solvent that inhibits micelle formation. Therefore, the resulting structures are formed by surface-induced direct adsorption of single polymer chains onto the substrate rather than by adsorption of whole micelles as described above for structure formation from a selective solvent. The pattern formation originates from the strongly different affinities of the building blocks towards the substrate. SFM showed that the P2VP exclusively adsorbs on the mica substrate almost in a molecular monolayer of 1 nm whereas the PS block arranges into ordered isolated clusters on top of the adsorbed P2VP layer. The structure of these surface-induced nanopatterns (SINPATs) is shown in Fig. 1.

X-ray photoelectron spectroscopy (XPS) measurements confirmed that P2VP is the adsorbing block since the N (1s) binding energies were higher for SINPAT than for thick films. In the case of SINPAT the N (1s) energy of every third to second 2VP unit was shifted about 1 eV to higher values indicating that these units are adsorbed directly on the substrate [33].

In order to equilibrate the surface patterns, samples were annealed for 24 h at 150 °C, which was significantly higher than the bulk glass temperature

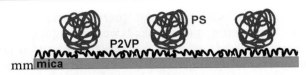

Fig. 1 Scheme of the structure of adsorbed PS-*b*-P2VP in ultra-thin films on mica. The P2VP adsorbs on the surface, whereas the PS dewets the adsorbed layer and forms isolated clusters

of both blocks. The SINPAT structure did not change during the annealing, demonstrating the thermodynamic stability of these patterns in contrast to the monomicellar film described above. Another characteristic aspect of SINPATs is a large periodicity comparing to the bulk. The radius of the adsorbed P2VP area corresponds to at least 5 times R_g of the unperturbed P2VP coil. This could only be explained by a transformation from a 3-dimensional coil to a 2-dimensional one when P2VP adsorbs strongly on mica. A transformation of the adsorbing block into a two-dimensional coil is entropically very unfavourable and can only be realized by a large gain in enthalpic energy due to the interactions between polymer and substrate.

Wetting of the P2VP layer by the PS block would be favourable in terms of the surface tension. However, the associated gain in enthalpy is not sufficient to facilitate the chain stretching. Therefore, the PS block dewets the adsorbed layer and aggregates to surface clusters in order to minimize the contact area with both the adsorbed P2VP block and air [10].

Typical PS-*b*-P2VP topographies obtained by SFM are presented in Fig. 2 [34]. Bright points correspond to elevations of PS aggregates.

These four images demonstrate how the degree of polymerization of the PS block affects the size and the periodicity of the islands. Comparison of Fig. 2 A (PS_{550}-*b*-$P2VP_{650}$) and B (PS_{550}-*b*-$P2VP_{1400}$) reveals that at constant length of the PS block, the increase of the degree of polymerization of P2VP enlarges the periodicity while the size of the islands remains the same (see Table 1).

While P2VP influences only the periodicity, the length of the PS block influences both periodicity and the dimensions of the islands as presented in Fig. 2C,D. With increasing the degree of polymerization of the nonadsorbing block diameter, height, periodicity of the islands and the aggregation numbers increase. Films formed by block copolymers with a relatively short PS block, for example PS_{70}-*b*-$P2VP_{300}$, were observed to lose the ability to form a regular pattern. This result is consistent with Monte Carlo simulations showing that beside the interaction potential, the length of the nonadsorbing block plays a crucial part in the formation of a regular surface pattern [35].

Theoretical analysis reveals that two different morphologies of surface micelles are possible. The type of structure on the surface depends on the N_A/N_B ratio of the lengths of the blocks, in which A corresponds to the adsorbing block and B to the nonadsorbing one. If the compatibility of the B block with air is poor, individual chains will collapse and for $N_A \ll N_B$ the collapsed

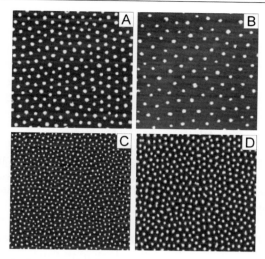

Fig. 2 SFM topography images of PS-b-P2VP SINPAT on mica. The scan size corresponds to 2.5×2.5 µm^2: **A** PS$_{550}$-b-P2VP$_{650}$, **B** PS$_{550}$-b-P2VP$_{1400}$, **C** PS$_{190}$-b-P2VP$_{190}$, **D** PS$_{340}$-b-P2VP$_{190}$

Table 1 Characteristic values of PS-islands for ultra-thin films formed by PS-b-P2VP block copolymers on mica

N$_{PS}$-b-N$_{P2VP}$	r [nm]	P [nm]	h [nm]	Q
190-b-190	25 ± 5	90 ± 10	4.3 ± 0.5	25 ± 10
330-b-140	65 ± 5	100 ± 10	5.3 ± 0.3	120 ± 20
340-b-190	70 ± 5	125 ± 15	6.1 ± 0.2	140 ± 20
550-b-660	85 ± 10	180 ± 20	7.5 ± 0.4	170 ± 30
570-b-490	90 ± 15	210 ± 20	7.5 ± 0.4	180 ± 50
550-b-1400	75 ± 10	250 ± 25	7.2 ± 0.5	120 ± 30
1350-b-400	130 ± 10	350 ± 35	10.9 ± 0.5	230 ± 30

r = average cluster diameter, P = periodicity, h = average height of the PS cluster, Q = aggregation number

blocks will form a homogeneous layer completely covering the surface. In the case that N_B is not so large with respect to N_A, the aggregation of the B chains can lead to lateral phase separation resulting in the formation of hemispherical islandlike surface micelles or wormlike surface aggregates (ribbons). These can be described as semi-cylinders lying on the surface (see Fig. 3) [36].

The free energy of either island or ribbon micelles can be written as follows [36, 37]:

$$F = F_{\text{surf}} + F_{\text{el}}^A + F_{\text{el}}^B \tag{1}$$

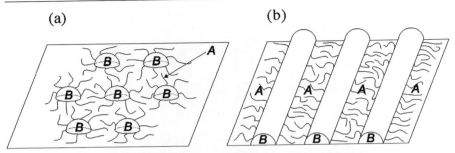

Fig. 3 Schematic representation of two possible surface morphologies. **a** Hexagonally arranged semi-spheres of B block (island morphology); **b** parallel oriented semi cylinders of B (ribbon morphology) (Reproduced with permission from [36]. Copyright 1999 American Chemical Society)

F_{el}^A and F_{el}^B represent the elastic energies of the A and B blocks, respectively. The first term, F_{surf}, is the surface free energy of the micelle. It minimizes the number of unfavourable contacts of monomer units of different blocks with each other and with the air, and can be written as follows:

$$F_{surf} = \gamma_1 S_1 + \gamma_2 S_2 + \gamma_3 S_3 \qquad (2)$$

The first two terms in Eq. 2 represent the interface energies between air and the blocks B and A, respectively. The third term accounts for the energy of the block A/block B interface. In this expression γ_i and S_i, $i = 1, 2, 3$ are the surface tension coefficients and the corresponding areas (see Fig. 4).

The form of the micelle was found to be close to a hemisphere for $\gamma_2 = \gamma_3$. Variation of the values of γ_2 and γ_3 changes the shape of the micelles but the dependence of their size on the length N_A and N_B of the blocks remains the same. By direct comparison of the free energies of brush, island and ribbon morphologies a phase diagram was constructed for two different values of γ_1, showing the transition between possible morphologies based on N_A-N_B variables (see Fig. 5).

For small values of γ_1 (Fig. 5A) a broad interval of N_A values exists for which the ribbon morphology is stable. By increasing γ_1 the region of ribbon stability becomes narrower (Fig. 5B). This indicates that for a short length of the nonadsorbing block, which shows a ribbon structure for low γ_1, at higher values of γ_1 a brush structure is the more favourable morphology in order to minimize the surface area. The detailed scaling analysis has been given in [36].

The theoretical results are in good agreement with the experimental observations for PS-b-P4VP block copolymers. In the experiments a well-defined ribbon morphology was obtained only for $N_S/N_{4VP} > 2$. An experimental phase diagram in a double logarithmic scale is shown in Fig. 6.

The triangle on the border line belongs to the PS_{350}-b-$P4VP_{260}$ polymer for which the film morphology consists of islands and ribbons side by side (see

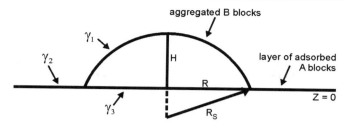

Fig. 4 Schematic representation of an island (ribbon) morphology formed by diblock copolymers on the surface $Z = 0$. Adsorbed A chains form a 2D melt on the surface. The island (ribbon) composed of B blocks is described as part of a sphere (cylinder) of radius R_S. H and R are the height and the radius of the surface micelle, respectively. γ_1, γ_2, γ_3 are the surface tension coefficients of the air/B, the air/A, and A/B interfaces, respectively (Reproduced with permission from [36]. Copyright 1999 American Chemical Society)

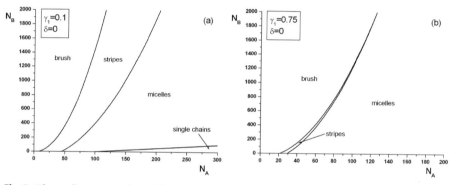

Fig. 5 Phase diagrams of A-B block copolymers with the A block adsorbed on a surface for $\delta = (\gamma_3 - \gamma_2)/\gamma_1 = 0$ while $\gamma_1 = 0.1$ (a) or $\gamma_1 = 0.75$ (b) (Reproduced with permission from [36]. Copyright 1999 American Chemical Society)

Fig. 7). Regarding other samples clearly polymers forming islands lie above the plot while the others are in the lower part [38].

Figure 8 shows the ribbon morphologies of four different PS-*b*-P4VP block copolymers. As in the case of PS-*b*-P2VP, the dimension of the ribbon strongly depends on the length of the nonadsorbing block. With increasing PS block length both the size of the ribbons and their spacing increase (see Table 2).

In contrast to PS-*b*-P4VP, PS-*b*-P2VP block copolymers exhibit only an island-brush transition. Since the exact size and the shape of the ribbon region of the phase diagram depend on the interfacial tension of the involved compounds, the ribbon region can be very small or even vanish. Therefore, the appearance of ribbons in the case of PS-*b*-P4VP is due to the differences in interfacial tension.

The topographic inhomogeneities of PS-*b*-P2VP SINPATs could be exploited in an etching process to transfer the polymer pattern into the underlying sub-

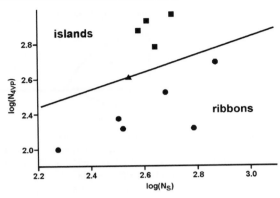

Fig. 6 Experimental morphology diagram for PS-*b*-P4VP SINPATs. *Squares* indicate polymers that form an island structure, *circles* polymers forming a ribbon structure, and the *triangle* the polymer that forms both island and ribbon structures side by side

Fig. 7 SFM micrograph of PS$_{350}$-*b*-P4VP$_{260}$ on mica. Scan size is $2.5 \times 2.5\ \mu m^2$

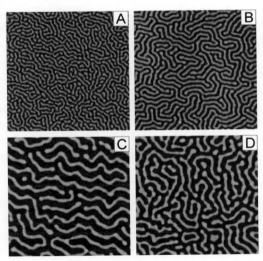

Fig. 8 SFM topography images of PS-*b*-P4VP SINPAT on mica. The scan size corresponds to $2.5 \times 2.5\ \mu m^2$: **a** PS$_{350}$-*b*-P2VP$_{260}$, **b** PS$_{380}$-*b*-P2VP$_{470}$, **c** PS$_{330}$-*b*-P2VP$_{130}$, **d** PS$_{610}$-*b*-P2VP$_{130}$

Table 2 Characteristic values of PS-ribbons for ultra-thin films formed by PS-*b*-P4VP block copolymers on mica

N_{PS}-*b*-N_{P4VP}	r [nm]	P [nm]	h [nm]
190-*b*-100	60 ± 5	72 ± 10	3.0 ± 0.4
320-*b*-150	80 ± 10	109 ± 10	4.1 ± 0.3
330-*b*-130	60 ± 10	105 ± 10	3.7 ± 0.4
480-*b*-210	75 ± 10	186 ± 20	4.8 ± 0.5
610-*b*-130	90 ± 10	140 ± 15	5.2 ± 0.5

r = average cluster diameter, P = periodicity, h = average height of the PS cluster; aggregation number is not given due to different length of ribbons

strate. Modern lithographic methods based on X-ray, ion beam or e-beam techniques allow fabrication of structures smaller than 100 nm [39]. However, these methods are not suitable when structuring of macroscopic areas is required. There is an obvious need for an intermediate process allowing easy patterning of macroscopic areas with a periodicity in the nanometer range. Based on concepts of macromolecular chemistry [39–41] much effort has been made to imitate the "bottom up" concept of nature. It has been shown that bottom up methods based on self-assembly of SINPAT, could be used for nanolithography [42]. By coating PS-*b*-P2VP SINPATs with a thin layer of Ti, the PS cluster height increased significantly more than expected from the average metal thickness, while the periodicity and the cluster size remained constant [43]. The initial growth of Ti was observed to be preferentially on the top of PS islands. In order to take advantage of the topographical inhomogeneities, samples were etched with an Ar ion plasma for various times. Figure 9 shows the resulting structures before and after etching of Ti-covered islands.

After 5 min the elevations gained their maximum height. At the same time the roughness between the elevations vanished completely, suggesting that the peaks are etched faster than the valleys. These results show that PS/Ti interaction could be exploited as etching masks for nanolithography. The SINPATs represent a new possibility to create < 100 nm islands which can act as a lithographic mask. Since the topological heterogeneities are related to the chemical composition, SINPATs seem to be good candidates for selective adsorption of other specimens and for further modification of the surface properties.

Not only di- but also triblock copolymers were used for the preparation of SINPATs. Ultra-thin film formation of PS-*b*-P2VP-*b*-PMMA and PS-*b*-PHEMA-*b*-PMMA triblock copolymers on silicon was reported by Böker et al. [44]. These films revealed regular surface patterns with stripe and island morphologies. The significant stretching of the adsorbing block resulted in rather large periodicities for relatively low molecular mass blocks. The lateral dimensions were in good agreement with the scaling laws derived by Potemkin et al. [36].

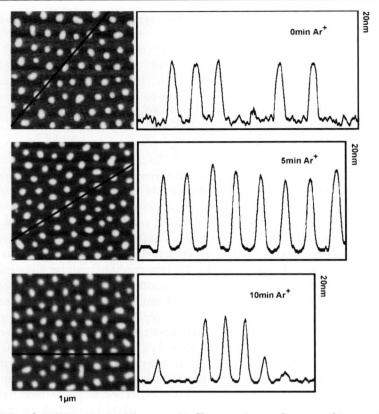

Fig. 9 PS_{340}-b-$P2VP_{190}$/4.5 nm Ti composite film on mica used as an etching mask. SFM micrographs and the cross section show the temporal evolution of the pattern. From *top* to *bottom:* 0 min, 5 min, and 10 min treatment with the plasma

5
Conclusions

In this review we summarized the research activity that has been performed on lateral phase segregation of PS-b-PVP block copolymers in ultra-thin films.

The use of selective solvents leads to micelle formation in solution. In toluene PS-b-P2VP forms micelles with P2VP cores and stretched PS coronas. By means of different techniques it was shown that on polar substrates a two-step adsorption process takes place. In solution single chains adsorb and form a brush. Subsequently this brush collapses upon sample withdrawal from the solution and whole micelles adsorb onto it. These laterally highly ordered patterns are not thermodynamically stable since they are obtained by rapid solvent evaporation.

Two-dimensional surface micelles were formed of poly(styrene)-*block*-poly(*N*-alkyl-4-vinylpyridinum iodide) by spreading a solution in a slightly selective solvent mixture onto the water/air interface by means of LB techniques. The resulting patterns showed a transition from a starfish conformation at low compression to a jellyfish conformation at high surface pressure. This transition does not depend on the degree of polymerization of the PVP chains but only on the length of the alkyl chain used for quaternization. These patterns could be transferred to solid substrates without structural changes.

Film preparation from ultra-diluted solutions of PS-*b*-P2VP and PS-*b*-P4VP in a nonselective solvent on mica resulted in highly ordered surface patterns formed due to the adsorption of single polymer chains in solution onto the substrate (SINPATs). The PVP block adsorbs strongly on mica in a two-dimensional coil whereas the PS block dewets the adsorbed layer forming isolated clusters on the top of the PVP layer. Three different morphologies of PS aggregates were found: hexagonally ordered islands, a ribbon morphology and a dense brush morphology. The occurrence of a specific morphology emerged depends on the block-length ratio N_S/N_{VP}, the chemical nature of the adsorbing block (P2VP or P4VP) and the degree of surface coverage. As one possible application the chemical heterogeneity of SINPATs was used as a template for Ti deposition. Furthermore Ti/SINPAT composites were successfully used as etch masks in order to transfer regular structures of nanometer size onto substrates.

Fabrication of highly ordered structures < 100 nm on surfaces via self-assembly of block copolymers opens up a new way to generate templates for various applications. However, there is still a great deal of research to be undertaken in order to understand the domain formation by lateral phase segregation of block copolymers in ultra-thin films. Understanding this process is crucial for the ability to design desired film morphologies by tailoring the chemical structure of the block copolymer.

References

1. Fleer GJ, Stuert MAC, Scheutjens JMH, Cosgrove T, Vincent B (1993) Polymers at Interfaces. Chapman and Hall, London
2. Frank CW, Rao V, Despotopoulou MM, Pease RFW, Hinsberg WD, Miller RD, Rabolt JF (1996) Science 273:912
3. Calvert PM (1996) Nature 384:311
4. Sperling LH (2001) Introduction to Physical Polymer Science. Wiley, New York
5. Milner ST (1991) Science 251:905
6. Lai PY, Binder KJ (1992) J Chem Phys 97:586
7. Siqueira DF, Köhler K, Stamm M (1995) Langmuir 11:3092
8. Grest GS, Murat M (1993) Macromolecules 26:3108
9. Dong H, Marko JF, Witten TA (1994) Macromolecules 27:6428 and references therein
10. Spatz JP, Sheiko SS, Möller M (1996) Adv Mater 8:513
11. Weber SE (1998) J Phys Chem B 102:2618
12. Bahadur P (2001) Curr Sci 80:1002

13. Riess G (2003) Prog Polym Sci 28:1107
14. Förster S, Plantenberg T (2002) Angew Chem Int Ed 41:688
15. Meiners JC, Ritzi A, Rafailovich MH, Sokolov J, Mlynek J, Krausch G (1995) Appl Phys A 61:519
16. Meiners JC, Quintel-Ritzi A, Mlynek J, Elbs H, Krausch G (1997) Macromolecules 30:4945
17. Li Z, Zhao W, Liu Y, Rafailovich MH, Sokolov J (1996) J Am Chem Soc 118:10892
18. Sohn BH, Yoo SI, Seo BW, Yun SH, Park SM (2001) J Am Chem Soc 123:12734
19. Hahn J, Weber SE (2003) Langmuir 19:3098
20. Hahn J, Weber SE (2004) Langmuir 20:1489
21. Zhu J, Eisenberg A, Lennox RB (1991) J Am Chem Soc 113:5583
22. Zhu J, Hanley S, Eisenberg A, Lennox RB (1992) Makromol Chem, Macromol Symp 53:211
23. Zhu J, Eisenberg A, Lennox RB (1992) Macromolecules 25:6556
24. Zhu J, Eisenberg A, Lennox RB (1992) Macromolecules 25:6547
25. Zhu J, Lennox RB, Eisenberg A (1992) J Phys Chem 96:4727
26. Zhu J, Lennox RB, Eisenberg A (1991) Langmuir 7:1579
27. Zhao ZLW, Quinn J, Rafailovich MH, Sokolov J, Lennox RB, Eisenberg A, Wu XZ, Kim MW, Sinha SK (1995) Langmuir 11:4785
28. Li S, Hanley S, Kahn I, Varshney SK, Eisenberg A, Lennox RB (1993) Langmuir 9:2243
29. Li S, Clarke CJ, Lennox RB, Eisenberg A (1998) Coll Surf A 133:191
30. Kumaki J, Nishikawa Y, Hashimoto T (1996) J Am Chem Soc 118:3321
31. Kumaki J, Hashimoto T (1998) J Am Chem Soc 120:423
32. Cox JK, Yu K, Constantine B, Eisenberg A, Lennox RB (1999) Langmuir 15:7714
33. Spatz JP, Möller M, Noeske M, Behm RJ, Pietralla M (1997) Macromolecules 30:3874
34. Eibeck P, Spatz JP, Potemkin II, Kramarenko EY, Khokhlov AR, Moeller M (1999) Polym Prepr 40:990
35. Spatz JP, Eibeck P, Mössmer S, Möller M, Kramarenko EY, Khalatur PG, Potemkin II, Khokhlov AR, Winkler RG, Reineker P (2000) Macromolecules 33:150
36. Potemkin II, Kramarenko EY, Khokhlov AR, Winkler RG, Reineker P, Eibeck P, Spatz JP, Möller M (1999) Langmuir 15:7290
37. Kramarenko EY, Potemkin II, Khokhlov AR, Winkler RG, Reineker P (1999) Macromolecules 32:3495
38. Eibeck P (1999) PhD Thesis, University of Ulm, Germany
39. Brodie I, Muray JJ (1992) The Physics of Micro/Nano-Fabrication. Plenum Press, New York
40. Kim E, Xia Y, Zhao XM, Whitesides GM (1997) Adv Mater 35:1510
41. Zhang L, Eisenberg A (1995) Science 268:1728
42. Spatz JP, Eibeck P, Mössmer S, Möller M, Herzog T, Ziemann P (1998) Adv Mater 10:849
43. Eibeck P, Spatz JP, Mössmer S, Möller M (1999) Nanostruc Mater 12:383
44. Böker A, Müller AHE, Krausch G (2001) Macromolecules 34:7477

Liquid Crystallinity in Block Copolymer Films for Controlling Polymeric Nanopatterns

Wim H. de Jeu (✉) · Yaëlle Séréro · Mahmoud Al-Hussein

FOM-Institute for Atomic and Molecular Physics (AMOLF), Kruislaan 407, 1098SJ Amsterdam, The Netherlands
dejeu@amolf.nl

1	Introduction	71
2	Liquid Crystalline/Amorphous Block Copolymers	73
3	Example 1. Smectic/Amorphous AB Diblock Copolymer	77
4	Example 2. Smectic/Amorphous A(A′B) "Triblock" Copolymer	80
5	Example 3. Fluorinated Smectic/Amorphous Diblock Copolymer	84
6	Conclusions and Outlook	87
	References	88

Abstract We review thin-film morphologies of hybrid liquid-crystalline/amorphous block copolymers. The microphase separation of the blocks and the smectic liquid crystalline ordering within one of the blocks are treated systematically in terms of the interaction parameters. The competition of the "tandem" interactions in terms of length scales and of surface anchoring can be used advantageously to control the orientation of block interfaces for nanopatterning.

Keywords Block copolymers · Nanostructures · Smectic liquid crystals · Thin films

Abbreviations
AFM   atomic force microscopy
GIXD   grazing-incidence X-ray diffraction
GISAXS grazing-incidence small-angle X-ray diffraction
LC   liquid crystal(line)
ODT   order-disorder transition
XR   X-ray reflectivity

1
Introduction

Several future technologies demand structures at length-scales below 1 μm (say 1–100 nm). In various cases such structures have to be ordered in a regular and tuneable fashion while in addition specific physical properties are needed. These requirements can often be met using the surfaces of polymers. Structural order over many length scales can be created by self-organizing polymer materials, controlled and possibly directed by molecular interactions and external constraints. Self-organization occurs from nanoscopic to macroscopic scales and includes phenomena like phase separation or microphase separation, adsorption and crystallization. Apart from potential applications, they are important for fundamental studies investigating the relation between nanostructure and resulting physical properties. In spite of considerable effort in this field, it is difficult to orient or align polymeric nanostructures. Thus, a major goal of present research is to arrange the polymeric features in regular well-ordered arrangements exhibiting multiple length scales [1]. Several competing factors are able to control the structure formation (chemical differences, conformational entropy, spatial constraints, external fields).

In this chapter we concentrate on block copolymers, which are well known to microphase separate in various structures that are scientifically interesting as well as technologically attractive [2, 3]. In thin films these structures can be further influenced by specific interactions between the various blocks and the limiting interfaces. However, in the context of nano-ordering, these possibilities of forming fairly well-defined mesoscopic structures are not sufficient. To obtain control over the ordering process, further handles are needed which can be provided by additional ordering principles in one of the blocks. Though again several mechanisms exist, we shall discuss here the possibilities of liquid crystalline (LC) ordering in one of the blocks, in particular smectic LC ordering. We restrict ourselves also to simple substrates, and refrain from the interesting field of "replication" of patterned surfaces, see for example [4–6].

In the next section the basic properties of block copolymers and liquid crystalline polymers will be summarized. Furthermore, we discuss the block microphase separation and its interaction with the LC ordering in one of the blocks in a systematic way. In the remaining sections a couple of examples (see Table 1) will be discussed of competing hierarchical ordering at surfaces, with emphasis on the possibilities for controlled nanopatterning.

Table 1 Compounds discussed and their properties

Polymer	M_n	$\Phi_{\text{LC-block}}$	DP	Phase behavior (°C)	Refs.
PS-b-PChEMA	57 000	0.51	1.1	g_{PS} 101 g_{PChEMA} 126 Sm-A 187 I	[43, 44]
PIBVE-b-PLC	9600	0.67	1.26	g_{PIBVE} −19 g_{LC} 11 Sm-C* 44 Sm-A 67 I	[47, 48]
PMMA-b-PF8H2A	45 400	0.79	1.2	Sm-A 85 I	[54, 55]

2
Liquid Crystalline/Amorphous Block Copolymers

The simplest case of a block copolymer is a diblock consisting of two covalently bonded polymers with chemically distinct repeat units A and B. If A and B are incompatible, below the order-disorder transition at T_{ODT} microphase separation is obtained into, for example, a spherical, cylindrical, or lamellar phase. The phase behavior depends on the relative volume fraction of A and B and on the magnitude of the product $\chi_{AB}N$, where χ_{AB} is the Flory–Huggins interaction parameter between the two polymers, and N the total degree of polymerization [7]. We can write

$$\chi_{AB} = \frac{Z}{k_B T}\left[\varepsilon_{AB} - \frac{1}{2}(\varepsilon_{AA} + \varepsilon_{BB})\right], \quad (1)$$

in which ε is the interaction energy and Z the number of nearest-neighbor contacts. Note that $\chi \sim 1/T$. In the case of roughly equal block sizes, the block copolymer microphase separates into a lamellar phase, the scaling of the lamellar period being $\sim N^{2/3}$. This result stems from the balance between the enthalpy gain of demixing A and B (Eq. 1) and the entropy cost of chain confinement within the layers [7–9]. For increasingly asymmetric volume fractions the lamellar structure makes way successively for a hexagonal cylindrical and a spherical micellar structure, very similar to that observed for low-mass surfactants. In fact the situation is somewhat more complicated due to the possible interference of intermediate morphologies like a cubic gyroid phase and a perforated lamellar phase.

In thin films of diblock copolymers, the interactions occurring at the air/film and film/substrate interfaces influence the microphase separation process, and can be used to control the orientation of the morphology. For example, a preferential interaction of one of the blocks with the boundaries ("surfactant"-like behavior) will favor uniformity of a lamellar phase on a macroscopic scale, with the lamellae of periodicity L parallel to the substrate [10, 11]. The resulting film thickness will belong to a discrete spectrum of allowed values D_n, depending on the boundary conditions [10, 12] given by

$$D_n = (n + \frac{1}{2})L \quad \text{or} \quad D_n = nL. \quad (2)$$

In the case where the actual thickness does not match this condition, terraces are formed with a height difference corresponding to L [13]. In a similar way the presence of interfaces will lead in other microphase-separated morphologies to some ordering of the cylinders or spheres. Note that structures of lamellae parallel to a substrate or cylinders perpendicular to a substrate are uniquely defined. However, this is no longer true when the lamellae are perpendicular or the cylinders parallel to the substrate. In the latter cases control of the azimuthal ordering in the plane of the film is needed to obtain uniform patterns. This is a crucial problem for the possible exploitation of such structures.

A combination of LC ordering and macromolecular properties can be obtained on a macroscopic scale in LC polymers [14, 15]. LC or mesogenic units can be built into a polymer in two different ways: as side groups attached to the polymer backbone (side-chain LC polymers) or built into the polymer backbone itself (main-chain LC polymers). The last class of materials can give strong fibbers, being best known under their trade names (Kevlar, Twaron). Side-chain LC polymers usually have an end group of the mesogenic units attached to the polymeric backbone via a flexible group that acts as a spacer. The resulting comb-shaped LC polymers are used as materials for the second-harmonic generation but also as protective coatings. Various types of LC phase (nematic, smectic) can be found, just as in the corresponding monomers, though in general the transition temperatures are shifted and the mesophases tend to be higher ordered than in the corresponding monomer systems. A special type of side-chain LC polymer is obtained if the mesogenic group is attached laterally to the polymer. In the absence of a spacer group this leads to stretching of the backbone into a rod-like shape: mesogenic jacketed LC polymers [16–18]. In this review we restrict ourselves to conventional comb-shaped LC polymers.

In the case of smectic-A (Sm-A) ordering the LC phase consists of stacks of liquid layers. In the context of nanostructuring this is the most relevant situation. The layer periodicity d can be close to the length of the side group (single layer), somewhat larger (interdigitation of the mesogenic groups) or about twice that value (double layers). The actual situation depends on the effective transverse dimension of the side group in relation to the spacing between the "anchoring" points at the polymeric backbone. In analogy to lamellar diblock microphase separation, the smectic ordering of a comb-shaped LC polymer can be considered as "nanophase" separation: the polymeric backbone separates from the mesogenic units [19, 20]. This analogy was nicely illustrated by Diele [20], who studied the smectic periodicity as a function of the number of side groups. Upon decreasing the density of attached side groups the smectic periodicity was found to increase. This can be understood if the mesogenic units always form the same type of smectic layering, forcing the backbone polymer to coil in between the layers (see Fig. 1a). With decreasing number of side groups, more backbone must be packed into this polymer sheet and its thickness increases. As a consequence also the smectic periodicity increases. One can push the analogy further by attributing an effective χ-parameter to the interaction backbone-mesogens. The situation mentioned then corresponds to strong phase separation; in other cases (weak phase separation) the backbone might be less strictly confined between the mesogenic layers. The Sm-A-nematic (T_{AN}) or smectic-isotropic (T_{Ai}) phase transition temperature can be considered as the relevant T_{ODT}.

Liquid Crystallinity in Block Copolymer Films for Polymeric Nanopatterns

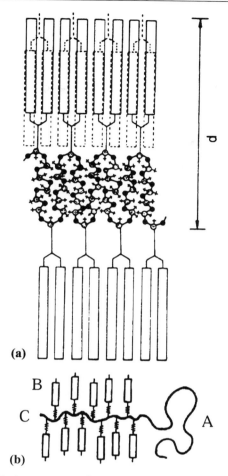

Fig. 1 a Smectic LC ordering as a nanophase separation process. The picture shows smectic layering when for only one out of five backbone positions a side chain is attached [20]. **b** Cartoon model of a LC/amorphous diblock copolymer

In hybrid amorphous/smectic block copolymers a hierarchical interplay arises between block copolymer ordering on the scale of tens of nm and smectic layering on the scale of nm. In addition to the enthalpy gain of demixing A and B and the entropy cost of chain stretching involved in "classical" microphase separation, several additional contributions to the free energy of block copolymers come into play. These include (i) the elastic energy of the LC phase and (ii) the orientational wetting of the LC phase at the internal micro-domain boundaries as well as (iii) at the external film surfaces. For strong block segregation the LC phase is confined by the amorphous block, similar to crystallization of block copolymers with a semi-crystalline block.

For weak block segregation the coupling may be more important and T_{AN} or T_{Ai} may act also as T_{ODT} for the block separation. On the other hand the LC phase behavior may be influenced by the confinement. In particular if the LC phase is the minority phase, in the case of spherical or cylindrical morphologies smectic ordering may be suppressed and nematic ordering favored. For lamellar block copolymer morphologies the smectic layers are usually oriented perpendicular to the block interfaces. The influence of molecular mass and polydispersity has hardly been investigated yet [21]. Some reviews in the field of LC/amorphous block copolymers are given in [22–26], a selection of original papers in [27–36].

In the spirit of the discussion so far, an amorphous/smectic LC diblock can be considered to be an A(CB) triblock copolymer, where A denotes the amorphous polymer and C and B the backbone and mesogens of the smectic polymer, respectively (Fig. 1b). If $T_{Ai} \ll T_{ODT}$, the Sm-A phase forms within a well microphase-separated block system and one would expect $\chi_{BC} \ll \chi_{AC}, \chi_{AB}$. Only when this condition is fulfilled the usual description as a diblock system is effectively correct. However, a different situation arises if the backbone in the smectic polymer is similar to the polymer in the amorphous block: $C \equiv A' \approx A$. Now little incompatibility exists between the polymer in the amorphous block and the backbone of the smectic LC block. In that situation $\chi_{AC} \approx 0$, and $\chi_{BC} \approx \chi_{AB} \gg 0$. A correct description of this situation would be A(A'B), and the interplay between the smectic ordering and the microphase separation is expected to become more important.

The competing hierarchical ordering in amorphous/smectic block copolymers provides a new handle to manipulate the structure in thin films. Let us consider as an example lamellar block structures. In bulk the smectic layers are usually orthogonal to the lamellae and separate the microphase segregated blocks. This can be understood from minimum contact interactions between the two polymer chains. As discussed above, in thin films the block lamellae in general like their interfaces to be parallel to the substrate. However, in such films the mesogenic units in a side-chain LC polymer prefer usually homeotropic anchoring to both the substrate and the air interface [37] (except for strongly polar end groups). For a smectic LC phase this would bring the smectic layers also parallel to the substrate. In that case a conflict arises between the directions of the smectic layering and the block lamellar ordering, that cannot both adjust to the same preferred boundary orientation and remain orthogonal to each other. The resulting frustration can be used advantageously for control purposes. In principle the film should be annealed at temperatures within the range of stability of the smectic phase. This requires T_{AN} or T_{Ai} to be larger than any glass transition in the system. Annealing at high temperatures well below T_{ODT} but above T_{AN} or T_{Ai} and subsequent cooling into the smectic phase is less suitable as it decouples the formation of the smectic layering from the initial block organization. Nevertheless, in such a situation smectic layering can already exist locally at the film interfaces.

In the following sections we shall illustrate these concepts by means of a few different examples. Similar considerations apply to the work of Ikkala and co-workers, who investigated systems with amphiphilic molecules attached via hydrogen bonding to one block of a diblock copolymer [38, 39]. In that situation the temperature at which the hydrogen bonding gets lost (around 80 °C) sets a limit to interactive annealing. The combination of X-ray reflectivity (XR) and atomic force microscopy (AFM) has proven to be most useful in gaining access to complementary real-space information of the surface and reciprocal-space knowledge of the interior of the films. As XR is sensitive to modulations of the average electron density along the surface normal [40], in this way a complete model of the thin film structure can be obtained, including possible differences between the surface and the interior of the film. Similar arguments apply to any lateral ordering of the film. Complementary to surface ordering probed by AFM, grazing-incidence X-ray diffraction (GIXD [41], including GISAXS at small azimuthal angles [42], can tell whether this ordering is continued through the whole film.

3
Example 1. Smectic/Amorphous AB Diblock Copolymer

In this first example, we investigate the lamellar thin film morphology of the approximately symmetric amorphous-smectic diblock copolymer PS-*b*-PChEMA. The molecular structure is shown in Fig. 2a while further molecular and phase information is given in Table 1. In the bulk the smectic layers ($d = 4.3$ nm) and the block lamellae ($L = 30$ nm) are orthogonal as anticipated [43]. Spin-coated films were annealed in the smectic phase around 170 °C, well above the glass transition of PS. XR measurements for films of different thickness D are displayed in Fig. 2b. A developing quasi-Bragg peak can be observed in thin films, with an associated periodicity of 4.2 nm that corresponds closely to the bulk smectic spacing of PChEMA. However, this observation is restricted to thin films with approximately $D \leq 2L$; for thicker films the situation is less clear. Thin films were investigated in some detail by Wong et al. [44]. The smectic layers are parallel to the surface, with the mesogenic side groups anchoring homeotropically, perpendicular to the surface. The same homeotropic anchoring is observed in XR measurements of PChEMA homopolymer thin films. The surface topography of a typical sample prepared under the same conditions and measured with AFM is given in Fig. 3a. The average thickness as measured is 33 nm, and two types of terrace with an average height difference of 5 nm can be seen. The latter value is comparable to the thickness of a smectic layer. This is consistent with the previous X-ray observation, and confirms that the smectic layers formed from the mesogenic side groups are parallel to the surface. In adjusting the thickness D to the boundary conditions, an integral number of smectic layers $D = nd$ now

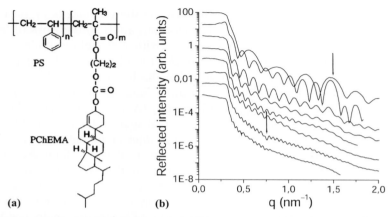

Fig. 2 **a** Structure of PS-*b*-PChEMA. **b** XR of films of various thickness, from *top* to *bottom*: 22, 45, 53.5, 55.5, 88, 91.5, 138, and 152 nm, respectively. The *upper arrow* indicates for thin films smectic layering parallel to the substrate; the *lower arrow* for thick films (weak) lamellar block orientation parallel to the substrate

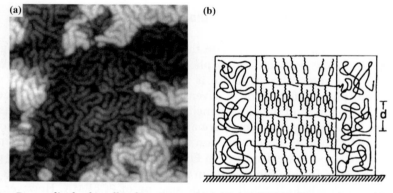

Fig. 3 **a** Perpendicular lamellae for a 36 nm thick PS-*b*-PChEMA film as observed by AFM (1×1 μm^2 viewing area). **b** Schematic representation of the perpendicular wetting [44]

takes the place of the quantization in L given by Eq. 2. The serpentine corrugations on the terraces in Fig. 3a have an average in-plane spatial period of 32 ± 4 nm, which is essentially the ABBA lamellar period.

The combination of XR and AFM results indicates that the generic ordering found in isotropic symmetric diblock films, where the lamellae are parallel to the surface, has been suppressed in these thin hybrid diblock films (Fig. 3b). The anchoring of the mesogenic groups dominates the wetting, and the hybrid diblock lamellae order in a direction perpendicular to the surface. PChEMA readily wets both the quartz and air interfaces and PS is usually repelled from hydrophilic substrates. Hence, intuition informed by theoretical descriptions of isotropic diblocks suggests that the contact interaction should dominate and lead to parallel wetting of the lamellae, with homeotropic PChEMA at both interfaces. For PS-*b*-PChEMA, however, homeotropic anchoring of the PChEMA side groups is inherently antagonistic to parallel wetting, because it necessarily generates defects and extra unfavorable segment-segment contact near PS-PChEMA interfaces. More important, the equilibrium bulk lamellar period defines a thickness for the PChEMA block at the interface, which is in general not commensurate with the thickness of an integral number of homeotropic smectic layers. This incommensurability is particularly strong for thin films in which D and L are of the same order but not equal. The frustration from both effects, however, can be avoided in perpendicular wetting of the lamellae, for which homeotropic anchoring is maintained without layer incommensurability, but at the cost of some unfavorable PS wetting at the substrate. The parallel ordering of lamellae, however, can be restored in asymmetric lamellar diblocks in which the contact area between PS and the substrate is increased [44]. In thicker films possibly a transition takes place from perpendicular lamellae at the substrate to parallel lamellae at the air interface, attributed to a decreased incommensurability. However, this point has not been investigated more extensively.

In spite of the interesting perpendicular morphology in thin films, the azimuthally random serpentine-like nature of the block lamellae prevents useful applications. Two possibilities have been explored to modify locally the substrate in order to create a preferential in-plane anchoring direction. The first is to play with the topography of the substrate. A sharp step on a silicon wafer will select a particular direction that may influence the orientation of the structure. Preliminary results for a film deposited on a saw-tooth patterned silicon wafer lead indeed to lamellae oriented in the step direction (see Fig. 4a). A second option consists of modifying on a nanometer scale the affinity of the substrate for the blocks. It is possible to graft alkyl chains on a silicon surface at selected positions determined by scratching the substrate in the presence of an appropriate solution of the grafting molecules [45]. Performing this scratching with an AFM cantilever tip, one can create a well-defined hydrophobic nanoline on the hydrophilic surface. As PS is a nonpolar polymer, it will preferably wet such a line. As shown in Fig. 4b this induces indeed some ordering of the lamellar features. However, in both cases the correlation lengths of the parallel oriented lamella is limited. In combination with the rather elaborate methods of obtaining the alignment, this is so far prohibitive for further applications.

In conclusion we have reached a morphology of stable lamellae oriented perpendicular to a nontreated substrate. Similar results have recently been reported by Hashimoto and co-workers using conventional lamellar block copolymers on a strongly corrugated substrate [46]. In both situations azimuthal ordering of the lamellae turned out to be difficult. We wanted to extend these experiments to a cylindrical morphology for which we should

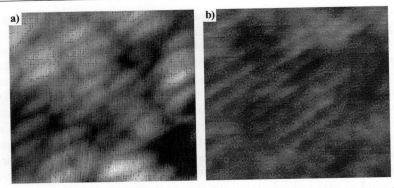

Fig. 4 AFM image of the alignment of perpendicular lamellae of a thin PS-b-PChEMA film along **a** a grating and **b** near a locally functionalized nano-scratch (see text, 0.35 × 0.35 μm^2)

expect the smectic layers to stabilize in thin film cylinders perpendicular to the substrate. However, though the analogous compound PS-b-PChEMA with 30–40% PS gives a cylindrical morphology, somewhat unexpectedly the smectic layers orient parallel to the block interfaces (cylinder axes) [43]. We attribute this anomalous behavior to the relatively low molecular mass of the particular block copolymers. Anyhow, it removed our "control handle" and made us turn to other systems (see Sect. 5).

4
Example 2. Smectic/Amorphous A(A′B) "Triblock" Copolymer

The next example consists of an amorphous polymer, which consists of a poly(isobutylvinylether) (PIBVE), chemically linked to a side-chain LC block with as a backbone the same vinyl ether. The structure is shown in Fig. 5a and further information is presented in Table 1. Note that the Sm-A phase region is at relatively low temperatures. Annealing was done close to T_{Ai}, subsequently the sample was quenched to room temperature for further analysis. As the amorphous block is only incompatible with the side-chain part of the LC block we expect the LC properties to be dominant. The system falls in the triblock class A(A′B) as described in Sect. 2. Nevertheless, two glass transitions have been detected in DSC, which is a direct proof of the immiscibility of the two blocks [47]. Films of the LC homopolymer indicate smectic layers parallel to the substrate with a periodicity of $d = 3.1$ nm. The period is equal to the stretched size of the side groups, which are homotropically oriented. This leads to a model of interdigitated side groups. A sample annealed in the isotropic phase indicates preferential interactions between the boundary layer and the substrate. XR shows that the film starts dewetting from the substrate except for a layer of 6.2 nm, corresponding to exactly two smectic layers.

The XR curve from a diblock film is given in Fig. 5b [48]. In addition to the Kiessig fringes several additional periodicities show up. Bragg-like features at $q_z \approx 7$ nm^{-1} with five higher orders indicate a lamellar period of $L \approx 9$ nm with long-range spatial order. This lamellar period is somewhat smaller than in the bulk (10.7 nm). The large intensity of the third order is due to superposition with the first-order Bragg peak from the smectic layering of about 3 nm parallel to the substrate. This indicates a rather special lamellar morphology in the diblock film, with both smectic and amorphous layers parallel to the substrate. The total thickness of 36 nm corresponds to four lamellar blocks. Fits to the reflectivity data indicate that the density profile through the film can be described by 12 layers of thickness of about 3 nm ($\pm 15\%$) while the interfacial roughness varies between 0.5 and 1 nm. Moreover, the density profile decreases monotonically from the substrate to the air/film interface leading to unrealistically low values for most of the layers. We are aware of one other report of parallel alignment of smectic layers and block lamellae [49], in which case it was attributed to peculiarities of the Sm-C* phase involved.

More insight about the diblock film structure is provided by the AFM data of Fig. 6. One can see terraces with a thickness of about 9 nm, the area of which decreases monotonically from the substrate to the air. Hence, the AFM image confirms the presence of lamellae parallel to the substrate. The decrease in area of the terraces on top of each other is consistent with the

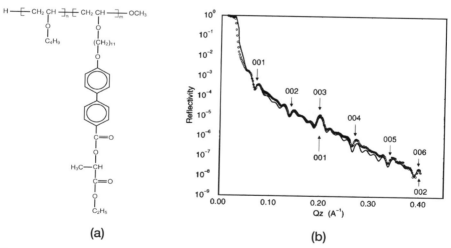

Fig. 5 a Structure of PIBVE-PLC. **b** X-ray reflectivity of a 36 nm thick film after annealing in the Sm-A phase at $T = 63\,°C$ (*circles*) with possible fitting curve (*solid line*) as discussed in the text. The Bragg positions of the diblock periodicity (*above the curve*) and of the smectic spacing (*below the curve*) are indicated [48]

monotonic decreasing density of the slabs in the XR model. It suggests that the organization of the lamellar structure starts from the substrate level. Reasonable fits of the XR data can be obtained by averaging the reflectivity of four different films of thickness of 1 to 4 lamellae, respectively, and realistic densities. We further notice in Fig. 6 instabilities during the microphase separation: the AFM picture shows intermediate layers of thickness of about 3 nm between adjacent terraces. These layers are dewetting the underneath terrace as indicated by their irregular contours. In fact, during the phase segregation process, islands and depressions of this thickness are successively formed, which are unstable at the air/film interface, leaving a few stable terraces of approximately 9 nm.

In the present film system the lamellar period must be composed of a smectic block of about 6 nm, and an amorphous block of approximately 3 nm, in order to be consistent with the bulk volume fractions of the blocks. Generally, such a parallel lamellar structure will be configurationally frustrated, because the smectic block size, as obtained by scaling the lamellar diblock period, may be incommensurate with the smectic periodicity [44]. However, for the present volume fractions, the smectic block size corresponds to twice the smectic periodicity. Several lamellar morphologies with block interfaces and smectic layers parallel are potentially possible. Various combinations of alternating amorphous blocks and an integral number of smectic layers can be obtained, with the mesogens oriented "up" and "down" with respect to the backbone. However, for neither of these cases, a lamellar period around 9 nm can be constructed. Moreover, such a situation leads to an unfavorable contact area between A and B components. Remember that the main segregation is due to the unfavorable χ_{AB} between the amorphous polymer

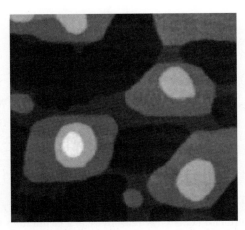

Fig. 6 AFM image of a dewetted PIBVE-PLC film (10×10 μm^2). The height difference between the terraces is about 9 nm. Dewetting layers of a thickness of about 3 nm can be seen adjacent to the terraces

A and the mesogenic units B and to $\chi_{A'B}$ between the backbone A' and its side groups B. These considerations led to a radically different proposal for the mesogen packing within the smectic layers. The mesogenic units are assumed to be densely packed, all pointing to the same direction (Fig. 7b). Alternating with the smectic layers, the PIBVE block will form the second sublayer. As the side groups point to the same direction, the polymer A and backbone A' face each other in a favorable way, because they are essentially of the same chemical composition and $\chi_{AA'} \approx 0$. In addition, phase segregation is ensured between the A, A' components at one side and the B component at the other side. In this situation, the smectic and amorphous block sizes correspond to about 6 and 3 nm, respectively, in agreement with the lamellar period. It seems that, in contrast to the bulk and homopolymer situations, a unidirectional conformation of the mesogens is the only way that allows a combination of the phase-separated structure and the smectic LC/substrate preferential interactions.

The difference in stability between the bulk structure of Fig. 7a and the film structure of Fig. 7b can be rationalized in terms of the influence of the surface [48]. First, the free energy contributions of the smectic ordering in the bulk and in the film are different because of the dissimilar packing configurations of the mesogens. The interdigitated smectic ordering of Fig. 7a is the natural morphology and is expected be more favorable than that of Fig. 7b, which requires a relatively strong confinement of the mesogens. However, this may be outweighed by the contribution from the block-block transition layer that contributes in the bulk (see Fig. 7a) unfavorably to the free energy of mixing of the A and B components. In the proposed film structure this is replaced by favorable AA' interactions. For the packing configuration of the amorphous block only minor changes are expected.

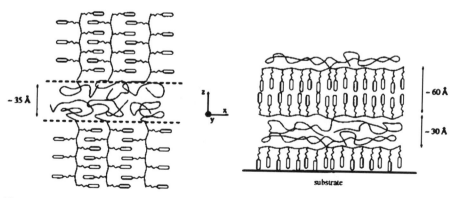

Fig. 7 Microscopic model of PIBVE-PLC for **a** the bulk orthogonal lamellar morphology and **b** the thin film parallel lamellar morphology

In conclusion the parallel lamellar structure of Fig. 7b seems to originate from the strong interaction between the mesogenic units and the substrate. The favorable internal diblock interactions will maintain the parallel lamellar structure, as long as the entropy cost due to the confinement of mesogens within the smectic layers is compensated for by the mesogens/substrate enthalpy bonus. The instability of the lamellar structure as observed in AFM (dewetting layers and terraces with decreasing area from the substrate to the air) indicates that this is not the case for all thicknesses. This point has not been investigated in further detail. Finally, we must conclude that though the situation of parallel block lamellae and smectic layers may be interesting from a physical point of view, it does not allow access to the layering in terms of nanostructuring. Unfortunately, samples with a probably useful cylindrical morphology were not available.

5
Example 3. Fluorinated Smectic/Amorphous Diblock Copolymer

Fluorinated polymers are of special interest because of the specific surface affinity of the fluorinated parts. This adds an extra element to the discussion so far. Coatings with controlled organization of the fluorinated parts are of interest as water/dirt repellent systems [50–53]. Fluorinated alkanes attached as side groups to a polymer can organize in smectic layers. Block copolymer ordering can be added as a third element to the smectic organization and the surface affinity of the fluorinated groups. In the example to be discussed [55], we move away from lamellar structures at about equal volume fractions and use the asymmetric compound PMMA-b-PF8H2A pictured in Fig. 8a (further properties in Table 1). Note that the backbone polymer is again rather similar (but not equal) to the amorphous polymer: in the A(A'B) picture we now expect $\chi_{AA'}$ to be small but not zero. Because of the strong difference with the fluorinated side chain we expect $\chi_{A'B} \approx \chi_{AB} \gg \chi_{AA'}$.

Small-angle X-ray scattering in the bulk shown in Fig. 8b indicates a hexagonal structure. Moreover, the two blocks are strongly microphase separated: $T_{ODT} > 300\,°C$. Upon cooling, a Sm-A phase appears within the fluorinated block below about 85 °C. The mesogens organize in double layers with a period $d \approx 3.3$ nm. The X-ray peaks corresponding to the microphase separated structure are not affected by the additional LC ordering. Single domain samples reveal that the smectic layers are perpendicular to the cylindrical block interfaces, in agreement with results for other fluorinated smectic/amorphous block copolymers [52, 56, 57]. While this situation is common from lamellar structures, it is not obvious for the other morphologies. In fact several cases have been reported of smectic layers parallel to the cylindrical block interfaces [43, 57].

Thin films of PMMA-b-PF8H2A were investigated with XR [55]. A typical result shown in Fig. 9a indicates rather rough interfaces (the Kiessig fringes

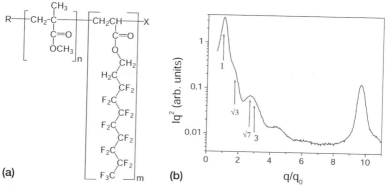

Fig. 8 a Structure of PMMA-*b*-PF8H2A; $n = 80$, $m = 74$. **b** Small-angle X-ray scattering in the bulk indicating hexagonal ordering with $L = 31.5$ nm and smectic layering with $d = 3.3$ nm

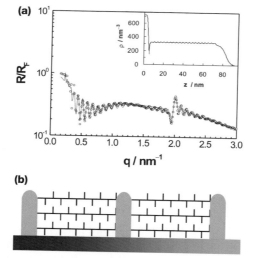

Fig. 9 a XR divided by the Fresnel reflectivity R_F and corresponding fit (*full line*) of a 82 nm film of PMMA-*b*-PF8H2A. **b** Associated model [55]

are not very well developed) and a smectic peak around $q = 2.01$ nm^{-1}. The corresponding spacing of 3.2 nm is very close to the bulk smectic spacing of 3.3 nm. This indicates that the smectic layers are aligned parallel to the substrate with the fluorinated chains anchoring homeotropically to the substrate. The best fit to the XR data indicates that the density profile through the film is composed of 21 smectic layers with a spacing of 3.2 nm sandwiched between different bottom and top layers (see inset of Fig. 9a). From this information we arrive at the preliminary model of Fig. 9b. Gronski and co-workers obtained somewhat similar results [34]. The latter authors describe in addition situations of random as well as regular positioning of the cylinders. Ordered arrays of perpendicular cylinders were recently also reported in thin films of a hybrid smectic/amorphous block copolymer with azobenzene mesogenic side-groups [59]. All these results indicate stabilization of a perpendicular orientation of the cylinders by smectic layering parallel to the substrate.

In general, for "classical" amorphous block copolymers one can expect cylinders perpendicular to the substrate if the possible gain in surface energy from preferential absorption of one of the blocks is more than compensated by the free energy change associated with the elastic distortions of the chain. This point was nicely demonstrated in the work of Russell and coworkers [60], who realized films with a perpendicular orientation of cylinders by making the substrate neutral to both blocks via coating with a random copolymer. Alternatively, perpendicular orientation of cylinders has been reached by applying an electric field [61] as well as by slow evaporation of the solvent after spin-coating [62, 63].

Figure 10a shows an AFM phase image of an annealed film at room temperature. Cylindrical PMMA domains (dark areas) can be seen, oriented normal to the surface and covering the whole surface of the film. The average diameter of the PMMA cylinders is about 18 nm and the average center-to-center distance between the cylindrical domains is approximately 35 nm. From the images there seems to be no long-range lateral order of the cylindrical domains. More information about the average in-plane ordering was obtained by grazing-incidence X-ray diffraction. In GIXD the X-ray beam shines on a surface at a small glancing angle α. If α is smaller than the critical angle α_c, total reflection occurs; now only an evanescent wave penetrates into the film. Using this wave as the "incident" beam, two-dimensional in-plane X-ray diffraction can be performed. The penetration depth of the X-rays depends on α and can be set to probe the full film. Hence, potentially GIXD allows us to quantify the internal structure of the blocks and to validate any model. In films of PMMA-b-PF8H2A a broad peak was found, corresponding to the liquid in-plane order of the fluorinated smectic block at an average distance of about 0.5 nm. This confirms the parallel orientation of the smectic layers with respect to the substrate as seen in XR. Any block structure would be at small angles and can only be observed with grazing-incidence small-angle X-ray scattering (GISAXS), the result of which is shown in Fig. 10b. The peak at $q = 0.207$ nm^{-1} corresponds to an average in-plane periodicity of 30.3 nm. This is very close to the interdomain period in the bulk (31 nm) and proves that the cylindrical domains are not only oriented normal to the air interface but span the entire film thickness. The peak can be well fitted by a Lorentzian function with a correlation length of about 40 nm (short-range liquid-like ordering).

In conclusion, AFM and grazing-incidence X-ray scattering essentially confirm the model of Fig. 9b. The cylindrical PMMA domains stand normal to the substrate and span the entire film. They are embedded in a matrix

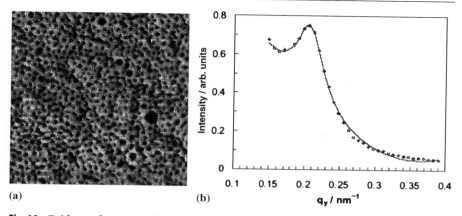

Fig. 10 Evidence for perpendicular cylinders in an annealed PMMA-*b*-PF8H2A film at room temperature. **a** AFM phase image in the tapping mode ($1 \times 1\ \mu m^2$). **b** GISAXS curve fitted to Lorentzian (*full line*)

of the PF8H2A phase with the side chains forming smectic bilayers parallel to the substrate. The structure can be understood in terms of the combined effect of the microphase separation, the smectic ordering and the orientational wettability of the side chains. At the air-film interface the fluorinated side chains orient toward air in order to lower the surface energy [64]. This would cause the smectic layers to orient parallel to the substrate. However, the smectic layers also tend to orient perpendicular to the microdomain interfaces in order to minimize the contact area between the two polymer blocks. Hence, to simultaneously maintain these preferential interactions the smectic layers and the cylindrical domains orient parallel and perpendicular to the substrate, respectively. Though the film has been annealed above the smectic-isotropic transition, evidently the essential elements leading to the smectic layering are already present locally at the interfaces.

6
Conclusions and Outlook

There is little doubt that supramolecular materials built from highly regular molecular nanostructures can possess interesting visco-elastic or electrical properties and may sensitively respond to various external fields (sensors). Furthermore, their nanostructured surfaces can exhibit highly selective interactions with molecules in the environment (biosensors). Nevertheless, we are only at the edge of coming to realistic applications. Controlled nanopatterning is a prerequisite for further progress. For various applications (nano-pore size filters, generation of a two-dimensional electric potential, medium for electrophoresis of DNA) simple periodic patterning is sufficient. Competing length scales provide a valuable tool to obtain and control the required nanopatterns.

In this review we have shown that thin films with periodic arrays of lamellae or cylinders perpendicular to a substrate can be nicely stabilized using the "tandem" interactions in hybrid smectic LC/amorphous block copolymers. Advantages over other methods [60–63] are: (i) absence of chemical treatment of the substrate; (ii) minimum number of steps in which the final result is accomplished; and (iii) robustness against annealing (within the smectic temperature range) because the structures are in thermodynamic equilibrium. Starting from a lamellar block morphology, the smectic LC layering stabilizes perpendicular block lamellae. However, control of the azimuthal orientation remains difficult so far [44, 46]. From this point of view stabilizing a cylindrical block morphology perpendicular to the substrate by parallel smectic layering is easier. A major recent achievement is stabilization by smectic layering of long-range order of the cylinders over large areas [59]. Possibilities of chemical variation of the liquid-crystal block have so far hardly been explored. Selective etching allows obtaining a variety of structures, in the case of cylinders either nano-holes or nano-pillars [65]. Applications of such possibilities are potentially numerous, but still require further investigations. In the mean time the physics involved is fascinating.

Acknowledgements The authors want to thank the members of the EU-POLYNANO network and Denitza Lambreva (AMOLF, Amsterdam) for fruitful discussions and cooperation over the past years. Gerard Wong and Daniel Sentenac are acknowledged for their contributions in the early stages. This work is supported in part by the European Community's Human Potential Programme under contract HPRN-CT-1999-00151 (POLYNANO) and is part of the research program of the "Stichting voor Fundamenteel Onderzoek der Materie (FOM)", which is financially supported by the "Nederlandse Organisatie voor Wetenschappelijk Onderzoek (NWO)". Y. Séréro and M. Al-Hussein acknowledge financial support provided through POLYNANO.

References

1. Muthukumar M, Ober CK, Thomas EL (1997) Science 277:1225
2. Park C, Thomas EL (2003) Adv Mater 15:585
3. Park M, Harrison C, Chaikin PM, Register RA, Adamson DH (1997) Science 276:1401
4. Rockford L, Mochrie SG, Russell TP (2001) Macromolecules 34:1487
5. Schaffer E, Thurn-Albrecht T, Russell TP, Steiner U (2000) Nature 403:874
6. Yang XM, Peters RD, Nealey PF, Solak HS, Cerrina F (2000) Macromolecules 33:9575
7. Bates FS, Fredrickson GH (1990) Annu Rev Phys Chem 41:525
8. Bates FS, Fredrickson GH (1999) Physics Today February:32
9. Hamley IW (1998) The Physics of Block Copolymers. Oxford University Press, New York
10. Russell TP, Coulon G, Deline VR, Miller DC (1989) Macromolecules 22:4600
11. Russell TP (1990) Mater Sci Rep 5:171
12. Coulon G, Russell TP, Deline VR, Green PF (1989) Macromolecules 22:2581

13. Coulon G, Collin B, Aussere D, Chatenay D, Russell TP (1990) J Phys (Paris) 51:2801
14. McArdle CB (ed) (1989) Side Chain Liquid Crystalline Polymers. Blackie, Glasgow
15. Plate NA (ed) (1993) Liquid-Crystal Polymers. Plenum, New York
16. Hardouin F, Mery S, Achard MF, Noirez L, Keller P (1991) J Phys (Paris) Lett 1:511
17. Pragliola S, Ober CK, Mather PT, Jeon HG (1999) Macrom Chem Phys 200:2338
18. Zhou QF, Zhu XL, Wen ZQ (1989) Macromolecules 22:491
19. Guillon D, Poeti G, Skoulios A, Fanelli E (1983) J Phys (Paris) Lett 44:L-491
20. Diele S, Oelsner S, Kuschel F, Hisgen B, Ringsdorf H, Zentel R (1987) Macrom Chem Phys 188:1993
21. Al-Hussein M, de Jeu WH, Vranichar L, Pispas S, Hajichristidis N, Itoh T, Watanabe J (2004) Macromolecules 67:6401
22. Chiellini E, Galli G, Angeloni AS, Laus M (1994) Trends Polym Sci 2:244
23. Walther M, Finkelmann H (1996) Progr Polym Sci 21:951
24. Poser S, Fischer H, Arnold M (1998) Prog Polym Sci 23:1337
25. Kato T (2002) Science 295:2414
26. Lee M, Cho B, Zin W (2001) Chem Rev 101:3669
27. Adams J, Gronski W (1989) Macrom Chem Rapid Commun 10:553
28. Bohnert R, Finkelmann H (1994) Macrom Chem Phys 195:689
29. Fischer H, Poser S, Arnold M, Frank W (1994) Macromolecules 27:7133
30. Yamada M, Uguchi T, Hirao A, Nakahama S, Watanabe J (1995) Macromolecules 28:50
31. Mao G, Wang J, Clingman SR, Ober CK, Chen JT, Thomas EL (1997) Macromolecules 30:2556
32. Sanger J, Gronski W, Maas S, Stuhn B, Heck B (1997) Macromolecules 30:6783
33. Galli G, Chiellini E, Laus M, Ferri D, Wolff D (1997) Macromolecules 30:3417
34. Figueiredo P, Geppert S, Brandsch R, Bar G, Thomann R, Spontak RJ, Gronski W, Samlenski R, Mueller-Buschbaun P (2001) Macromolecules 34:171
35. Anthamatten M, Hammond PT (2001) Macromolecules 34:8574
36. Ansari IA, Castelletto V, Mykhaylyk T, Hamley IW, Lu ZB, Itoh Y, Imrie CT (2003) Macromolecules 36:8898
37. Jérome B, Commandeur J, de Jeu WH (1997) Liq Cryst 22:685
38. Ikkala O, ten Brinke G (2002) Science 295:2407
39. Ruokolainen J, Makinen R, Torkkeli M, Makela T, Serimaa R, ten Brinke G, Ikkala O (1998) Science 280:557
40. Tolan M (1999) X-ray Scattering from Soft-Matter Thin Films. Springer, Heidelberg
41. Als-Nielsen J, Jacquemain D, Kjaer K, Leveiller F, Lahav M, Leiserowitz L (1994) Phys Rep 246:251
42. Madsen A, Konovalov O, Robert A, Gruebel G (2001) Phys Rev E 64:61406
43. Fischer H, Poser S, Arnold M (1995) Liq Cryst 18:503
44. Wong GCL, Commandeur J, Fischer H, de Jeu WH (1996) Phys Rev Lett 26:5221
45. Niederhauser TL, Jiang G, Lua Y-Y, Dorff MJ, Woolley AT, Asplund MC, Berges DA, Lindford MR (2001) Langmuir 17:5889
46. Sivaniah E, Hayashi Y, Iino M, Hashimoto T, Fukunaga K (2003) Macromolecules 36:5894
47. Omenat A, Hikmet RAM, Lub J, van der Sluis P (1996) Macromolecules 29:6730
48. Sentenac D, Demirel A, Lub J, de Jeu WH (1999) Macromolecules 32:3235
49. Zheng W-Y, Albalak RJ, Hammond PT (1998) Macromolecules 31:2686
50. Krupers M, Slangen P-J, Moeller M (1998) Macromolecules 31:2552
51. Genzer J, Sivaniah E, Kramer EJ, Wang J, Korner H, Xiang M, Char K, Ober CK, Sambasivan S, Fischer DA (2000) Macromolecules 33:1882
52. Li X, Ruzzi L, Chiellini E, Galli G, Ober CK, Hexemer A, Kramer EJ, Fischer DA (2002) Macromolecules 35:8078

53. Schmidt DL, Coburn CE, DeKoven BM, Potter GE, Meyers GF, Fisher DA (1994) Nature 368:39
54. Krupers M, Moeller M (1997) Macrom Chem Phys 198:2163
55. Al-Hussein M, Séréro Y, Konovalov O, Mourran A, Moeller M, de Jeu WH (2005) Macromolecules 38 (in press)
56. Osuji C, Zhang Y, Mao G, Ober CK, Thomas EL (1999) Macromolecules 32:7703
57. Wang J, Mao G, Ober CK, Kramer EJ (1997) Macromolecules 30:1906
58. Anthamatten M, Hammond PT (1999) 32:8066
59. Iyoda T, Kamata K, Watanabe K, Watanabe R, Yoshida H (2004) E-polymers, Macro-04 (Paris) Session 221
60. Mansky P, Liu Y, Huang E, Russell TP, Hawker C (1997) Science 275:1458
61. Thurn-Albrecht T, DeRouchey J, Russell TP, Jaeger HM (2000) Macromolecules 33:3250
62. Lin ZQ, Kim DH, Wu XD, Boosahda L, Stone D, LaRose L, Russell TP (2002) Adv Mater 14:1373
63. Kim SH, Misner MJ, Xu T, Kimura M, Russell TP (2004) Adv Mater 16:226
64. Wu SH (1982) Polymer Interfaces and Adhesion. Marcel Dekker, New York
65. Thurn-Albrecht T, Steiner R, DeRouchey J, Stafford CM, Huang E, Bal M, Tuominen M, Hawker CJ, Russell TP (2000) Adv Mater 12:787

Surface Nano- and Microstructuring with Organometallic Polymers

Igor Korczagin · Rob G. H. Lammertink · Mark A. Hempenius · Steffi Golze · G. Julius Vancso (✉)

University of Twente, MESA+ Institute for Nanotechnology, P.O. Box 217, 7500 AE Enschede, The Netherlands
g.j.vancso@tnw.utwente.nl

1	Introduction	92
2	Poly(ferrocenyldimethylsilane) as a Reactive Ion Etch Barrier	94
3	Lithographic Applications of Poly(ferrocenylsilane) Homopolymers	97
3.1	Printing of Organometallic Polymers by Soft Lithography	97
3.2	Directed Dewetting	99
3.3	Capillary Force Lithography	100
4	Lithographic Applications of Poly(ferrocenylsilane) Block Copolymers	102
4.1	Structure Formation via Block Copolymer Self-Assembly	102
4.2	Block Copolymer Thin Films	104
4.3	Self-Assembling Resists	106
5	Guided Deposition of Poly(ferrocenylsilane) Polyions	108
6	Conclusions	113
	References	114

Abstract This paper gives an overview of the use of poly(ferrocenylsilane)s in the surface patterning of silicon substrates. Due to the presence of iron and silicon in their main chain, poly(ferrocenylsilane)s show a very high resistance to reactive ion etching, allowing one to transfer polymer patterns directly onto the substrate. Methods for introducing etch-resistant polymer patterns on substrate surfaces include soft lithography approaches such as microcontact printing, directed dewetting, and capillary force lithography. Next to top-down methods, self-assembly strategies are discussed. Phase separation in thin films of asymmetric organic–organometallic block copolymers leads to the formation of nanoperiodic organometallic patterns. The use of such thin films as nanolithographic templates is demonstrated. Surface patterning can also be realized using electrostatic self-assembly of organometallic polyions. Layer-by-layer deposition of poly(ferrocenylsilane) polyanions and polycations on chemically patterned substrates allows one to guide the growth of multilayer thin films and to produce patterned organometallic coatings.

Keywords Block copolymers · Etch resistance · Nanolithography · Organometallic polymers · Thin films

1 Introduction

Macromolecules featuring inorganic elements or organometallic units in the main chain are of considerable interest as they may combine potentially useful chemical, electrochemical, optical, and other interesting characteristics with the processability of polymers [1, 2]. Poly(ferrocenylsilane)s (PFSs), composed of alternating ferrocene and silane units in the main chain, belong to the class of organometallic polymers. The presence of iron and silicon in the PFS backbone adds a distinctive functionality to this class of materials [3]. Poly(ferrocenylsilane)s were found to be effective resists in reactive ion etching processes due to the formation of an etch-resistant iron/silicon oxide layer in oxygen plasmas [4], resulting in several lithographic applications of PFSs. Patterns on micron and sub-micron scales were obtained using PFSs as ink in various soft lithographic techniques [5], and nanopatterning was realized by means of block copolymer lithography. Block copolymers featuring poly(ferrocenylsilane) blocks form nanoperiodic microdomain structures upon phase separation [6]. In thin films of such block copolymers, e.g., poly(isoprene-*block*-ferrocenyldimethylsilane), the high resistance of the organometallic phase to reactive ion etching compared to the organic phase was used to form nanopatterned surfaces. These patterns were transferred onto silicon or silicon nitride substrates in a one-step etching process [7]. Ferrocenylsilane block copolymers can even be employed to pattern thin metal films, as was demonstrated by the use of ferrocenylsilane-styrene block copolymers as templates in the fabrication of nanometer-sized cobalt magnetic dot arrays [8]. Furthermore, phase-separated block copolymer thin films containing PFS domains were found to be efficient precursors to nanoperiodic arrays of iron oxide particles that are active catalysts in the growth of carbon nanotubes [9, 10]. The presence of ferrocene units in the main chain renders PFS electrochemically active [11]. Reversible redox-induced morphology and volume/thickness changes were observed in self-assembled poly(ferrocenylsilane) monolayers on gold [12, 13]. These effects are intimately related to the changes in solubility and conformation of the PFS macromolecules upon oxidation and reduction. Thus, surface-immobilized PFSs constitute an interesting electrochemically addressable stimulus-responsive system [14]. The redox activity of PFS was also employed in the preparation of a redox- and solvent-tunable photonic crystal. The material used in this device was composed of silica microspheres in a matrix of crosslinked PFSs. The geometry and optical properties of the crystal could be changed using the chemomechanical response of PFSs [15].

Pyrolysis of PFS polymers yields nanocomposites containing magnetic Fe clusters with tunable magnetic properties [16]. Finally, block copolymers containing PFS blocks self-assemble, generating remarkable nanoarchitectures such as cylindrical micelles in selective solvents [17, 18]. Consequently, these block copolymers are interesting candidates for a range of potential applications from the fabrication of nanostructured magnetic materials to electronic and photonic applications.

Poly(ferrocenylsilane)s were first obtained by Rosenberg in a condensation polymerization of an appropriate biscyclopentadienide anion with iron(II)chloride [19]. This resulted in oligomers with degrees of polymerization of up to 10. An important step in the development of these polymers was the discovery by Manners et al. that strained, silicon-bridged[1]ferrocenophanes undergo thermal ring-opening polymerization, producing high-molar-mass PFSs [20]. Furthermore, sila[1]ferrocenophanes were found to be polymerizable in solution, by using anionic initiators [21] or transition metal catalysts [22, 23], or in the solid state using a ^{60}C γ-ray source [24] (Scheme 1).

The macromolecular characteristics of the resulting polymers depend on the substituents at silicon. Polymerization of asymmetrically substituted sila[1]ferrocenophanes [25] yields amorphous polymers while symmetrically substituted ferrocenophanes yield semicrystalline materials [26]. Additionally, the glass transition temperature of PFSs can be tuned by varying the size and type of the substituents [27–29]. Several types of iron-containing macromolecules have been synthesized by polymerizing ferrocenophanes with varying numbers and types of bridging atoms [30] and substituents on the cyclopentadienyl rings [31]. The synthesis of water-soluble PFS polyanions [32–34] and polycations [35, 36] and their layer-by-layer deposition [36, 37] to form multilayer thin films was also reported.

The discovery of the anionic ring-opening polymerization of silicon-bridged[1]ferrocenophanes enabled the synthesis of well-defined, near-monodisperse poly(ferrocenylsilane) homo- and block copolymers. PFS was combined with polystyrene [38], polyisoprene [39], poly(dimethylsiloxane) [40], poly(ethylene oxide) [41], poly(ferrocenylphenylphosphine) [42], poly(aminoalkyl methacryate) [43] and recently with poly(methyl methacrylate) [44, 45] blocks.

In the following sections, methods for introducing poly(ferrocenylsilane) patterns on silicon substrates are discussed. These methods include soft lithography approaches such as microcontact printing and capillary force lithography and the self-organization of organic–organometallic block copolymers in thin films, allowing one to form nanoperiodic organometallic patterns, which

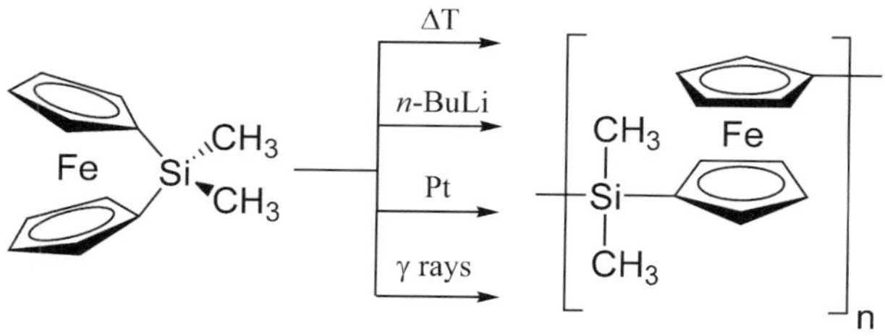

Scheme 1 Ring-opening polymerization of strained dimethylsila[1]ferrocenophane

can serve as nanolithographic templates. Finally, the electrostatic self-assembly of poly(ferrocenylsilane) polyanions and polycations will be discussed as a means to area-selectively introduce ultrathin poly(ferrocenylsilane) films on chemically patterned substrates.

2
Poly(ferrocenyldimethylsilane) as a Reactive Ion Etch Barrier

Oxygen and oxygen-containing plasmas are most commonly employed to modify polymer surfaces [46]. Reactive ion etching can be divided into two etching processes; namely, chemical and physical [47]. Chemical etching refers to the chemical reactions that take place between the active plasma species and the surface of interest. After absorption of the reactive species, the reaction takes place, and the product desorbs from the surface. Physical etching occurs due to the bombardment of positive ions. The positive ions are accelerated towards the surface and break bonds upon impact, which results in physical etching. The balance between the two etching mechanisms depends on many variables, such as gas pressure and composition, reactor design, and temperature [48, 49].

Organometallic compounds are known to act as etching barriers in oxygen plasmas. Contrary to organic compounds, products of chemical etching with oxygen plasmas are non-volatile and therefore do not desorb from the surface. This is the fundamental reason for the low etch rates found for inorganic species when using oxygen plasmas. When poly(ferrocenylsilane) films are exposed to oxygen reactive ion etching conditions, a thin Fe/Si-oxide layer is formed on top of the film, as was established by XPS [4].

From Auger electron spectroscopy (AES), using argon ion-sputtering depth profiling, the composition in the depth of the PFS film was investigated. In the AES spectra (Fig. 1), the first sputtering cycle corresponds to the sample surface, and the increase of the Si signal around cycle 90 indicates the underlying substrate. A thin oxide-rich layer of approximately 10 nm is present at the surface. Compared to the film interior, relatively little carbon and a significant amount of oxygen are present at the surface oxide layer. Interestingly, more silicon is removed from the surface compared to iron in the oxygen plasma treatment.

Radio-frequency discharges in low-pressure fluorocarbon gases are often used for etching silicon, silicon oxide, and silicon nitride [50, 51]. Gases commonly employed in reactive ion etching are CF_4, CHF_3, C_2F_6, SF_6, etc. [52]. By adjusting the composition of the gas, the nature of the plasma can be dramatically changed [53]. Although the process of etching with such plasmas is very complex, and diverse variants of the process exist, some general observations can be highlighted. The C/F atomic ratio of the feed gas is a crucial parameter for the nature of the plasma [54]. If the C/F atomic ratio of the

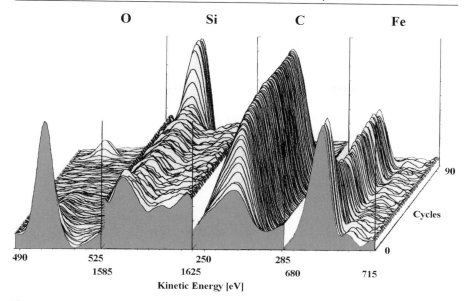

Fig. 1 Auger electron spectroscopy depth profile of an oxygen-plasma-treated film of poly(ferrocenyldimethylsilane). The front of the image corresponds to the exposed free surface. Reproduced with permission from [4]. Copyright 2001, American Chemical Society

feed-in gas increases, the concentration of CF and CF_2 radicals will be higher at the expense of F, and the plasma becomes "polymerizing" and hence protective rather than "etching" due to deposition of a thin fluorocarbon polymer film on the substrate [55, 56]. Even a very thin fluorocarbon film deposited on the substrate results in significant reduction of the SiO_2 etch rate. To suppress this so-called etch stop, the fluorocarbon gases have been diluted with other gases like oxygen [57–59]. Oxygen atoms act as scavengers for carbon, and consequently a relative higher [F] (the etching component) rather than $[CF_x]$ (the polymerizing component) is obtained. Figure 2 shows an example of a structured silicon wafer produced by exposure to CF_4/O_2 reactive ion etching (RIE), using a PFS mask prepared by capillary force lithography [5]. Under the employed conditions, the etch rate contrast between PFS resist and silicon substrate is on the order of 10 : 1.

In order to fabricate structures with higher aspect ratios, one has to decrease sputtering action, which might affect the polymeric mask and increase the rate of chemical etching of the substrate. This has to be achieved without compromising the anisotropy of the etching process. By using SF_6 as the etch gas in a cryogenic reactive ion etcher, where the physical and chemical etching parameters could be varied independently, we could optimize conditions for the highest etch rate contrast [60]. The substrate temperature was kept at – 110 °C during processing. As a result, the rate of chemical etching for Si increased and the rate of physical sputtering was drastically reduced. Using this process we obtained etch rates of 3000 nm/min into Si and around 5 nm/min in the PFS layer (etch rate contrast of 600 : 1). A typical cross-section of an etched stripe pattern is shown in Fig. 3. Such a high etch rate contrast enables the fabrication of high-aspect-ratio structures. Silicon nanopillars with aspect ratios of 10 (see Fig. 4) were obtained with ease.

As the potential of poly(ferrocenyldimethylsilane) as an etch barrier has been demonstrated, we now focus on pathways to generate patterns of these

Fig. 2 SEM image of a patterned silicon substrate. Lines of poly(ferrocenylmethylphenyl-silane) were introduced by capillary force lithography, followed by CF_4/O_2 reactive ion etching (10 min). The polymer mask was subsequently removed using nitric acid

Fig. 3 Cross-sectional SEM image of a patterned silicon substrate. By means of micromolding in capillaries (MIMIC), Si was decorated with poly(ferrocenyldimethylsilane) lines, followed by etching with SF_6 (3 min) in a cryogenic reactive ion etcher. The organometallic polymer residue was removed with nitric acid

Fig. 4 SEM image of silicon nanopillars obtained using cryogenic RIE and a poly(ferrocenyldimethylsilane) mask. The substrate was etched with SF_6 (2 min)

polymers. Masking layers with a high etch resistance are potentially useful for very thin resist layer applications, since reducing resist film thickness is a viable method to prevent pattern collapse when fabricating high-aspect-ratio structures.

3
Lithographic Applications of Poly(ferrocenylsilane) Homopolymers

The fabrication of new functional submicron and nanoscale devices with chemical or true three-dimensional patterns requires new complementary soft-lithography approaches. Polymers come naturally as ideal "inks" or building blocks for soft lithography because of their defined architecture, wide range of chemical functionalities that can be incorporated, and ease of processing on micro and nano scales. Optical systems, and microelectronic devices such as transistors and light-emitting diodes, have been realized with polymers and soft lithography. Among the most promising polymer-based microfabrication strategies are nanoimprint lithography, microcontact printing, micro-fluid-contact printing, lift-up, micromolding in capillaries, replica molding, solvent-assisted micromolding, and its variations (for references see [5]). Polymer patterns can also be formed on chemically heterogeneous surfaces prepared by microcontact printing of self-assembled monolayers and by performing photolithography in the optical near field of an elastomeric mask. However, one of the drawbacks is that most polymers possess poor etch resistivity. Thus, the choice of macromolecular "inks" that can be shaped or transferred with the stamp, to a silicon wafer, for example, and used as a "single-step" resist is rather limited. Poly(ferrocenylsilane)s constitute a group of polymers that combine both macromolecular properties and etch resistivity and are ideal materials for one-step resists. This section describes the fabrication of microstructures of poly(ferrocenylmethylphenylsilane) (PFMPS) and their use as etch resist. The asymmetrically substituted PFS was selected, as symmetrically substituted polymers (such as poly(ferrocenyldimethylsilane)) crystallize, giving heterogeneous, semicrystalline films. Clearly, for homopolymer etch resist applications, structurally homogeneous films are needed. Two soft lithography approaches for pattern formation are presented. One employs solvent-assisted polymer dewetting, and the other relies on the concept of capillary force lithography.

3.1
Printing of Organometallic Polymers by Soft Lithography

The printing of etch-resistant polymers by soft lithography methods is an attractive option for the fabrication of lithographic masks. Microcontact printing is a versatile technique to generate chemically distinct patterns on solid substrates. An elastomer stamp, usually made from poly(dimethylsiloxane) (PDMS), is used to transfer the "ink" to the surface of the substrate by contact printing [61]. Usually, small molecules are employed as inks that upon contact will react with the substrate. When the stamp is removed, a pattern of ink molecules, dictated by the stamp, remains on the surface of the substrate. Recently, approaches where the ink consists of a macromolecule have been attracting attention [62, 63].

The limited use of macromolecular inks can be attributed to poor wetting characteristics of PDMS surfaces by many polymers. Stamps prepared from PDMS have a very low surface tension (around 20 mN/m), compared to common polymers [64]. This is why attempts to use poly(ferrocenyldimethylsilane) as an ink in classical microcontact printing (μCP) were not successful. Wettability of the PDMS stamps used for μCP can be improved by treating the stamp with an oxygen plasma prior to inking. Exposure to an O_2-based plasma oxidizes the surface of the stamp and increases the surface free energy of PDMS [65]. However, the thin silica-like layer that is formed during plasma treatment [66] thermally expands and generates mechanical stress on the stamp upon cooling. The relaxation of this stress induces buckling of the brittle surface layer, and a periodic wavy structure with an orientation perpendicular to the pattern of the stamp is introduced [67]. This fine structure can also be replicated by the PFS used as an ink, and a pattern on two independent length scales is formed (Fig. 5) [68, 69].

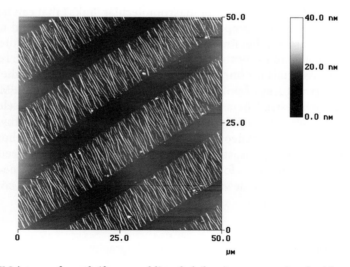

Fig. 5 AFM image of a poly(ferrocenyldimethylsilane) pattern printed with an oxygen-plasma-treated PDMS stamp. The corrugated structure is a result of the stress generated by the difference in thermal expansion between the PDMS matrix and the oxidized surface. Reprinted with permission from [69]. Copyright 2004, John Wiley & Sons, Inc.

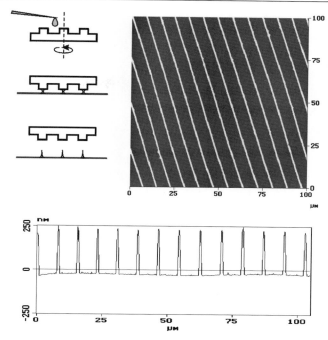

Fig. 6 Solvent-assisted dewetting of poly(ferrocenylmethylphenylsilane) (PFMPS) confined between a PDMS stamp and silicon substrate. Schematic diagram (*left*) and AFM height image (*right*) with section analysis after 10 minutes of CF_4/O_2-RIE treatment. Reprinted with permission from [5]. Copyright 2003, American Chemical Society

When using less aggressive cleaning and oxidation techniques, it is also possible to eliminate the corrugation formation. Cleaning the PDMS stamp in a mild ozone/UV environment will still render the PDMS surface hydrophilic [70], but will not result in extensive surface heating and therefore thermal-expansion-related stress. Patterns revealed after using these stamps have a spacing that corresponds to the periodicity of the stamp structures. This suggests that during printing, the polymer solution dewets between the stamp and Si surface, forming continuous lines in the middle of the protruding stamp contact areas (Fig. 6). Similar results were observed by Braun and co-workers using a hexagonally structured stamp [71]. As shown in Fig. 6, these lines with a width on the order of 1 µm can be etched into the underlying silicon substrate.

3.2
Directed Dewetting

A chemically patterned substrate, obtained by microcontact printing, may serve as a template that directs the adsorption of macromolecules, as is demonstrated in Fig. 7. A 1*H*,1*H*,2*H*,2*H*-perfluorodecyltrichlorosilane self-

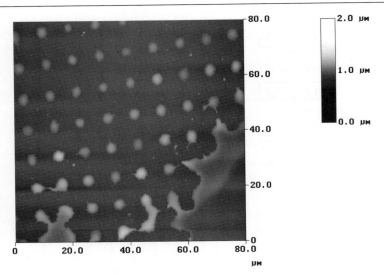

Fig. 7 AFM image of a microcontact printed silicon substrate, partially dewetted by a poly(ferrocenyldimethylsilane) film. The circular features, corresponding to bare silicon, are wetted by PFS. Reprinted with permission from [69]. Copyright 2004, John Wiley & Sons, Inc.

assembled monolayer was printed on a silicon substrate. The circular regions correspond to bare silicon. Poly(ferrocenyldimethylsilane) was then spin-coated onto this hydrophylically/hydrophobically patterned substrate, which directed the dewetting of the polymer film. PFS preferentially wetted the bare silicon circles.

3.3
Capillary Force Lithography

The principles of capillary force lithography [72, 73] and a representative pattern of poly(ferrocenylmethylphenylsilane) (PFMPS) stripes on Si obtained by this technique, are shown in Fig. 8. A PDMS mold was placed in contact with a thin PFMPS film (thickness 27 nm) and, subsequently, the temperature was raised above the T_g (74 °C) of the polymer. The polymer, initially confined in a thin film, is squeezed out from areas of contact between stamp and substrate. It diffuses into the grooves where structures are formed along the vertical walls of the stamp due to capillary rise. Polymer structures, which are approximately 110 nm high and 500 nm wide, were fabricated. Section analysis of the AFM height images revealed a meniscus of the capillary rise (note the different scales for the vertical and horizontal directions). The structures were developed by CF_4/O_2-RIE. After resist removal, a patterned substrate as shown earlier in Fig. 2 was obtained.

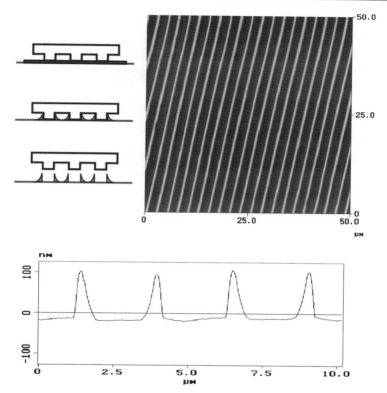

Fig. 8 Capillary force lithography of a thin PFMPS film with a 2 × 3 μm PDMS stamp. Schematic diagram (*left*) and AFM height image with section analysis. The height of the resulting features is 110 nm and the initial thin film thickness was 27 nm. Reprinted with permission from [5]. Copyright 2003, American Chemical Society

Very thin polymer films do not provide enough material to fill the grooves of the stamp completely. As a result, two polymeric lines are formed per groove. This is clearly seen in AFM profiles and was also confirmed by SEM. Increasing the thickness of the initial polymer film results in thicker double lines, which eventually merge when the film thickness exceeds approximately 140 nm. This allows one to tune the lateral dimensions of the polymer lines (Fig. 9).

Figure 10 displays profiles of microstructures obtained with PFMPS before and after etching, showing that approximately 300 nm of silicon was removed in a 10 min treatment. The remaining resist (polymer with oxide layer) still present at the top of the silicon structures can easily be stripped in HNO_3.

Soft lithography-based approaches are versatile and cost-effective methods to generate patterns on the micrometer lengthscale. Submicrometer structures are also accessible provided that PDMS stamps with corresponding geometries are available.

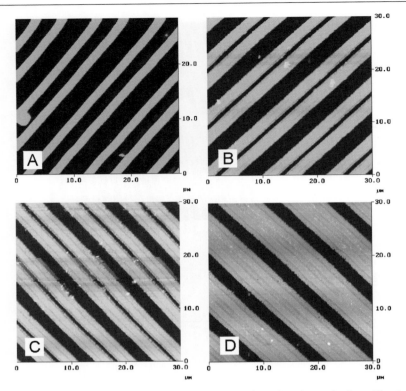

Fig. 9 Patterns in silicon after etching (CF_4/O_2-RIE) and resist stripping. The initial polymer film thickness was **A** 30 nm, **B** 80 nm, **C** 100 nm, and **D** 150 nm. Reprinted with permission from [5]. Copyright 2003, American Chemical Society

4
Lithographic Applications of Poly(ferrocenylsilane) Block Copolymers

Issues regarding the self-assembly of organic–organometallic block copolymers in thin films and applications of the resulting nanoperiodic patterns are discussed in this section.

4.1
Structure Formation via Block Copolymer Self-Assembly

The self-organization of block copolymers constitutes a versatile means of producing ordered periodic structures with phase-separated microdomain sizes on the order of tens of nanometers. The morphology of microdomains formed by diblock copolymers in the bulk has been intensively researched and is by now a relatively well-understood area [74, 75].

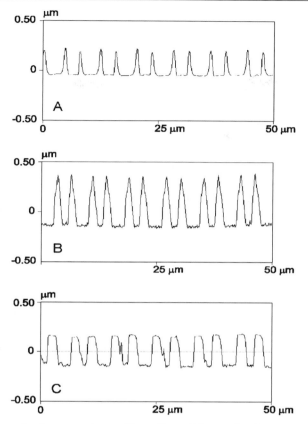

Fig. 10 Pattern development in capillary force lithography (3 × 5 μm stamp) with poly(ferrocenylmethylphenylsilane). AFM section analysis of: **A** PFS mask, **B** structures after etching (CF$_4$/O$_2$-RIE, 10 min), **C** Si structures after removal of the resist in HNO$_3$. Reprinted with permission from [5]. Copyright 2003, American Chemical Society

In neat diblocks, three "classical" ordered microphases are usually distinguished (Fig. 11). These include alternating lamellae, hexagonally packed cylinders, and body-centered-cubic packed spheres. In addition, some other, more complex microstructures may appear, especially near the order–disorder transition.

A series of styrene-ferrocenyldimethylsilane block copolymers form periodic structures on a lengthscale comparable to the size of the polymers [6] (Fig. 12). Pure diblock copolymers phase-separate into an equilibrium structure that depends on the corresponding χ-parameter [77], the length of the block copolymer, and the volume fraction of the two phases. Knowledge of the χ-parameter will allow one to target a specific morphology by adjusting the molar mass and composition of the diblock copolymer accordingly (Scheme 2).

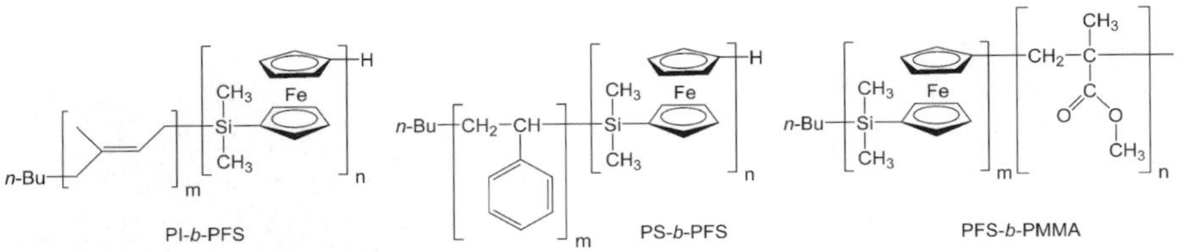

Scheme 2 Examples of organic-ferrocenylsilane block copolymers

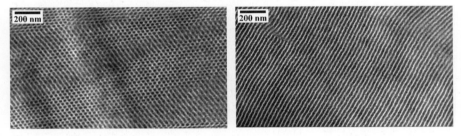

Fig. 11 The classical morphologies in block copolymer systems

Fig. 12 Bright-field TEM micrograph of a cylinder-forming and a lamellar-forming PS-b-PFS diblock copolymer. Reproduced with permission from [6]. Copyright 1999, Wiley Periodicals, Inc., A Wiley Company

4.2
Block Copolymer Thin Films

In thin films, the presence of a substrate and surface can induce orientation of the structure and can result in changes in domain dimensions or in phase transitions due to preferential segregation of one of the blocks at the substrate or the surface. If one block has a higher affinity for the substrate, it will exhibit preferential wetting, resulting in an orientation of the phase-separated domains parallel to the substrate [78]. Similar wetting will occur at the free surface, i.e., the block with the lower surface free energy will enrich the surface. In relatively thick block copolymer films, a mismatch between the film thickness and the bulk lattice spacing can be distributed over many layers. As the film thickness is decreased to a value equal to a few domain periods, the frustration due to the mismatch becomes more significant and can be released by the formation of islands or holes [79] (Fig. 13).

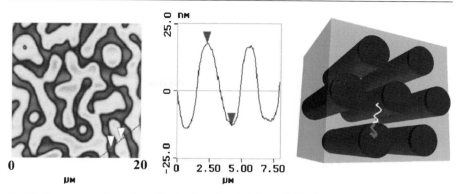

Fig. 13 Representation of a cylinder-forming (PS-*b*-PFS) block copolymer (*right*). When confined in a thin film with a thickness of 2× the lattice spacing, islands and holes are formed (*left*) due to an incommensurate film thickness

Fig. 14 Tapping mode AFM height images of PS-*b*-PFS block copolymer thin films of different initial thickness: **A** 30 nm, **B** 35 nm, **C** 40 nm. Each scan size is 1 μm^2. The organic matrix phase was selectively removed prior to AFM imaging by means of an O_2-RIE treatment. Reprinted with permission from [80]. Copyright 2001, American Chemical Society

Film thickness constraints can also be exploited to control the orientation of the nanodomains. More specifically, when the thickness of the film approaches the lattice constant of the block copolymer, the orientation of the nanodomains is sometimes altered with respect to the substrate. Simply stated, the stress generated by the incommensurate film thickness can be relieved by a change in domain orientation. This is further illustrated by an example in Fig. 14. A styrene-ferrocenyldimethylsilane block copolymer (PS-*b*-PFS), consisting of a styrene fraction of 0.73 forms a cylindrical structure in the bulk [6]. Within a thin film, the pattern that is formed by the block copolymer also depends on the film thickness [80].

Alternatively, the morphology of a thin diblock film can be altered by a subtle change in composition. It was demonstrated for a series of isoprene-ferrocenyldimethylsilane block copolymers (PI-*b*-PFS) that the structure

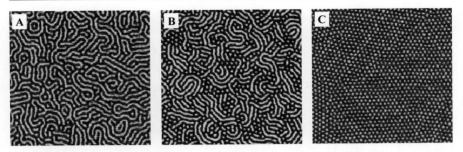

Fig. 15 Tapping mode AFM images of 30 nm thin PI-*b*-PFS copolymer films of different compositions: **A** PI volume fraction of 0.72, **B** PI fraction of 0.76, **C** PI fraction of 0.80. Each scan size is 1 μm^2. Reprinted with permission from [81]. Copyright 2000, American Chemical Society

within a 30 nm thin film could be significantly changed by a slight compositional difference, although the bulk structures all had the same morphology [81]. This is demonstrated in Fig. 15.

4.3
Self-Assembling Resists

Block copolymers can be employed as templates to direct the deposition of inorganic nanostructures. Park et al. [82] used an OsO$_4$-stained microphase-separated thin film of poly(styrene-*block*-butadiene) that produced holes upon RIE in silicon nitride substrates. The etch ratio between the two phases, stained butadiene and styrene, was only about 1 : 2. Möller et al. discussed the use of poly(styrene-*block*-2-vinylpyridine), to prepare masks for nanolithography by loading the PVP domains with gold particles [83] or by selective growth of Ti on top of the PS domains [84].

Thin films of organic–organometallic block copolymers self-assemble to form lateral regions that have a significantly different etching behavior. Furthermore, the two phases already contain all the elements necessary to generate large etching contrast, without the need for staining or loading. The organometallic-rich areas enclose regions of high resistance against removal by oxygen and fluorocarbon plasmas, whereas the organic rich phase is quickly removed. This opens up the possibility of transferring the pattern generated by block copolymer self-assembly in a one-step etching process onto the underlying substrate [7].

A nanostructured silicon surface obtained after etching, as measured via AFM, is shown in Fig. 16. The organic constituents of the block copolymer have been selectively removed by the action of the oxygen plasma, leaving the oxidized metal-containing phase behind. Such self-assembling nanolithographic templates were recently employed to obtain high-density magnetic arrays in cobalt substrates (Fig. 17) [8].

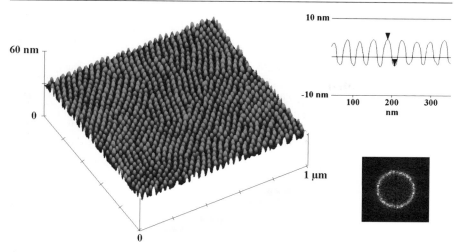

Fig. 16 Tapping mode AFM height image of an etched PI-*b*-PFS organic–organometallic diblock copolymer film. The dots are a result of the block copolymer phase separation. Reprinted with permission from [7]. Copyright 2000, American Chemical Society

Patterns resulting from block copolymer phase separation show good short-range order but usually lack long-range order, which could limit their use in some applications. Combining block copolymer self-assembly with long-range ordering methods would allow nanostructures to be fabricated in precise positions on a substrate. Graphoepitaxy is a method that allows ordered arrays of nanostructures to be formed by spin casting a block copolymer over surfaces patterned with shallow grooves [85]. Figure 18 shows a remarkably well-ordered pattern obtained by spin-coating a PS-*b*-PFS block copolymer on the appropriately structured silicon wafer [86]. It also shows how the long-range order of the pattern changes with the width of the grooves. Within a 500-nm-wide groove, about three rows of close-packed PFS features are aligned parallel to the sidewall (Fig. 18a). In the 320 nm grooves, some regions show a close-packed pattern extending across the groove, but there is a significant number of defects in the pattern (Fig. 18b). However, the alignment is nearly perfect in 240-nm-wide grooves, in which the groove width is comparable to the typical polymer grain size. The features have a sixfold symmetry, and the superposed fast Fourier transforms of the images of several grooves (Fig. 18c) show that the pattern in each groove has the same orientation. The rare domain-packing defects are apparently generated from the edge roughness of the grooves. The ordered block copolymer domain patterns are then transferred into an underlying silica film using a single etching step to create a well-ordered hierarchical structure consisting of arrays of silica pillars with 20 nm feature sizes and aspect ratios greater than 3.

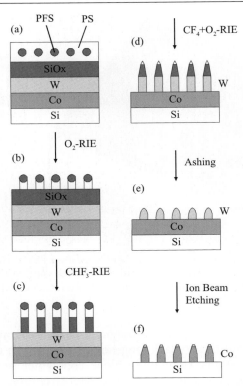

Fig. 17 Fabrication process of cobalt dot arrays via block copolymer lithography. **A** A block copolymer thin film on a multilayer of silica, tungsten, and cobalt. **B** The block copolymer lithographic mask is formed through a O_2-RIE process. The PFS domains are partly oxidized. **C** The silica film is patterned using CHF_3-RIE. **D** The tungsten hard mask is patterned using CF_4/O_2-RIE. **E** Removal of silica and residual polymer by high-pressure CHF_3-RIE. **F** The cobalt dot array is formed using ion beam etching. Reproduced with permission from [8]. Copyright 2001, Wiley-VCH Verlag GmbH

5
Guided Deposition of Poly(ferrocenylsilane) Polyions

Water-soluble poly(ferrocenylsilane) polycations, belonging to the rare class of main chain organometallic polyelectrolytes, have been reported by us and others [35, 36, 87, 88]. These compounds are of interest because they combine the unusual properties of poly(ferrocenylsilane)s with the processability of polyelectrolyte solutions—for example, enabling one to make use of ionic interactions to deposit these polymers onto substrates. Polyelectrolytes can be employed in layer-by-layer self-assembly processes to form ultrathin multilayer films with controlled thickness and composition [89, 90].

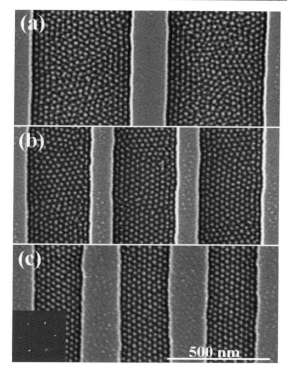

Fig. 18 Scanning electron micrographs of annealed (48 h) and plasma-treated PS-b-PFS films spun at 3500 rpm on silica gratings with **A** 500-nm-wide grooves; **B** 320-nm-wide grooves; **C** 240-nm-wide grooves. The inset is the Fourier transform of the plan-view pattern showing sixfold symmetry. Reprinted with permission from [86]. Copyright 2002, American Institute of Physics

The incorporation of metals in multilayer thin films significantly extends the scope of useful characteristics associated with these films. By employing, for instance, polymeric Ru(II) complexes as polycationic species and poly(sodium acrylate) as polyanions in the layer-by-layer deposition process, efficient light-emitting solid-state devices could be fabricated [91]. In another example, a ferrocene-containing redox-active polycation was combined with an enzyme to produce electrocatalytically active enzyme/mediator multilayer structures [92]. Multilayers composed of poly(4-vinylpyridine) complexed with $[Os(bpy)_2Cl]^{+/2+}$ and poly(sodium 4-styrenesulfonate), for example, were used to accomplish the electrocatalytic reduction of nitrite [93].

Although cationic poly(ferrocenylsilane)s have been used in combination with commercially available organic polyanions to fabricate heterostructured multilayer films [36, 37], fully organometallic multilayers were not reported until recently due to the lack of availability of anionic organometallic polyions [94, 95]. Multilayer structures composed of poly(ferrocenylsilane) polyanions and polycations are of interest as redox-active thin films. Besides forming continuous organometallic multilayer thin films, we explored the layer-by-layer deposition of poly(ferrocenylsilane) polyions onto, for example, hydrophilically/hydrophobically modified substrates, with the aim of building two-dimensionally patterned organometallic multilayers. In general, surfaces modified with microscopically patterned conducting [96], luminescent [97], or redox-active polymer [98] films have potential use in microelectronic and optoelectronic devices and microsensor arrays. Patterned organometallic multilayers may be useful as etch barriers in reactive ion etch processes [4].

The synthesis of PFS polyions was described in detail elsewhere [32, 33, 36]. We employ a poly(ferrocenylsilane) featuring chloropropylmethylsilane repeat units as an organometallic main chain that already has reactive pendant groups in place for further functionalization (Scheme 3). Poly(ferrocenyl(3-chloropropyl)methylsilane) was readily accessible by transition-metal catalyzed ring-opening polymerization [22, 23] of the corresponding (3-chloropropyl)methylsilyl[1]ferrocenophane [36]. By means of halogen exchange, poly(ferrocenyl(3-chloropropyl)methylsilane) can be con-

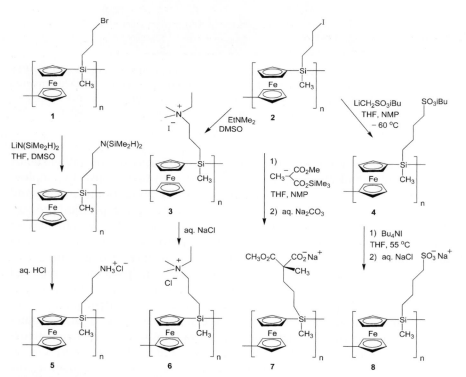

Scheme 3 Examples of poly(ferrocenylsilane) polycations and polyanions

verted quantitatively into its bromopropyl (1) or iodopropyl (2) analogues, which are particularly suitable for functionalization by nucleophilic substitution. Ionic functionalities were introduced by side-group modification of 1 and 2. Scheme 3 shows examples of weak and strong PFS polyelectrolytes.

UV/Vis absorption spectroscopy was used to monitor the electrostatic self-assembly of the organometallic polyions 5/7 and 6/8. The increase in absorption at $\lambda_{max} = 216$ nm shows a linear dependence on the number of bilayers deposited on quartz slides. Information on the increase of film thickness with the number of bilayers was obtained from spectroscopic ellipsometry [92]. Thickness measurements on the films, built up on silicon wafers, were carried out after each bilayer deposition. The fitted multilayer thickness increased linearly with the number of deposited bilayers, in accordance with the UV/Vis absorption spectroscopy results. For the strong polyions 6 and 8, deposited under salt-free conditions, a thickness contribution of 0.6 nm per bilayer was found. In the presence of NaCl, markedly thicker multilayers were obtained (4.5 nm/bilayer).

Poly(ferrocenylsilane)s are redox-active materials, showing fully reversible electrochemical oxidation and reduction [27–29]. A typical voltammogram shows two oxidation waves, indicating intermetallic coupling between neighboring iron centers in the polymer chain. The first oxidation wave was attributed to oxidation of ferrocene centers having neutral neighboring units. In the second wave, at higher potentials, oxidation of the remaining ferrocene centers, predominantly in positions next to oxidized units, is completed [99–101]. Thus, the charge ratio between the two oxidation peaks is approximately 1 : 1. The redox behavior of the multilayer thin films, fabricated on gold electrodes, was studied. Stable multilayers were obtained by first adsorbing a sodium 3-mercapto-1-propanesulfonate monolayer on Au, producing a negatively charged surface ideally suited for polyion adsorption [92]. Cyclic voltammograms of multilayer thin films of polyions 5/7 were recorded for samples having an increasing number of bilayers (1–7) to monitor the electrochemical response as the surface concentration of redox sites increased. The CVs of thin films composed of 1, 3, and 6 bilayers (Fig. 19) show the two oxidation and reduction waves typical of poly(ferrocenylsilane)s. Integration of the voltammetric peaks allows one to calculate the charge involved in the redox processes. From this the surface coverage Γ of ferrocene units can be obtained, using the relation $\Gamma = Q/n \cdot F \cdot A$, where Q is the charge, n is the number of exchanged electrons ($n = 1$ in this case), F is Faraday's constant (96 485 C mol^{-1}), and A is the electrode surface area employed in the measurements (0.44 cm^2) [102, 103]. Using this relation, one organometallic bilayer of polyions 5/7, deposited under salt-free conditions, was found to correspond to a ferrocene surface coverage of 0.45 ferrocene units/nm^2. The surface coverage Γ (ferrocene units/nm^2) increased linearly with the number of bilayers.

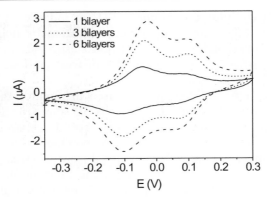

Fig. 19 Cyclic voltammogram of 1, 3, and 6 organometallic bilayers (5/7) deposited on a gold electrode featuring a monolayer of sodium 3-mercapto-1-propanesulfonate taken at a scan rate of $v = 30$ mV/s. Reprinted with permission from [94]. Copyright 2002, American Chemical Society

In addition to producing continuous fully organometallic multilayer thin films, one can form patterned organometallic multilayer structures by confining the deposition of the poly(ferrocenylsilane) polyelectrolytes to selected areas on substrates, which broadens the applicability of such multilayers [104]. The selective deposition of polyelectrolytes on hydrophilically/hydrophobically patterned gold substrates has been described [105, 106]. In this case, patterned self-assembled monolayers consisting of, for example, methyl-terminated and oligo(ethylene glycol)-terminated alkanethiols were introduced on gold substrates using microcontact printing [57–59]. Areas covered with oligo(ethylene glycol)-terminated alkanethiols were found to prevent adsorption of polyelectrolytes. Here, as a demonstration, a gold substrate was patterned with 5 μm-wide methyl-terminated alkanethiol lines, separated by 3 μm, by microcontact printing of 1-octadecanethiol. The uncovered areas were subsequently filled in with 11-mercapto-1-undecanol, resulting in a hydrophilically/hydrophobically patterned substrate. AFM height and friction force images of these patterned self-assembled monolayers (Fig. 20, top images) show minimal height contrast but a large contrast in friction force, with the hydroxyl-terminated lines corresponding to the high-friction areas. The patterned substrate was then coated with 12 bilayers of polyions 5/7 and again examined by contact mode AFM. Clearly, after deposition, the height contrast increased, and the contrast in friction force was reversed, which shows that the multilayers grow selectively on the broad, methyl-terminated stripes (Fig. 20, lower images) [94].

The resistivity of the hydroxyl-terminated areas to polyion deposition was demonstrated by first forming a monolayer of 11-mercapto-1-undecanol on a gold substrate, which was then processed in a similar manner as the patterned substrates, and subsequently analyzed by XPS. Fe 2p signals, in-

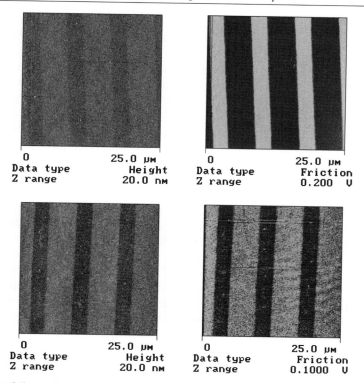

Fig. 20 Multilayer deposition on a hydrophilically/hydrophobically patterned gold substrate. Upper AFM images: height (*left*) and friction force (*right*) images of patterned methyl- and hydroxyl- alkanethiol self-assembled monolayers. Adsorption of poly(ferrocenylsilane) polyions (5/7, 12 bilayers) occurs selectively on the broad methyl-terminated stripes (lower AFM images). Reprinted with permission from [94]. Copyright 2002, American Chemical Society

dicating adsorbed polyions, were absent in the survey scan. The selective adsorption of the polyions on the methyl-terminated regions of the surface is most likely driven by favorable hydrophobic interactions [107] between these areas and the hydrophobic poly(ferrocenylsilane) backbone, minimizing the interfacial free energy of the system. Such favorable secondary interactions with the hydrophilic regions, which are hydrated under the processing conditions [108], are excluded.

6
Conclusions

Metal-containing polymers appear to be valuable candidates for developing highly etch-resistant resists. Masking layers with a high etch resistance are potentially useful for very thin resist layer applications, since reducing resist film thickness is a viable method to prevent pattern collapse when fabricating high-aspect-ratio structures.

Advances in printing techniques of polymers may also contribute to the progress in cost-effective pattern replication procedures. The ability to replicate and print polymer patterns on micron and submicron lengthscales opens up a route to economically fabricate etch-resistant patterns for applications where metallic contaminations are not relevant. Examples of such applications include the fabrication of photomasks, microfluidic devices, optical components, data storage arrays, etc.

Self-assembly of organometallic block copolymers provides access to nanolithographic masks. A broad range of nanoscale morphologies, which can be developed in one step using reactive ion etching, is available. These patterns can be confined on a surface, and their long-range order can be further improved using techniques such as graphoepitaxy [109].

Poly(ferrocenylsilane) polyions are readily processed to novel, fully organometallic multilayer thin films. Furthermore, these organometallic polyions, featuring a hydrophobic backbone, enable one to make use of both electrostatic and hydrophobic interactions to fabricate self-organized patterns on templated substrates, with potential applications as aqueous processable ultrathin etch resists.

Acknowledgements This work was supported in part by the European Community's Human Potential Programme under contract HPRN-CT-1999-00151, [PolyNano].

References

1. Abd-El-Aziz AS (2004) Overview of organoiron polymers. In: Abd-El-Aziz AS, Carraher CE, Pittman CU, Sheats JE, Zeldin M (eds) Macromolecules containing metal and metal-like elements, vol 2. Wiley, New York, p 1–27
2. Abd-El-Aziz AS (2002) Macromol Rapid Commun 23:995
3. Manners I (2004) Synthetic metal-containing polymers. Wiley, New York
4. Lammertink RGH, Hempenius MA, Chan VZH, Thomas EL, Vancso GJ (2001) Chem Mater 13:429
5. Korczagin I, Golze S, Hempenius MA, Vancso GJ (2003) Chem Mater 15:3663
6. Lammertink RGH, Hempenius MA, Thomas EL, Vancso GJ (1999) J Polym Sci Part B: Polym Phys 37:1009
7. Lammertink RGH, Hempenius MA, van den Enk JE, Chan VZH, Thomas EL, Vancso GJ (2000) Adv Mater 12:98
8. Cheng JY, Ross CA, Chan VZH, Thomas EL, Lammertink RGH, Vancso GJ (2001) Adv Mater 13:1174
9. Hinderling C, Keles Y, Stöckli T, Knapp HF, de los Arcos T, Oelhafen P, Korczagin I, Hempenius MA, Vancso GJ, Pugin R, Heinzelmann H (2004) Adv Mater 16:876
10. Lastella S, Jung YJ, Yang HC, Vajtai R, Ajayan PM, Ryu CY, Rider DA, Manners I (2004) J Mater Chem 14:1791
11. Nguyen MT, Diaz AF, Dementev VV, Pannell KH (1993) Chem Mater 5:1389

12. Péter M, Hempenius MA, Lammertink RGH, Vancso GJ (2001) Macromol Symp 167, 285
13. Péter M, Hempenius MA, Kooij ES, Jenkins TA, Roser SJ, Knoll W, Vancso GJ (2004) Langmuir 20:891
14. Zou S, Ma Y, Hempenius MA, Schönherr H, Vancso GJ (2004) Langmuir 20:6278
15. Arsenault AC, Míguez H, Kitaev V, Ozin GA, Manners I (2003) Adv Mater 15:503
16. MacLachlan MJ, Ginzburg M, Coombs N, Coyle TW, Raju NP, Greedan JE, Ozin GA, Manners I (2000) Science 287:1460
17. Massey J, Power KN, Manners I, Winnik MA (1998) J Am Chem Soc 120:9533
18. Cao L, Massey JA, Winnik MA, Manners I, Riethmuller S, Banhart F, Spatz JP, Möller M (2003) Adv Funct Mater 13:271
19. Rosenberg H, Rausch MD (1962) US Patent 3 060 215
20. Foucher DA, Tang BZ, Manners I (1992) J Am Chem Soc 114:6246
21. Rulkens R, Lough AJ, Manners I (1994) J Am Chem Soc 116:797
22. Ni YZ, Rulkens R, Pudelski JK, Manners I (1995) Macromol Rapid Commun 16:637
23. Gómez-Elipe P, Macdonald PM, Manners I (1997) Angew Chem Int Ed 36:762
24. Rasburn J, Petersen R, Jahr T, Rulkens R, Manners I, Vancso GJ (1995) Chem Mater 7:871
25. Foucher D, Ziembinski R, Petersen R, Pudelski J, Edwards M, Ni YZ, Massey J, Jaeger CR, Vancso GJ, Manners I (1994) Macromolecules 27:3992
26. Foucher DA, Ziembinski R, Tang BZ, Macdonald PM, Massey J, Jaeger CR, Vancso GJ, Manners I (1993) Macromolecules 26:2878
27. Manners I (2003) Macromol Symp 196:57
28. Kulbaba K, Manners I (2001) Macromol Rapid Commun 22:711
29. Manners I (1999) Chem Commun 857
30. Nelson JM, Rengel H, Manners I (1993) J Am Chem Soc 115:7035
31. Nelson JM, Lough AJ, Manners I (1994) Angew Chem Int Ed 33:989
32. Hempenius MA, Vancso GJ (2002) Macromolecules 35:2445
33. Hempenius MA, Brito FF, Vancso GJ (2003) Macromolecules 36:6683
34. Wang Z, Lough A, Manners I (2002) Macromolecules 35:7669
35. Power-Billard KN, Manners I (2000) Macromolecules 33:26
36. Hempenius MA, Robins NS, Lammertink RGH, Vancso GJ (2001) Macromol Rapid Commun 22:30
37. Ginzburg M, Galloro J, Jäkle F, Power-Billard KN, Yang S, Sokolov I, Lam CNC, Neumann AW, Manners I, Ozin GA (2000) Langmuir 16:9609
38. Ni YZ, Rulkens R, Manners I (1996) J Am Chem Soc 118:4102
39. Lammertink RGH, Hempenius MA, Vancso GJ (2000) Langmuir 16:6245
40. Rulkens R, Ni YZ, Manners I (1994) J Am Chem Soc 116:12121
41. Resendes R, Massey J, Dorn H, Winnik MA, Manners I (2000) Macromolecules 33:8
42. Wang XS, Winnik MA, Manners I (2002) Macromolecules 35:9146
43. Wang XS, Winnik MA, Manners I (2002) Macromol Rapid Commun 23:210
44. Korczagin I, Hempenius MA, Vancso GJ (2004) Macromolecules 37:1686
45. Kloninger C, Rehahn M (2004) Macromolecules 37:1720
46. d'Agostino R (1990) Plasma Deposition, Treatment, and Etching of Polymers. Academic Press, New York
47. Manos DM, Flamm DL (1988) Plasma etching: an introduction. Academic Press, New York
48. van Roosmalen AJ, Baggerman JAG, Brader SJH (1991) Dry etching for VSLI. Plenum, New York
49. Flamm DL, Donnelly VM (1981) Plasma Chem Plasma Proc 1:317

50. Knizikevicius R, Galdikas A, Grigonis A (2002) Vacuum 66:39
51. Kokkoris G, Gogolides E, Boudouvis AG (2002) J Appl Phys 91:2697
52. Jansen H, Gardeniers H, de Boer M, Elwenspoek M, Fluitman J (1996) J Micromech Microeng 6:14
53. Oehrlein GS, Williams HR (1987) J Appl Phys 62:662
54. Coburn JW, Kay E (1979) Solid State Technol 22:117
55. Strobel M, Corn S, Lyons CS, Korba GA (1987) J Polym Sci A 25:1295
56. Strobel M, Thomas PA, Lyons CS (1987) J Polym Sci A 25:3343
57. Mogab CJ, Adams AC, Flamm DL (1978) J Appl Phys 49:3796
58. Donnelly VM, Flamm DL, Dautremont-Smith WC, Werder DJ (1984) J Appl Phys 55:242
59. Egitto FD, Matienzo LJ, Schreyer HB (1992) J Vac Sci Technol A 10:3060
60. Legtenberg R, Jansen H, de Boer M, Elwenspoek M (1995) J Electrochem Soc 142:2020
61. Xia YN, Whitesides GM (1998) Angew Chem Int Ed 37:551
62. Li HW, Huck WTS (2002) Curr Opin Solid State Mat Sci 6:3
63. Granlund T, Nyberg T, Roman LS, Svensson M, Inganas O (2000) Adv Mater 12:269
64. Clarson SJ, Semlyen JA (1993) Siloxane polymers. Prentice-Hall, New York
65. Delamarche E, Geissler M, Bernard A, Wolf H, Michel B, Hilborn J, Donzel C (2001) Adv Mater 13:1164
66. Hillborg H, Gedde UW (1998) Polymer 39:1991
67. Chua DBH, Ng HT, Li SFY (2000) Appl Phys Lett 76:721
68. Bowden N, Brittain S, Evans AG, Hutchinson JW, Whitesides GM (1998) Nature 393:146
69. Lammertink RGH, Korczagin I, Hempenius MA, Vancso GJ (2004) Metal-containing polymers for high-performance resist applications. In: Abd-El-Aziz AS, Carraher CE, Pittman CU, Sheats JE, Zeldin M (eds) Macromolecules containing metal and metal-like elements, vol 2. Wiley, New York, p 115–133
70. Efimenko K, Wallace WE, Genzer JJ (2002) J Colloid Interface Sci 254:306
71. Wang MT, Braun HG, Kratzmuller T, Meyer E (2001) Adv Mater 13:1312
72. Suh KY, Kim YS, Lee HH (2001) Adv Mater 13:1386
73. Suh KY, Lee HH (2002) Adv Funct Mater 12:405
74. Bates FS, Rosedale JH, Fredrickson GH (1990) J Chem Phys 92:6255
75. Hamley IW Nanotechnology 14:R39
76. Park C, Yoon J, Thomas EL (2003) Polymer 44:6725
77. Eitouni HB, Balsara NP, Hahn H, Pople JA, Hempenius MA (2002) Macromolecules 35:7765
78. Coulon G, Russell TP, Deline VR, Green PF (1989) Macromolecules 22:2581
79. Coulon G, Ausserre D, Russell TP (1990) Journal de Physique 51:777
80. Lammertink RGH, Hempenius MA, Vancso GJ, Shin K, Rafailovich MH, Sokolov J (2001) Macromolecules 34:942
81. Lammertink RGH, Hempenius MA, Vancso GJ (2000) Langmuir 16:6245
82. Park M, Harrison C, Chaikin PM, Register RA, Adamson DH (1997) Science 276:1401
83. Spatz JP, Herzog T, Mößmer S, Ziemann P, Möller M (1999) Adv Mater 11:149
84. Spatz JP, Eibeck P, Mößmer S, Möller M, Herzog T, Ziemann P (1998) Adv Mater 10:849
85. Segalman RA, Yokoyama H, Kramer EJ (2001) Adv Mater 13:1152
86. Cheng JY, Ross CA, Thomas EL, Smith HI, Vancso GJ (2002) Appl Phys Lett 81:3657
87. Neuse EW, Khan FBD (1986) Macromolecules 19:269
88. Kelch S, Rehahn M (1999) Macromolecules 32:5818

89. Decher G (1997) Science 277:1232
90. Bertrand P, Jonas A, Laschewsky A, Legras R (2000) Macromol Rapid Commun 21:319
91. Wu A, Yoo D, Lee JK, Rubner MF (1999) J Am Chem Soc 121:4883
92. Hodak J, Etchenique R, Calvo EJ, Singhal K, Bartlett PN (1997) Langmuir 13:2708
93. Sun J, Sun Y, Zou S, Zhang X, Sun C, Wang Y, Shen J (1999) Macromol Chem Phys 200:840
94. Hempenius MA, Robins NS, Péter M, Kooij ES, Vancso GJ (2002) Langmuir 18:7629
95. Halfyard J, Galloro J, Ginzburg M, Wang Z, Coombs N, Manners I, Ozin GA (2002) Chem Commun 1746
96. Beh WS, Kim IT, Qin D, Xia YN, Whitesides GM (1999) Adv Mater 11:1038
97. Clark SL, Handy ES, Rubner MF, Hammond PT (1999) Adv Mater 11:1031
98. Schanze KS, Bergstedt TS, Hauser BT, Cavalaheiro CSP (2000) Langmuir 16:795
99. Rulkens R, Lough AJ, Manners I, Lovelace SR, Grant C, Geiger WE (1996) J Am Chem Soc 118:12683
100. Foucher D, Ziembinski R, Petersen R, Pudelski J, Edwards M, Ni YZ, Massey J, Jaeger CR, Vancso GJ, Manners I (1994) Macromolecules 27:3992
101. Foucher DA, Honeyman CH, Nelson JM, Tang BZ, Manners I (1993) Angew Chem Int Ed 32:1709
102. Finklea HO (1996) Electrochemistry of organized monolayers of thiols and related molecules on electrodes. In: Bard AJ, Rubinstein I (eds) Electroanalytical chemistry, vol 19. Dekker, New York, p 109
103. Murray RW (1984) Chemically modified electrodes. In: Bard AJ (ed) Electroanalytical chemistry, vol 13. Dekker, New York, p 191
104. Decher G, Lehr B, Lowack K, Lvov Y, Schmitt J (1994) Biosens Bioelectron 9:677
105. Hammond PT, Whitesides GM (1995) Macromolecules 28:7569
106. Clark SL, Hammond PT (1998) Adv Mater 10:1515
107. Clark SL, Hammond PT (2000) Langmuir 16:10206
108. Sprik M, Delamarche E, Michel B, Rothlisberger U, Klein ML, Wolf H, Ringsdorf H (1994) Langmuir 10:4116
109. Cheng JY, Mayes AM, Ross CA (2004) Nature Materials 3:823

Genetic Engineering of Protein-Based Polymers: The Example of Elastinlike Polymers

J. Carlos Rodríguez-Cabello (✉) · Javier Reguera · Alessandra Girotti · F. Javier Arias · Matilde Alonso

BIOFORGE research group, Dpto. Física de la Materia Condensada, E.T.S.I.I., Universidad de Valladolid, Paseo del Cauce s/n, 47011 Valladolid, Spain
cabello@eis.uva.es

1	Introduction	120
1.1	The Present and Future Global Challenges of Polymer Science	120
1.2	Biological Macromolecules: The Lesson from Nature	122
2	Genetic Engineering of Protein-Based Polymers: The "Gutenberg Method" in Polymer Design and Production	123
3	State of the Art in GEPBPs	128
4	Silklike Polymers	128
5	Elastinlike Polymers: A Privileged Family of GEPBPs	135
5.1	Introducing ELPs	135
5.2	Smart and Self-assembling Properties of ELPs	136
5.3	Basic Molecular Designs: Thermal Responsiveness	137
5.4	Introducing Further Chemical Functions in the Monomer: pH-responding ELPs and the ΔT_t Mechanism	142
5.5	Self-Assembling Capabilities of ELPs	146
5.6	Further Chemical Functionalization of the Monomer: Photoresponding ELPs and the Amplified ΔT_t Mechanism	150
5.7	The Outstanding Biocompatibility of Elastinlike Polymers: The Third Pillar for Extraordinary Biomaterial Designs	155
5.7.1	ELPs for Drug Delivery Purposes: Different Strategies for Molecular Designs	156
5.7.2	ELPs for Tissue Engineering: Introducing Tailored Biofunctionality	158
6	Conclusions	162
	References	163

Abstract In spite of the enormous possibilities of macromolecules as key elements in developing advanced materials with increased functionality and complexity, the success in this development is often limited by the randomness associated with polymer synthesis and the exponential increase in technical difficulties caused by the attempt to reach a sufficiently high degree of complexity in the molecular design. This paper describes a new approach in the design of complex and highly functional macromolecules, the genetic engineering of protein-based macromolecules. The exploitation of the efficient machinery of protein synthesis in living cells opens a path to obtain extremely well-defined and complex macromolecules.

Different molecular designs are presented, with increasing degree of complexity, showing how the controlled increase in their complexity yields (multi)functional materials with more select and sophisticated properties. The simplest designs show interesting properties already, but the adequate introduction of given chemical functions along the polymer chain presents an opportunity to expand the range of properties to enhanced smart behavior and self-assembly. Finally, examples are given where those molecular designs further incorporate selected bioactivities in order to develop materials for the most cutting-edge applications in the field of biomedicine and nano(bio)technology.

Keywords Elastinlike polymers · Genetic engineering · Protein-based polymers · Self-assembly · Smart polymers

1
Introduction

1.1
The Present and Future Global Challenges of Polymer Science

In recent decades, polymer science has definitively shown that macromolecules can be excellent candidates to create highly functional materials. With the availability of thousands of different monomers and the possibilities opened by their different combinations, polymer science has succeeded on many occasions when a material was needed for a particular application, from the simplest uses as bulk commodities to the most sophisticated and special biomedical, engineering, or nanotechnological ones. Very few other technical developments in history have shown both the rapid development and deep societal impact of polymer science. Currently, the number of different technologies enabled by the existence of the adequate polymer is amazing, and the crucial role of polymer science in the current stage of societal development and well-being is beyond question.

Up till now, when a new development was required from polymer science, it has been possible to design and obtain a new polymer fitting that particular requirement. The most challenging tasks for polymer science were rather of a more logistical nature than scientific. For example, a reduction in the number of different polymers used in practice to cover the whole range of consumer demands (in order to simplify and make more profitable their manufacture) has been and still is the cause of important research efforts. Additionally, environmental and other related matters, such as sustainability, have also been addressed, but they have always remained in the background and have not significantly limited the development of polymer science.

However, this situation started to change a couple decades ago. At that time, the concepts of self-assembly and hierarchical organization, as well as others such as "smartness," began to awake extended interest within the polymer science community and boosted expectations for new applications. The deeper knowledge on the physical-chemical basis of the high functionality of those pioneer polymers triggered a rapid scaleup in the complexity of new designs as well as the need for controlling their composition.

However, all methodologies of polymer synthesis are characterized by an unavoidable component of randomness and lack of control. This is especially true for the classical radical, cationic, or ring-opening polymerizations, where, even in the simplest polymers, it is not possible to control parameters such as the degree of polymerization. We are used to considering this as a mean value bearing a statistical meaning. Generally, the information given by the mean molecular weight needs to be completed by a polydispersity index in order to quantify how broad or narrow is the molecular weight distribution of our polymer. In the case of copolymers, we are also used to dealing with random copolymers, although, with some effort, we can prepare alternating or block copolymers.

However, the design of highly functional polymers unavoidably means the design of complex molecules and a tight control in their synthesis.

New discoveries in the area of catalysis and controlled polymerizations with the work of Matyjaszewski, Hawker, Waymouth, Coates, Deming, and many more [1–5] have allowed us to keep pace, but with the impression that the demands are growing faster than the achievements. This could be a signal that in the future conventional polymer science could reach one of its most critical limits; polymer chemistry could be overwhelmed by the demands of the new polymer designs. Additionally, in the existing technologies of polymer synthesis, unavoidably, there is an exponential relation between the cost (money and time) needed to synthesize a polymer and the complexity of its primary structure. But that is a matter not only of costs. As the complexity increases, the synthesis methods and protocols become less and less robust and more and more difficult to scale up, preventing, to a large extent, its commercial exploitation. Presently, although we already have the knowledge needed to design advanced polymers envisaged as possessors of extraordinary properties, the frustrating fact is that they are very difficult to synthesize in practice (Fig. 1).

In addition, perhaps this is not the only limiting condition in the current state of development of polymer science. The field could be facing an additional crucial problem in the middle or long term. Most synthesis methodologies and the polymers we currently produce are based exclusively on petroleum-derived chemicals. It is estimated that more than 200 million Tm of crude are used yearly as raw material to produce plastics and rubbers, while an equivalent amount of oil is burned to generate the energy needed for their synthesis. As a source of materials, oil is not renewable. Although there is no consensus about the level of oil reserves, it is clear that this resource is not infinite and that its price will likely continue to increase if we keep our increasing rate of demand. Additionally, perhaps we do not have to wait until the imminent exhaustion of oil reserves to reduce oil's use as a source of en-

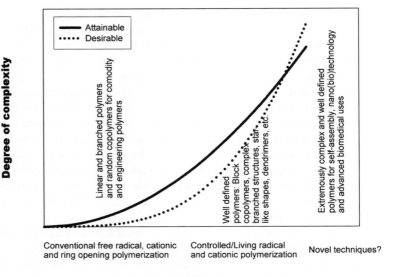

Fig. 1 Evolution of polymer synthesis methodologies and of demand by polymer designers

ergy and plastics. The growing evidence that the increase in the atmospheric CO_2 level is causing a palpable modification of the global climate [6] could lead, in the middle or long term, to abandon, or at least reduce drastically, oil as our main source of raw materials for plastics.

The above-described scenario is, obviously, unpleasant. However, as has happened before, when a technology arrives at a bottleneck, there could exist an alternative way to break through the impasse. This paper is devoted to gathering arguments in favor of one of those possible alternative routes: the genetic engineering of protein-based polymers (GEPBPs). By this approach, evidences on the possibility of obtaining very complex and highly functional polymers, well beyond the reach of the present chemical methods of synthesis and from exclusively renewable sources, will be presented.

1.2
Biological Macromolecules: The Lesson from Nature

Biology discovered long ago that macromolecules are the best option for obtaining highly functional materials. Novel concepts in materials science such as hierarchical organization, mesoscale self-assembly, or smartness are common to many natural macromolecules such as proteins, nucleic acids, and polysaccharides (or combinations thereof). In fact, the slow but implacable process of natural selection has produced materials showing a level of functionality that is really much higher than the level we have reached in our synthetic materials. One of the best (and nicest) examples is proteins. Proteins in living cells show an amazing set of capabilities in terms of functionality. From structural proteins, all of them showing acute self-assembly capabilities, to extraordinary enzymes, with their superior catalytic performance and highly efficient molecular machines (flagellar rotary motor, etc.), examples abound. Natural proteins are usually large and very complex molecules containing diverse specific functional groups to generate and direct self-assembly and function. Nature makes use also of different physical processes that allow for directed and controlled organization from the molecular to the macroscopic level. As a whole, both local organization through functional chemical groups and the physical properties giving rise to order up to the highest scales provide the properties and functions that biological systems require for their efficient functioning.

Nevertheless, all of this amazing functionality displayed by natural proteins seems to be based on a simple fact: a complex and completely defined primary structure. In living cells, protein biosynthesis is carried out with an absolute control of the amino acid sequence, from the first amino acid to the last with a complete absence of randomness. In fact, the need for this absolute control is dramatically clear in some genetic disorders in which the lack or a substitution of a single amino acid in the whole protein leads to a complete loss of the original function, which can have dramatic consequences in some cases such as falciform anemia (sickle cell anemia), phenylketonuria, and cystic fibrosis [7].

Herein lies the lesson that, if we want to create really functional materials, we must find a way to synthesize complex and completely defined macromolecules. This task, which completely overwhelms our most sophisticated chemical methods, is taking place incessantly in all living cells. One more characteristic of protein biosynthesis deserves mention. The protein biosynthesis machinery is extraordinarily flexible. Ribosomes are able to process and produce practically any amino acid sequence stored in the holders of information called genes, so its flexibility is nearly absolute. Therefore, for practical purposes, it is interesting to realize that if one controls the information that genes deliver to the machinery, then one completely controls the biosynthesis process itself.

2
Genetic Engineering of Protein-Based Polymers: The "Gutenberg Method" in Polymer Design and Production

Due to current developments in molecular biology, we have for the first time the ability to create almost any DNA duplex codifying any amino acid sequence at will. We also have the chance to introduce this synthetic gene in the genetic content of a microorganism, plant, or other organisms and induce the production of its codified protein-based polymer (PBP) as a recombinant protein [8–16]. Therefore, as we now have all the required technology, the use of genetically modified cells as cellular factories to produce sophisticated polymers is extremely tempting. This approach has many advantages.

First, as our knowledge of the protein-function relationship continues to grow, GEPBPs will be able to show any function or property, simple or complex, present in natural proteins. In this sense, this method opens the opportunity for exploiting the huge amounts of resources, in terms of functionality, hoarded and refined in a very efficient manner over the long course of natural selection. GEPBPs easily make use of the vast amounts of functional wealth present in the hundreds of thousands of different proteins in living organisms in the widest sense, from the smallest prions or viruses to highly complex animals.

On the other hand, as we can construct the codifying gene base by base following our own original designs and without being restricted to genes of fragments found in living organisms, we can design and produce GEPBPs to obtain materials, systems, and devices exhibiting a function not displayed in living organisms but of a particular technological interest [17].

Third, from the point of view of a polymer chemist, the degree of control and complexity attained by genetic engineering is clearly superior to those achieved by any present chemical synthesis technologies. GEPBPs are strictly monodisperse, while they can be obtained from a few hundred daltons to more than 200 kDa, and these limits are continuously expanding [18]. Among other things, this has opened the possibility of studying, in a simple and highly precise manner, the dependence of different material properties on the molecular weight (MW) [19, 20], and this knowledge also opens the possibility of finely tuning those properties in the designed materials. In addition, although monodispersity is not an important requirement for bulk polymers—it is even desirable in some cases—it clearly enhances the chances of success in designing materials with self-assembling and smart behavior [21].

Although, as discussed above, the increase in complexity of conventional polymers unavoidably means an exponential increase in time and cost of production, this relationship is not fulfilled by GEPBPs. Paradoxically, experience constantly shows that the enzymes and all other techniques of molecular biology that are used for the construction of synthetic genes as well as all the molecular machinery implicated in protein biosynthesis work better with complex GEPBPs than with simple and highly repetitive GEPBPs. Biological systems are adapted to build complex natural proteins, so they feel more comfortable in an environment of complexity. Therefore, for GEPBPs, there is not a clear and direct relationship between production yield and polymer complexity. In practice, usually complexity is more feasible than simplicity. In addition, the cost of production of GEPBPs is not related to their complexity. By this approach, the most costly task in terms of time and money is gene construction. However, once the genetically modified (micro)organism is obtained, the fast, robust, and cheap GEPBP production readily compensates the costs associated with the molecular biology steps.

In addition, in contrast to conventional polymers, where the raw materials are the monomers, the raw materials employed in GEPBP biosynthesis are not the amino acids themselves. Recall that protein synthesis in living cells is inserted within a dense and complex metabolic network, by which many simple, renewable, and cheap sources of carbon and nitrogen can be finally converted into the needed amino acids and, finally, to the desired GEPBP.

Fourth, the number of different combinations attainable by combining the 20 natural amino acids is virtually infinite, so the number of different GEPBPs that can be obtained seems to be more than enough.

Somehow, this situation recalls the time when Johannes Gutenberg began building his press (in 1436). At that time, rather than writing books one by one, Gutenberg found that the time spent in building the movable type and the press, even to print high-quality and complex texts, was rapidly compensated by the reduced time in printing many identical copies. Therefore, perhaps we are now in a position to apply this concept to polymer production (the "Gutenberg Method"). If we want to obtain several identical batches of a sophisticated and complex polymer, we should not direct our main effort to building the polymer itself but to building the gene that codifies it. Then, polymer production can be done by expressing the gene in a cellular factory. Thus, these cells play the role of the press in book printing.

Although this list does not pretend to be exhaustive, the final advantage mentioned here stresses environmental considerations. GEPBPs are made from biomass, and their production involves only renewable biomass and environmentally clean processes from raw materials to waste. In addition, no petroleum-based chemicals are used. GEPBPs are, evidently, biodegradable, and water is used as the exclusive solvent in most GEPBPs produced to date. GEPBPs are obtained by an easily scalable technology, fermentation, that uses moderate amounts of energy and temperatures. Additionally, a main goal in the production of GEPBPs is their production in genetically modified plants. In this way, there is even no need for fermentation facilities, which reduce significantly the productions costs, while this could be a way to help revitalize the agriculture sectors in many countries.

It is not easy to imagine clear disadvantages of GEPBPs vs. petroleum-based polymers because even the differential in production costs is decreasing rapidly on one hand due to the progressive increase in bioproduction yields and the possibilities opened by using genetically modified plants instead of microorganisms, and the continuous increase in oil prices on the other hand. Perhaps, as polymer scientists, the first thing to come to mind would be that conventional polymer science has produced thousands of different useful monomers. Therefore, the possibilities opened by this high number of petroleum-based monomers, in terms of availability of function, seems to be overwhelming if we consider that, in designing GEPBPs, we must restrict ourselves to just the 20 natural amino acids. However, this reasoning could also be fallacious if we paid attention to nature once more. It is unquestionable that no synthetic material matches the exquisite and very special functionality of enzymes or biological molecular machines, but let us set aside sophistication for now and restrict our comparison to simple mechanical properties.

We find in biology extraordinary proteins that show surprising mechanical properties. Indeed, we find proteins that match and clearly outperform the mechanical properties of our best petroleum-based polymers. For example, some kinds of spider silks, such as the *Nephila clavipes* dragline, show a superior strength [22, 23]. An *N. clavipes* dragline silk shows a Young's modulus, tensile strength, and stress at break of the same order of Kevlar, which is a benchmark of modern polymer fiber technology but absorbs almost one order of magnitude more energy than Kevlar when breaking [22–24]. In fact, their mechanical properties can be considered above those of steel itself. Its absorbed energy at breaking point is almost two orders of magnitude higher, while its tensile strength is almost six times higher and the stresses at breaking point are equivalent [22–24]. Additionally, although the Young's modulus of steel is about three times higher than the spider-silk modulus, this last material has a much lower density. Its ratio of tensile strength to density is perhaps five times better than steel. Therefore, at equal mass, the spider silk behaves much better than steel. In conclusion, spider-silk fibers are nearly as strong as several of the current synthetic fibers and can outperform them in many applications in which total energy absorption is important.

Spider silks deserve additional commentary. Again, this example shows as that Nature never gives up to complexity, as if complexity were an intrinsic part of natural materials, and this is so even in these apparently simple materials that Nature has designed just to reach a given mechanical performance. Spider silks show a highly efficient self-healing behavior that is now under intense scrutiny due to its evident technological potential [25].

Dragline spider silks are not the only impressive example. Among elastic protein fibers, Nature shows us examples covering a wide range of elastomeric properties. Again, we find other kinds of spider silks, such as flagelliform silks, that show elastomeric behavior with the ability to withstand high levels of elastic strain; such silks can be extended up to $\sim$ 200% without breaking, but they also show a high rate of energy dissipation [22, 26]. This is well known in flying insects that collide with spider webs; the insects, in spite of their high kinetic energy, very rarely are able to break through the webs. On the contrary, this impact energy is absorbed without catapulting the insect out of the web [22, 26]. In addition, once trapped, they find that breaking the web is a very exhausting and hopeless task.

In contrast, other elastic proteins show precisely the opposite property, i.e., they dissipate a negligible amount of energy in a stress-strain cycle or, equivalently, they show a resilience value near 100% (100% of the elastic energy stored in the deformed sample is restored when released). This is so for resilin, the main elastic protein of jumping insects [27, 28], and the abducting of the swimming bivalves. Also, elastin has been claimed to show and almost ideal elasticity [29]. All these elastic proteins are characterized by high resilience, large strain, and low stiffness [27].

The nearly ideal elasticity of some proteins or some of their functional domains has been identified recently as being a central part of a universal foundation of protein function: the coupled hydrophobic and elastic consilient mechanisms. This has been nicely described by Urry [30], who made a profound study of the Gibbs free energy of hydrophobic hydration and the coupled hydrophobic and elastic consilient mechanisms in specially designed protein-based polymers. This mechanism has been postulated as being the universal principle of functioning of biological protein-based machines and has been identified with biology's vital force (*élan vital*). The model for protein function based on this mechanism has already been postulated for key molecular machines of the cell, such as the complex III in the mitochondrial electron transport chain that produces a proton gradient, the F1 motor of the ATP synthase that uses the proton gradient to produce ATP, and the myosin II motor of muscle contraction that uses ATP to generate motion [30, 31].

The list of proteins with superior mechanical performance can also include keratins. This protein shows a superior impact resistance with a Young's modulus of 2.50 GPa [22]; not for nothing is it the main component of hoofs, beaks, and horns. Again, this protein shows multifunctional character and complexity because keratin is also the main component of feathers, a prodigy of rigidity and lightness.

Although this list could be extended ad infinitum with many other fascinating examples, such as collagen and others, just one more example will be mentioned: mussel adhesives. Mussel adhesive proteins are remarkable materials that display an extraordinary ability to adhere to almost any kind of natural or artificial substrate, and, in addition, they do so in extreme conditions. The environments where these proteins show their functionality are underwater (in salty water, for instance) and standing continuous and changing stresses (waves, tides, underwater flows, etc.). No artificial adhesive is able to work, even minimally, under those circumstances. It is important to emphasize that this kind of environment is not much different than the one found, for example, inside living tissues. For that reason, recent investigations from groups coming from quite diverse areas of expertise have made substantial progress in the identification of the genes and proteins that are involved in adhesive formation. These discoveries have led to the development of recombinant proteins and synthetic polypeptides that are able to reproduce the properties of mussel adhesives for applications in medicine and biotechnology [27].

In summary, the above examples show that a reduced set of 20 amino acids as exclusive primary source to build polymers could be enough to design materials with extraordinary properties, even in the less complex sense of bulk materials. It could be even extended with the recent progress in the development of methods for incorporating nonnatural amino acids into recombinant proteins that can be an alternative strategy for extending GEPBPs with diverse chemical, physical, and biological properties [32]. Therefore, the properties of GEPBPs span a broad range in all directions, from the simplest mechanical properties to the most complex, smart, and self-assembling characteristics. Practically all the properties displayed by petroleum-based polymers are within this range. Thus from the technological point of view, the possibility of obtaining many different materials with a wide range of properties that outperform existing polymers, are obtained by only one common basic technology, and in addition show clear environmental advantages, is a highly interesting scenario.

3
State of the Art in GEPBPs

Presently, genetic engineering of PBPs is still in its early infancy. The radically different approach in the methodology used to produce these polymers has resulted in the fact that, even now, a limited number of research groups and companies have made the effort to make this transition. Among these pioneer groups, the main interest has been mainly concentrated in two major polymer families: spider-silk-like polymers and elastin-like polymers ("ELPs"), although some other interesting protein polymers have also been researched. Those include coiled-coil motifs and their related leucine zippers [33–36], β-sheet-forming polymers [37], poly(alylglicine) [38], and homopolypeptides such as poly(glutamic acid) [39].

The different strategies and methodologies for gene construction, iterative, random, and recursive ligations, have been summarized recently [13, 40, 41].

4
Silklike Polymers

Silks are fibrous proteins produced by spiders and insects such as the silk worm (*Bombyx mori*). There are an astonishing variety in different mechanical properties and compositions of the different silks naturally produced. Many spiders and insects have a varied tool kit of task-specific silks with divergent mechanical properties [42–49]. Those silks seem to have evolved to match a very particular need for the creature that produces them. Furthermore, although some spiders may use silk sparingly, most make rather elaborate nests, traps, and cocoons typically using more than one type of finely tuned and specialized silk. Those different silks are produced by a wide and diverse range of glands, ducts, and spigots. However, in spite of the extraordinary physical properties of spider silks as well as the enormous variety there is only limited information on the composition of the various silks produced by different spiders. Among the different types of spider silks, draglines from the golden orb weaver *Nephila clavipes* and the garden cross spider *Araneus diadematus* are most intensely studied.

Based on DNA analysis it could be shown that all spider silk proteins are chains of iterated peptide motifs ("repeating units"). The small peptide motifs can be grouped into four major categories: GPGXX (with X often representing Q), alanine-rich stretches [An or (GA)n], GGX, and spacers. A fifth category is represented by nonrepetitive (NR) regions at the amino and carboxyl termini of the proteins, often representing polypeptide chains of 100 or more amino acids [48–56].

On the basis of several studies, the major categories of peptide motifs in spider silk proteins have been assigned structural roles [57–61]. It has been suggested that the GPGXX motif is involved in a β-turn spiral, probably providing elasticity, based on structures of comparable proteins [62–65]. If elasticity is due to GPGXX β-spirals, then this motif should be found in the more elastic silks. Flagelliform silks, which show the highest elasticity with more than 200%, consist of contiguous repeats of this motif at least 43 times in each repeating unit. Alanine-rich motifs typically contain 6–9 alanine residues and have been found to form crystalline β-sheet stacks leading to tensile strength [23, 57, 58]. The major and minor ampullate silks are both very strong, and at least one protein in each silk (there are always pairs) contains the $(A)_n$ or $(GA)_n$ motif. Interestingly, this motif is not found in flagelliform silks. A glycine-rich 3₁-helix is adopted by the GGX motif forming an amorphous matrix that connects crystalline regions and that provides elasticity [49, 66, 67]. The postulated GGX motif is widely distributed and this motif can be found in major and minor ampullate and flagelliform silks. Several groups have suggested that the GPGXX and GGX motifs might be involved in forming an amorphous matrix, which would provide the elasticity of the fiber. The spacers contain charged groups and separate the iterated peptide motifs into clusters. NR termini are common to all sequenced major and minor ampullate and flagelliform silks belonging to the Araneoidea family with highly conserved carboxyl-terminal sequences [53, 68, 69].

Regarding genetically engineered silklike polymers (GESLPs), they have been mainly restricted to those designed on the repetition of the sequences [GGAGQGGYGGLGSQ-GAGRGGLGGQGGAG] and [GPGGYGGPGQQGPGGY APGQQPSGPGS] from the silk produced by the *N. clavipes* major ampullate glands 1 and 2, respectively. Some modifications of those base sequences have also been explored. In the first instance, some of them were used to control the degree of crystallinity as a way to improve the processability of those polymers. However, some other modifications have been

Table 1 Selected examples of GEPBPs

Polymer structure	Type	Properties/Uses	Ref.
(GVGVP)$_n$	ELP	Model polymer, prevention of adhesions	[73]
(GVGVAP)$_n$	ELP	Model polymer	[73]
(GVGIP)$_n$	ELP	Model polymer	[74]
(GGIP)$_n$	ELP	Model polymer	[75]
(GGAP)$_n$	ELP	Model polymer	[75]
(GVGVP GVGVP GKGVP GVGVP GVGVP GVGVP)$_n$	ELP	Model polymer, drug delivery	[76]
(GVGVP GVGFP GKGFP GVGVP GVGVP GVGVP)$_n$	ELP	Model polymer, drug delivery	[76]
(GVGVP GVGFP GKGVP GVGVP GVGFP GFGFP)$_n$	ELP	Model polymer, drug delivery	[76]
(GVGVP GVGFP GKGFP GVGVP GVGFP GVGFP)$_n$	ELP	Model polymer, drug delivery	[76]
(GVGVP GVGFP GKGFP GVGVP GVGFP GFGFP)$_n$	ELP	Model polymer, drug delivery	[76]
(GVGVP GVGVP GEGVP GVGVP GVGVP GVGVP)$_n$	ELP	Model polymer, drug delivery	[31]
(GVGVP GVGFP GEGFP GVGVP GVGVP GVGVP)$_n$	ELP	Model polymer, drug delivery	[31]
(GVGVP GVGFP GEGVP GVGVP GVGFP GFGFP)$_n$	ELP	Model polymer, drug delivery	[31]
(GVGVP GVGFP GEGFP GVGVP GVGFP GVGFP)$_n$	ELP	Model polymer, drug delivery	[31]
(GVGVP GVGFP GEGFP GVGVP GVGFP GFGFP)$_n$	ELP	Model polymer, drug delivery	[31]
[(GVGVP)$_2$ GGGVP GAGVP (GVGVP)$_3$ GGGVP GAGVP GGGVP]$_n$	ELP	Model polymer	[77]
[(GVGVP) (GAGVP) (GGGVP)$_7$ GAGVP]$_n$	ELP	Model polymer	[77]
[VPGKG(VPGVG)$_m$]$_n$	ELP	Model polymer	[78]
[VPGQG(VPGVG)$_6$]$_n$	ELP	Model polymer	[79]
[(GVGVP) (GAGVP GGGVP)$_7$ GAGVP]$_4$-(GVGVP)$_{60}$	ELP	Model polymer	[40]

Table 1 (continued)

Polymer structure	Type	Properties/Uses	Ref.
[(GVGVP)$_2$ GGGVP GAGVP (GVGVP)$_3$ GGGVP GAGVP GGGVP]-doxorubicin	ELP-Protein	Purification of recombinant proteins and cancer therapy	[80]
[(GVGVP)$_2$ GGGVP GAGVP (GVGVP)$_3$ GGGVP GAGVP GGGVP]$_n$-Trx (*Trx: thioredoxin*)	ELP-Protein	Purification of proteins, and capture release of proteins by stimuli responsive switches	[81]
[(GVGVP)$_2$ GGGVP GAGVP (GVGVP)$_3$ GGGVP GAGVP GGGVP]$_n$-CAT (*CAT: acetyltransferase*)	ELP-Protein	Purification of proteins	[82]
[(GVGVP)$_2$ GGGVP GAGVP (GVGVP)$_3$ GGGVP GAGVP GGGVP]$_n$-BFP (*BFP: blue fluorescent protein*)	ELP-Protein	Purification of proteins	[82]
[(GVGVP)$_2$ GGGVP GAGVP (GVGVP)$_3$ GGGVP GAGVP GGGVP]$_n$-CalM (*CalM: calmodulin*)	ELP-Protein	Purification of proteins	[82]
[(GA)$_n$GE]$_m$	Silklike	High performance fibers and biomaterials	[83]
[(GEEIQIGHIPREDVDYHLYP)(GVPGI)$_m$]$_n$	ELP	Tissue engineering	[84]
[LD(GEEIQIGHIPREDVDYHLYP)G((VPGIG)$_2$VPGKG(VPGIG)$_2$)$_4$VP]$_n$	ELP	Tissue engineering	[85]
[(AG)$_3$PEG]$_n$	alanylglycine-rich	Biomaterials	[86]
[(AG)$_4$ PEG]$_n$	alanylglycine-rich	Biomaterials	[86]
[E$_{17}$D]$_n$	Rodlike-polymer	Biomaterials	[39]
MRGSHHHHHHGSDDDDKWA-Helix-IGDHVAPRDTSMGGC (Helix: SGDLENEVAQLEREVRSLEDEAAELEQKVSR LKNEIEDLKAE)	Leucine zipper	Biomaterials and reversible hydrogels	[87]

Table 1 (continued)

Polymer structure	Type	Properties/Uses	Ref.
MRGSHHHHHHGSDDDKASYRDPMG-[(AG)$_3$PEG]$_{10}$-ARMPTSW	alanylglycine-rich	Biomaterials	[87]
MRGSHHHHHHGSDDDDKWA-Helix-IGKHVAPRDTSYRDPMG-[(AG)$_3$PEG]$_{10}$- ARMPTSD-Helix- IGDHVAPRDTSMGGC	Leucine zipper-alanylglycine-rich	Biomaterials and reversible hydrogels	[87]
[(PGVGV)$_2$(PGEGV)(PGVGV)$_2$]$_n$	ELP	Model polymer	[19]
[(VPGIG)$_2$(VPGKG)(VPGIG)$_2$(EEIQIGHIPREDVDYHLYP)(VPGIG)$_2$(VPGKG) (VPGIG)$_2$(GVGVAP)$_3$]$_n$	ELP	Tissue Engineering	[88]
[(GVGVP)$_4$GKGVP(GVGVP)$_3$(GAGAGS)$_4$]$_n$	Silk-ELP	High performance fibres and biomaterials	[89]
[(GVGVP)$_4$GEGVP(GVGVP)$_3$GAGAGS]$_n$	Silk-ELP	High performance fibres and biomaterials	[90]
[(GVGVP)$_8$GAGAGS]$_n$	Silk-ELP	High performance fibres and biomaterials	[90]
[(GAGAGS)$_9$GAAVTGRGDSPASAAGY]	Pronectin F^{TM}	Tissue engineering	[70]
[(GAGAGS)$_2$(GVGVP)$_4$(GKGVP)(GVGVP)$_3$]$_n$	Silk-ELP	High performance fibres and biomaterials	[91]
[(GAGAGS)$_3$(GVGVP)$_4$(GKGVP)(GVGVP)$_3$]$_n$	Silk-ELP	High performance fibres and biomaterials	[91]
[(GAGAGS)$_4$(GVGVP)$_4$(GKGVP)(GVGVP)$_3$]$_n$	Silk-ELP	High performance fibres and biomaterials	[91]
[(GAGAGS)$_4$ (GVGVP)$_8$]$_n$	Silk-ELP	High performance fibres and biomaterials	[91]
[(GAGAGS)$_8$(GVGVP)$_{16}$]$_n$	Silk-ELP	High performance fibres and biomaterials	[91]
[(VPGVG)$_4$(VPGKG)]$_n$	ELP	Model Polymer	[12]
[(VPGVG)$_4$(VPGIG)]$_n$	ELP	Model Polymer	[12]
[(GPGGSGPGGY)$_2$GPGGK]$_n$	Silk	High performance fibres and biomaterials	[92]
[VPGEG(IPGAG)$_4$]$_{14}$-[VPGFG(IPGVG)$_4$]$_{14}$	ELP (Block Copolymer)	Drug Delivery	[93]

Table 1 (continued)

Polymer structure	Type	Properties/Uses	Ref.
[(APGGVPGGAPGG)$_2$]$_x$-[VPAVG(IPAVG)$_4$]$_{16}$	ELP (Block Copolymer)	Drug Delivery	[93]
[VPAVG(IPAVG)$_4$]$_{16}$-[VPGVG(IPGVGVPGVG)$_2$]$_{19}$-[VPAVG(IPAVG)$_4$]$_{16}$	ELP (Block Copolymer)	Drug Delivery	[93]
[VPAVG(IPAVG)$_4$]$_{16}$-[VPGEG(VPGVG)$_4$]$_{30}$-[VPAVG(IPAVG)$_4$]$_{16}$	ELP (Block Copolymer)	Drug Delivery	[93]
[VPAVG(IPAVG)$_4$]$_{16}$-[VPGEG(VPGVG)$_4$]$_{48}$-[VPAVG(IPAVG)$_4$]$_{16}$	ELP (Block Copolymer)	Drug Delivery	[93]
[VPAVG(IPAVG)$_4$]$_{16}$-[(APGGVPGGAPGG)$_2$]$_{22}$-[VPAVG(IPAVG)$_4$]$_{16}$	ELP (Block Copolymer)	Drug Delivery	[93]
[VPAVG(IPAVG)$_4$]$_{16}$-[(VPGMG)$_5$]$_x$-[VPAVG(IPAVG)$_4$]$_{16}$	ELP (Block Copolymer)	Drug Delivery	[93]
{VPAVG[(IPAVG)$_4$(VPAVG)]$_{16}$IPAVG} - VPGVG[(VPGVG)$_2$VPGEG[(VPGVG)$_2$]$_{48}$VPGVG - {VPAVG[(IPAVG)$_4$(VPAVG)]$_{16}$IPAVG}	ELP (Block Copolymer)	Drug Delivery	[94]
{VPAVG[(IPAVG)$_4$(VPAVG)]$_{16}$IPAVG} - VPGVG VPGVG - {VPAVG[(IPAVG)$_4$(VPAVG)]$_{16}$IPAVG}	ELP (Block Copolymer)	Drug Delivery	[94]
(AG)$_6$(VPGVG)(AG)$_7$	Silk-ELP	High performance fibres and biomaterials	[95]
(AG)$_5$(VPGVG)$_2$(AG)$_5$	Silk-ELP	High performance fibres and biomaterials	[95]
(AG)$_n$	Silk	High performance fibres and biomaterials	[95]
(GGAGSGYGGGYGHGYGSDGG(GAGAGS)$_3$)$_4$	Silkworm	High performance fibres and biomaterials	[96]
AS(A)$_{18}$TS(GVGAGYGAGAGYGVGAGYGAGVGYG AGAGY)TS	Silkworm	High performance fibres and biomaterials	[96]

Table 1 (continued)

Polymer structure	Type	Properties/Uses	Ref.
[(GPGGYGPGQQ)$_2$GPSGPGSA$_8$]$_n$	Silk	High performance fibres and biomaterials	[23]
[GGAGQGGYGGLGQGAGRGGLGGQ(GA)$_2$A$_5$]$_n$	Silk	High performance fibres and biomaterials	[23]
[A GQG GYS GLS SQG)(A GQG GYG GLG SQH A GRG GLG GQG A GAAAAAAGG) (A GQG GLG SQH A GQG A GAAAAAA GG) (A GQG GYG GLG SQH A GRG GQG A GAAAAA GG)]$_n$	Silk	High performance fibres and biomaterials	[97]
[(GPGGY GPGQQ GPGGY GPGQQ GPGGY GPQGG GPSGPGS AAAAAA)(GPGQQ GPGGY GPGQQ GPGGY GPQGG GPSGPGS AAAAAAA)(GPGGY GPGQQ GPGGY GPQGG GPSGPGS AAAAAAAA)]$_n$	Silk	High performance fibres and biomaterials	[97]
(AEAEAKAKAEAEAKAK)$_9$	β-sheet fibres	High performance fibres and biomaterials	[37]
(GGP)$_n$-collagen	Fusion polymer	Tissue engineering	[98]
(VPGVG)$_{78}$-OPH (organophosphorus hydrolase:OPH)	ELP-protein	Biotechnology	[99]
[GGAGQGGYGGLGSQGAGRGGLGGQGGAG]	Silk	High performance fibres and biomaterials	[15]
[GPGGYGGPGQQGPGGYAPGQQPSGPGS]	Silk	High performance fibres and biomaterials	[15]

added to further functionalize the polymers, such as the incorporation of RGD cell attachment sequences (Pronectin) [70] or the creation of block copolymers combining silk and elastin motifs [71, 72]. Some of the representative examples of GEPBPs produced to date have been summarized in Table 1.

5
Elastinlike Polymers: A Privileged Family of GEPBPs

5.1
Introducing ELPs

The ELP family has shown a versatile and ample range of interesting properties that go well beyond their simple mechanical performance. Certainly, ELPs show a set of properties that places them in an excellent position towards designing advanced polymers for many different applications, including the most cutting edge biomedical uses, for which ELPs are particularly well suited, as will be discussed later. In addition, the deepening understanding of their function in terms of their molecular composition and behavior is shedding light on one of the most interesting basic problems still faced in modern science, the understanding of protein folding and function in living organisms.

The basic structure of ELPs is a repeating sequence having its origin in the repeating sequences found in the mammalian elastic protein, elastin. Regarding their properties, some of their main characteristics are derived from the natural protein they are based on. For example, the cross-linked matrices of these polymers retain most of the striking mechanical properties of elastin [100], i.e., an almost ideal elasticity with Young's modulus, elongation at break, etc. in the range of natural elastin and an outstanding resistance to fatigue [85, 101].

Interestingly, this mechanical performance is accompanied by an extraordinary biocompatibility, although, however, the most striking properties are perhaps their acute smart and self-assembling nature. These properties are based on a molecular transition of the polymer chain in the presence of water when their temperature is increased above a certain level. This transition, called the "inverse temperature transition" (ITT), has become the key issue in the development of new peptide-based polymers as molecular machines and materials. The understanding of the macroscopic properties of these materials in terms of the molecular processes taking place around the ITT has established a basis for their functional and rational design [102].

All these aspects of the ELP family will be presented below in the context of the present state of the art and the foreseeable future outcomes.

5.2
Smart and Self-assembling Properties of ELPs

The most numerous members of the ELP family are those based on the pentapeptide VPGVG (or its permutations). A wide variety of polymers have been (bio)synthesized with a general formula (VPGXG), where X represents any natural or modified amino acid [103–105] with the exception of L-proline. All the polymers with that general formula that can be found in the literature are functional, i.e., all show a sharp smart behavior. However, the achievement of functional ELPs by the substitution of any of the other amino acids in the pentamer is not so straightforward. For example, the first glycine cannot be substituted by any other natural amino acid different from L-alanine [105].

The model poly(VPGVG), whose amino acid side chains are simple aliphatic chains without further functionalization, shows an acute thermoresponsive behavior associated to the existence of the ITT.

All of the functional ELPs exhibit this reversible phase transitional behavior [105]. In aqueous solution and below a certain transition temperature (T_t), the free polymer chains remain disordered, random coils in solution [106] that are fully hydrated, mainly by hydrophobic hydration. This hydration is characterized by the existence of ordered clathratelike water structures surrounding the apolar moieties of the polymer [107–109] with a structure somehow similar to that described for crystalline gas hydrates [109, 110], although showing a more heterogeneous structure with structures varying in perfection and stability [108]. In contrast, above T_t, the chain hydrophobically folds and assembles to form a phase-separated state of 63% water and 37% polymer by weight [111] in which the polymer chains adopt a dynamic, regular, nonrandom structure, called β-spiral, involving type II β-turns as the main secondary feature, and stabilized by intraspiral, interturn, and interspiral hydrophobic contacts [105]. This is the product of the ITT. In this folded and associated state, the chain loses essentially all of the ordered water structures of hydrophobic hydration [107]. During the initial stages of polymer dehydration, hydrophobic association of β-spirals takes on fibrillar form. This process starts from the formation of filaments composed of three-stranded dynamic polypeptide β-spirals that grow to a several-hundred-nanometer particle before settling into a visible phase-separated state [105, 112]. This folding is completely reversible upon lowering again the sample temperature below T_t [105].

Although, generally speaking, the phenomenology shown by these ELPs resembles that found in amphiphilic LCST polymers, such as poly-(N-isopropylacrylamides) (PNIPAM), the presence of an ordered state in ELPs above the transition temperature, which is not present in the LCST polymers, has prevented the use of LCST as a descriptive term for the ITT of ELPs [74].

5.3
Basic Molecular Designs: Thermal Responsiveness

Poly(VPGVG) (or its permutations) can be considered one of the simplest ELPs. The nonexistence of further functionalization, apart from the hydrophobic nature of valine and proline side chains, gives rise to a straightforward thermal response as shown in Fig. 2. As mentioned above, the transition can be easily followed either by turbidity measurements or by calorimetric methods, measuring the heat flow during the transition. The first method is characterized by a turbidity profile showing a sharp step. T_t is considered the temperature showing a 50% change in the relative turbidity change. In contrast, DSC measurements are always characterized by a broad peak, expanding 20 °C or more. In this case, T_t can be considered either as the onset or peak temperature. Usually, T_t values obtained by these methods differ among each other. Different factors cause such differences. The first one is the dynamic nature of the DSC and its associated thermal lags; those thermal lags being, of course, higher for higher heating rates. However, those thermal lags can be eliminated using different heating (or cooling) rates and obtaining an extrapolated T_t value to a heating rate equal to zero [113]. Figure 2a clearly shows the influence of this parameter; the DSC peak temperature for a 10 °C/min heating rate is several degrees higher than the turbidity T_t.

Another factor that can cause T_t differences between the two techniques is the different polymer concentrations. It is well known that polymer folding is a cooperative process that is facilitated by the presence of other polymer chains and, accordingly, T_t can be several degrees higher for low concentrations [20, 105]. There is a strong dependence of T_t on concentration in the range of 0.01 to 5–10 mg/mL. Above this concentration, T_t does not show further significant changes with increasing concentrations up to a limit of 150–200 mg/mL. Above this value, we find deficiently hydrated polymer chains and, due to the heterogeneity of the hydrophobic hydration structures, in water deficiency states only the strongest structures are formed, which leads to a new increase in T_t as the polymer concentration increases [108]. Typical concentrations for turbidity experiments are in the range of 2–5 mg/mL, while those for DSC usually are in the range of 50–150 mg/mL, so further differences in T_t caused by concentration effects could be possible.

In addition, T_t also depends on the MW. T_t decreases as the MW increases [19, 20, 101]. Furthermore, the presence of other ions, such us those of the buffer, and molecules also changes the T_t value. In conclusion, all these factors make the comparison of T_t values among not only different techniques but also different authors a delicate matter.

The endothermic peak found in a DSC heating run is in fact the net result of a complex process containing different thermal contributions. Once

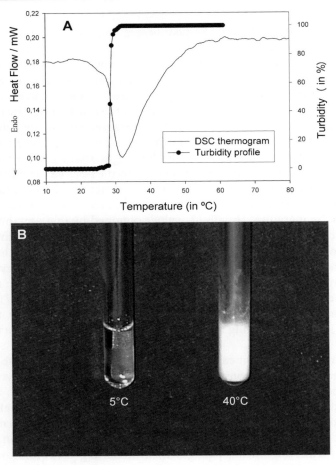

Fig. 2 A Turbidity profile as a function of temperature for a poly(VPGVG) 5-mg/L sample dissolved in water and DSC thermogam of a 50-mg/L water solution of the same polymer (heating rate 5 °C/min). **B** photographs of a water solution (5 mg/mL) of this poly(VPGVG) *below* (5 °C) and *above* (40 °C) its T_t

a poly(VPGVG) solution reaches its T_t, there is first a destruction of the ordered hydrophobic hydration structures surrounding the polymer chain. This is further accompanied by an ordering of the polymer chain into the β-spiral structure. In turn, these β-spirals further establish interchain hydrophobic contacts (Van der Waals cohesive interactions) that caused the formation of nano- and microaggregates segregating from the solution. The first process must be considered endothermic while the second one must be exothermic. Although both events take place simultaneously, they are very different in nature. In particular, it is reasonable to consider that both phenomena occur with different kinetics. In effect, previous kinetic studies made on poly(VPGVG) showed that the process of phase separation is faster than the process of redissolution [114]. This difference creates a chance to split the different contributions of the ITT. This has been recently achieved for the first time using temperature-modulated DSC (TMDSC) [115]. TMDSC is an improved DSC measurement that is able to separate thermally overlapping phenomena with different time dependences by using a heating program containing an alternating function of the temperature, such as a sinus, superimposed on the constant heating rate (v) [116–120]. In principle, TMDSC will provide a clear split of two overlapping phenomena when, under the particular dynamical conditions, one is reversible and the other is not. Therefore, by this experimental approach, both phenomena could be split by finding a frequency for the periodic component low enough for the faster phenomenon to follow the oscillating temperature changes ("reversing") while high enough to impede this alternating behavior of the slower one ("nonreversing"). This approach has been used to study the ITT of three different ELPs chemically synthesized poly(VPGVG), recombinant (VPGVG)$_{251}$, and recombinant (IPGVG)$_{320}$ [115]. Figure 3a shows an example of the TMDSC thermogram found for (VPGVG)$_{251}$, while Fig. 3b shows the results of its analysis. Under those experimental conditions, the endothermic total curve (ΔH_{tot} = -10.40 Jg^{-1}, $T_t = 27.72\,°C$) is composed by a nonreversing endothermic component ($\Delta H_{non-rev} = -13.98$ Jg^{-1}, $T_t = 27.63\,°C$) and a reversing exotherm ($\Delta H_{rev} = 3.33$ Jg^{-1}, $T_t = 27.30\,°C$).

A detailed analysis has been carried out to study the dependence of the reversing and nonreversing components as a function of v and amplitude (A) and period (P) of the alternating component. For the total contribution, the changes in v (0.5 to 1.5 °C/min), A (0.1 to 1 °C), and P (0.1 to 1.0 min) did not significantly affect the enthalpy and T_t values, which are similar to those obtained by DSC. Also the reversing and nonreversing components were not affected by changes in v and A. However, P exhibits a strong influence on the enthalpy values of both components.

ΔH_{rev} is plotted in Fig. 4 as a function of P for the three polymers. In all cases, at low frequencies (high P), the reversing component shows an endothermic peak with an enthalpy comparable to the one shown by the endothermic peak of the nonreversing component. Thus, at these high P, the chain-folding and dehydration contributions were not well separated. However, as P decreases, ΔH_{rev} undergoes a substantial increase. At $P = 0.8$–1 min, the reversing component turns into a positive exothermic peak which reaches a maximum at $P = 0.5$–0.6 min (P_M). Parallelly, $\Delta H_{non-rev}$ suffers an equivalent decrease. Therefore, as P decreases, the reversing component is being enriched in the exothermic component (chain folding), while the non-reversing is being enriched in the endothermic contribution (dehydration). The ΔH_{rev}, $\Delta H_{non-rev}$, ΔH_{tot} values found at P_M can be seen in Table 2. Further decrease in P results in a progressive reduction in ΔH_{rev} to zero and

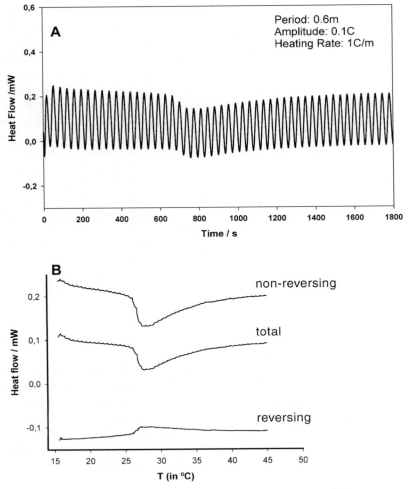

Fig. 3 A Heat flow vs. time in a TMDSC analysis of a 125-mg mL^{-1} water solution of (VPGVG)$_{251}$. **B** Reversing, nonreversing, and total thermograms. Reproduced with permission from Elsevier

an increase in $\Delta H_{\text{non-rev}}$ to the total enthalpy as a result of the complete overlap of both phenomena in the nonreversing component.

The maximum splitting was found at approximately the same P_M regardless of the polymer. Additionally, a comparison of the data found for (VPGVG)$_{251}$ and (IPGVG)$_{320}$ indicates that the reversing component at maximum is higher for (IPGVG)$_{320}$. Due to the higher hydrophobicity of I as compared to V, its chain folding has to show a higher exothermic ΔH_{rev} (Table 2). Therefore, ΔH_{rev} values could then be used as a quantitative measurement of the amino acid hydrophobicity. Additionally, the increased hydrophobicity of

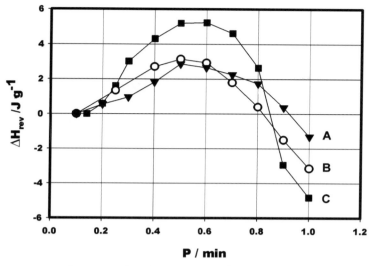

Fig. 4 ΔH_{rev} as a function of P for 125 mg mL^{-1} water solution of **A** synthesized poly(VPGVG), **B** recombinant (VPGVG)$_{251}$, and **C** recombinant (IPGVG)$_{320}$ ($\nu = 1\,°C\,min^{-1}$, and $A = 0.1\,°C$). Reproduced with permission from Elsevier

(GVGIP)$_{320}$ would also induce a higher extension of hydrophobic hydration, so its higher endothermic $\Delta H_{non-rev}$ is also reasonable.

There are no significant differences when comparing data from (VPGVG)$_{251}$ and poly(VPGVG) (Table 2). Since the only difference between these two polymers is their MW dispersity, their TMDSC results are practically the same, which would imply that the reversing and nonreversing TMDSC components depend mainly on the mean hydrophobicity of the monomer.

Therefore, TMDSC has been demonstrated to be an effective method to split the overlapping phenomena present in the ITT of elastic protein-based polymers. By tuning the frequency of the periodic component, a maximum split can be achieved that shows an exothermic contribution arising from the Van der Waals contacts attending chain folding and assembly, and an endothermic contribution associated with loss of hydrophobic hydration, the

Table 2 Enthalpy values of the reversing, non-reversing and total components found at P_M

Polymer	$\Delta H_{rev}/Jg^{-1}$	$\Delta H_{non-rev}/Jg^{-1}$	$\Delta H_{tot}/Jg^{-1}$	P_M/min
(IPGVG)$_{320}$	5.61	−22.82	−17.21	0.6
(VPGVG)$_{251}$	3.14	−11.34	−7.50	0.5
Poly(VPGVG)	2.96	−11.11	−8.79	0.5

former being about one fourth of the latter, in absolute values. To the best of our knowledge, TMDSC is the only method currently available to separate both contributions. Accordingly, its utility for evaluating the hydrophobicity of the full compliment of naturally occurring amino acids and relevant modifications thereof is clear, and its relevance to hydrophobic folding of polymers and natural proteins is noteworthy.

5.4
Introducing Further Chemical Functions in the Monomer: pH-responding ELPs and the ΔT_t Mechanism

In all ELPs, T_t depends on the mean polarity of the polymer, increasing as the hydrophobicity decreases. This is the origin of the so-called "ΔT_t mechanism" [105]; i.e., if a chemical group that can be present in two different states of polarity exists in the polymer chain, and these states are reversibly convertible by the action of an external stimulus, the polymer will show two different T_t values. This T_t shift ("ΔT_t") opens a working temperature window in which the polymer isothermally and reversibly switches between the folded and unfolded states following the changes in the environmental stimulus. This ΔT_t mechanism has been exploited to obtain many elastinlike smart derivatives [105, 121–124].

This mechanism is also exploited in the following model pH-responding polymer: $[(VPGVG)_2\text{-}VPGEG\text{-}(VPGVG)_2]_n$. In this ELP, the γ-carboxylic group of the glutamic acid (E) suffers strong polarity changes between its protonated and deprotonated states as a consequence of pH changes around its effective pK_a.

Figure 5 shows the folded chain content as a function of T at two different pHs for a genetically engineered polymer with the above general formula ($n = 45$). At pH = 2.5, in the protonated state, the T_t shown by the polymer is 28 °C. Below this temperature the polymer is unfolded and dissolved, while above it the polymer folds and segregates from the solution. However, at pH = 8.0 the increase in the polarity of the γ-carboxyl groups, as they lose their protons, becoming carboxylate, is enough to cause T_t to rise to values above 85 °C, opening a working temperature window wider than 50 °C. Therefore, at temperatures above 28 °C the polymer would fold at low pHs and unfold at neutral or basic pHs. In addition, this fact reveals the extraordinary efficiency of ELPs as compared to other pH-responding polymers since this huge ΔT_t is achieved with just 4 E residues per 100 amino acids in the polymer backbone. This is of practical importance in using these polymers to design molecular machines and nanodevices such as nanopumps or nanovalves because just a low number of protons is needed to trigger the two states of the system.

The materialization of an electric charge in a side chain of a given ELP due to acid-basic equilibrium has been considered in the literature as a highly ef-

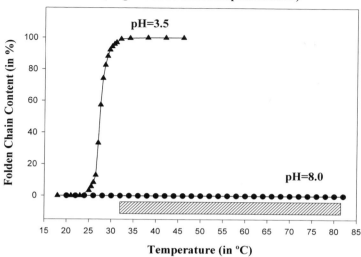

Fig. 5 Turbidity temperature profiles of a model genetically engineered pH responding ELP (see [19] for details on bioproduction of this polymer). *Box at bottom*: window of working temperatures. Experimental conditions are given in plot

ficient way to achieve high ΔT_t. In the number of ELPs designed and studied to date, the capability of the free carboxyl or amino groups of aspartic acid, glutamic acid, or lysine to drive those T_t shifts is only surpassed by the ΔT_t caused by the phosphorylation of serine [31].

Contrary to what happens with polydisperse synthetic polymers, the exquisite control on the molecular architecture and the strict monodisperse MW attained by genetic engineering make easy the study of the dependence of the different polymer properties vs. MW.

This has been done in the $[(VPGVG)_2\text{-}VPGEG\text{-}(VPGVG)_2]_n$ series for pH-responding ELPs. A set of different monodisperse versions of polymers has been bioproduced, with n = 5, 9, 15, 30, and 45. These were set to study the effects of MW on the properties of their ITT and its dependence on pH. As a result, the transition temperature decreased and the transition enthalpy increased as MW increased, especially for the lowest MWs. This can be qualitatively seen in Fig. 6, where a series of DSC thermograms has been plotted for a given polymer concentration and pH.

Quantitatively, these dependences can be seen in Fig. 7, in which enthalpy and true T_t values have been plotted vs. MW. True T_t is the term used to describe the T_t value obtained by extrapolation to zero heating rate ($\nu = 0$) of the DSC peak temperature.

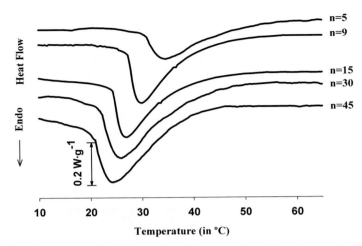

Fig. 6 DSC thermograms of 50 mg mL^{-1} phosphate buffered (0.1 M, pH 2.5) water solutions of studied polymers. Their polymerization degree (n) is shown on the *right-hand side* of the plot. Heating rate 10 °C min^{-1}. Reproduced with permission from American Chemical Society

Moreover, we have observed that the pK_a of the free carboxyl of the glutamic side chain also depends on MW. This striking fact can be seen in Fig. 8, where T_t has been followed as a function of pH for the different MWs.

As shown in that figure, the pH at which T_t starts to increase, following the first deprotonations of the free carboxyl groups, is lower for lower MWs. With the help of the enthalpy values found at different pHs and MWs, it has been possible to estimate the apparent pK_a (pK_a') of this free carboxyl group as a function of MW [19] (Fig. 9).

That behavior would imply that for higher MWs this carboxyl group is less acidic and shows a greater tendency to remain in the protonated state, and this despite the fact that the surroundings of this carboxyl are equivalent in all MWs. This striking behavior could be partially explained by the influence of the polar chain-end groups, as this influence is higher for lower MWs. However, the exclusive effect of the end-chain polarity seems insufficient to account for the strong influence reported. We believe that a large part of the effect of MW on the ITT is caused by the inter- and intrachain cooperativity of the hydrophobic self-assembly taking place during the ITT [106]. In this sense, it is reasonable to think that short chains do not show an efficient cooperation so their self-assembly is hindered, while for high MWs the inter- and intrachain cooperativity during folding is more efficient, which, to some degree, forces the carboxyl group to be in the protonated (less polar) state.

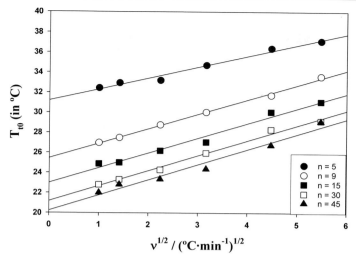

Fig. 7 Dependence of T_t on square root of heating rate for studied polymers. The corresponding polymerization degree (n) is indicated in plot. *Lines*: least square linear regressions of data for each n. Phosphate-buffered samples (0.1 M, pH 2.5). Reproduced with permission from American Chemical Society

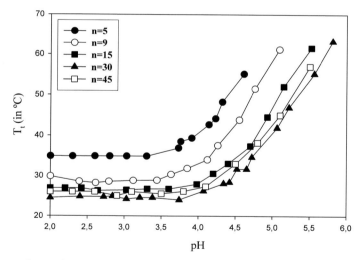

Fig. 8 Dependence of T_t on pH for studied polymers (as indicated in plot). 0.1 M phosphate-buffered samples. Reproduced with permission from American Chemical Society

As shown in Fig. 9, the pH at which T_t starts to increase, following the first deprotonations of the free carboxyl groups, is lower for lower MWs. With the help of the enthalpy values found at different pHs and MWs, it has been

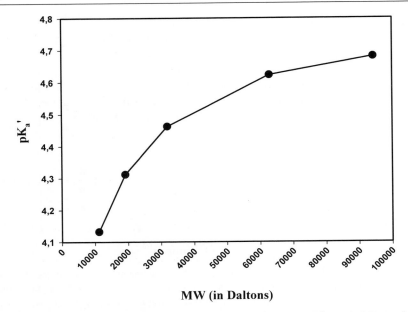

Fig. 9 Dependence of pK_a' for the γ-carboxyl group of glutamic acid on MW. Reproduced with permission from American Chemical Society

possible to estimate the apparent pK_a (pK_a') of this free carboxyl group as a function of MW [19] (Fig. 9).

That behavior would imply that for higher MWs this carboxyl group is less acidic and shows a greater tendency to remain in the protonated state, and this despite the fact that the surroundings of this carboxyl are equivalent in all MWs. This striking behavior could be partially explained by the influence of the polar chain-end groups, as this influence is higher for lower MWs. However, the exclusive effect of the end-chain polarity seems insufficient to account for the strong influence reported. We believe that a large part of the effect of MW on the ITT is caused by the inter- and intrachain cooperativity of the hydrophobic self-assembly taking place during the ITT [106]. In this sense, it is reasonable to think that short chains would not show an efficient cooperation so their self-assembly is hindered, while for high MWs the inter- and intrachain cooperativity during folding is more efficient, which, to some degree, forces the carboxyl group to be in the protonated (less polar) state.

5.5
Self-Assembling Capabilities of ELPs

In relation to self-assembling, natural elastin suffers a self-aggregation process in its natural environment. Elastin is produced from a water-soluble precursor, tropoelastin, which spontaneously aggregates yielding fibrilar structures that are finally stabilized by enzymatic interchain cross links. This produces the well-known insoluble and elastic elastin fibers that can be found in abundance in the skin, lungs, arteries, and, in general, those parts of the body undergoing repeated cycles of stress-strain.

The self-assembling ability of elastin seems to reside in certain relatively short amino acid sequences, as has been recently probed by Yang et al. [124] working in recombinant ELPs. Some of these polypeptides have shown that, above their T_t, they are able to form nanofibrils that further organize into hexagonally close-packed arrangements when the polymer was deposited onto a hydrophobic substrate [124].

However, in ELPs, this tendency to self-assemble in nanofibers can be expanded to other topologies and nanostructured features [93, 125, 126]. Taking advantage of the opportunities and potential given by genetic engineering in designing new polymers, the growing understanding in the molecular behavior of ELPs and the enormous wealth of experimental and theoretical experience gained in recent decades on the self-assembling characteristics of different types of block copolymers, different self-assembling properties are starting to be unveiled within the ELP family. For example, Reguera et al. have shown that the ELP previously shown as a pH-responding polymer, [(VPGVG)$_2$(VPGEG)(VPGVG)$_2$]$_{15}$, was able to form polymer sheets showing self-assembled nanopores [126] (Fig. 10).

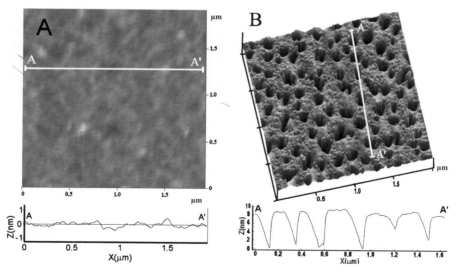

Fig. 10 Tapping Mode AFM image of [(VPGVG)$_2$-(VPGEG)-(VPGVG)$_2$]$_{15}$ deposited from a water solution on a Si hydrophobic substrate. Sample conditions: **A** 10 mg mL^{-1} in 0.02 M HCl water solution (acid solution); and, **B** 10 mg mL^{-1} in 0.02 M NaOH water solution (basic solution). Adapted from [126]. Reproduced with permission from American Chemical Society

An AFM study of the topology of polymer spin-coated depositions of Glu-containing ELPs, from acid and basic solutions, on a Si hydrophobic substrate at temperatures below T_t has shown that in acidic conditions, the polymer deposition just shows a flat surface without particular topological features (Fig. 10a).

However, from basic solutions the polymer deposition clearly shows an aperiodic pattern of nanopores ($\sim$ 70 nm width and separated by about 150 nm) (Fig. 10b). This different behavior as a function of pH has been explained in terms of the different polarity shown by the free γ-carboxyl group of glutamic acid. In the carboxylate form, this moiety shows a markedly higher polarity than the other polymer domains and the substrate itself. Under this condition, the charged carboxylates impede any hydrophobic contact with their surroundings, which is the predominant way of assembling for this kind of polymer. These charged domains, along with their hydration sphere, are then segregated from the hydrophobic surroundings, giving rise to nanopore formation (Fig. 11).

The self-association of ELPs is starting to be employed to develop different applications. For example, Molina el al. [127] have tested self-assembled nano- and microparticles of poly(VPAVG), another version of ELP, as carriers of the model drug dexamethasone phosphate in order to develop injectable systems for controlled drug release. In these particles, the drug is entrapped while the particles self-assemble as the temperature rises above its T_t.

In another remarkable example, Chilkoti et al. have developed nanostructured surfaces by combining ELPs and dip-pen nanolithography that show reversible changes in their physicochemical properties in response to changes in their environmental conditions. In particular, these systems are able to capture and release proteins on nanopatterned surfaces by using the self-assembling characteristics of ELPs in an effort to develop advanced biomaterials, regenerable biosensors, and microfluidic bioanalytical devices [127–130].

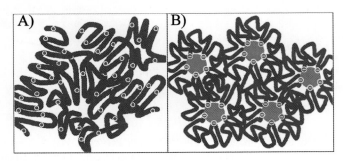

Fig. 11 Schematic cartoon of polymer distribution on hydrophobic substrate. **A** In a acid medium. **B** In a basic medium. Counterions have been not drawn for clarity. Adapted from J Am Chem Soc. Reproduced with permission from American Chemical Society

However, the exploitation of the huge potential of ELPs in producing self-assembling polymers is still very poor. In recent decades, the development achieved in the design of self-assembling polymers, especially block copolymers, has been enormous in spite of the difficulties found in the synthesis of these polymers [21]. The different blocks show different compositions and physicochemical properties, so in an adequate environment those blocks segregate in various immiscible phases that with the combination of adequate external fields are able to self-organize into different highly interesting nanostructures [21]. Among the different physicochemical properties that can be used to trigger phase segregation among the different blocks is their hydrophobic-hydrophilic nature. This opens an interesting possibility for using ELP blocks to construct self-assembling block copolymers. The tendency of these ELPs to show controlled hydrophobic association can be exploited to obtain advanced multiblock copolymers with the advantage given by three salient facts. First, the hydrophobic association of ELPs can be externally controlled since it is associated to the ITT, and this is stimulus triggered (temperature, pH canges, etc.). Second, currently and as a consequence of extensive and deep work on tens of different model ELPs of the type (VPGXG)$_n$ carried out by Urry's group in recent decades, there is a deep and quantitative body of knowledge on the degree of hydrophobicity of the different amino acid side chains [30, 31, 131]. We now have a precise classification of the hydrophobicity of amino acids. The parameter used to precisely quantify the hydrophobic character is based on the direct experimental measurement of the Gibbs free energy of hydrophobic association. Therefore, for the first time, the hydrophobic character has been evaluated from the origin of the hydrophobicity itself and not from its indirect effects, such as the distribution coefficient between solvents, etc. The available data include the 20 natural amino acids and some derivatives. For those amino acids with polarizable side chains, such as glutamic acid, lysine, or phosphorilated and unphosphorilated serine, this datum has been evaluated in the two states. These values have been summarized in Fig. 12.

As can be observed, the hydrophobic character of the different amino acids covers a broad range between the most hydrophobic, tryptophan, to the most hydrophilic, phosphorilated serine. The energy gap between these two extremes is as high as 15 kcal per mol of VPGXG. A gradual transition between those extreme values can be used to adjust the hydrophobicity of the designed blocks with unprecedented precision.

The third relevant fact is the unparalleled capacity to achieve complex and completely controlled PBPs given by genetic engineering. The block length, hydrophobicity, composition, and position can be engineered at will with absolute precision. Additionally, genetically engineered elastinlike block copolymers can easily incorporate any other structural feature of interest for self-assembly and function such as β-sheet-forming domains, leucine zippers, binding of domains to different substrates, and any biofunctionality

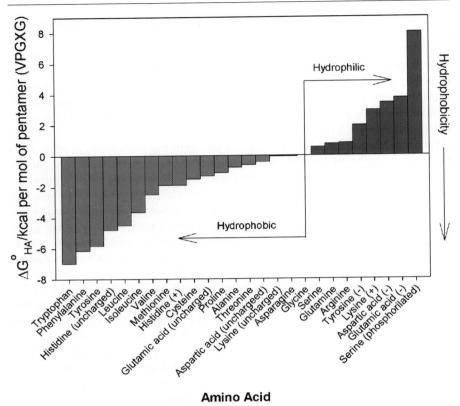

Fig. 12 Hydrophobicity scale of 20 natural amino acids in their different polarizated state. Adapted from data taken from [30, 31, 131]

imparted by bioactive peptides (cell attachment sequences, etc.). All three of these characteristics will certainly open new ways of creating advanced multiblock copolymers with applications spreading to many technological fields.

5.6
Further Chemical Functionalization of the Monomer: Photoresponding ELPs and the Amplified ΔT_t Mechanism

The range of stimuli that can exploit the ΔT_t mechanism is not limited to those chemical reactions taking place on natural amino acid side chains. It is possible to modify certain side chains to achieve systems with extended properties. A good example of this are photoresponding ELPs, which bear photochromic side chains either coupled to functionalized side chains in the previously formed polymer (chemically or genetically engineered) or by using nonnatural amino acids that were already photochromic prior to chemical polymerization.

The first example corresponds to this last kind. The polymer is an azobenzene derivative of poly(VPGVG), the copolymer poly[f_V(VPGVG), f_X(VPGXG)] (X,L-p-(phenylazo)-phenylalanine; f_V and f_X are mole fractions). The p-phenylazobenzene group suffers a photo-induced cis-trans isomerization. Dark adaptation or irradiation with visible light around 420 nm induces the presence of the trans isomer, the most unpolar isomer. In contrast, UV irradiation (at around 348 nm) causes the appearance of high quantities of the cis isomer, which is slightly more polar than the trans isomer. Although the polarity change is not high, it is enough to obtain functional polymers due to the sensitivity and efficiency of ELPs. Figure 13 shows the photoresponse of one of these polymers with $f_X = 0.15$. That mole fraction represents only 3 L-p-(phenylazo)phenylalanine groups per 100 amino acids in the polymer chain. In spite of the low polarity change and the exiguous presence of chromophores, the existence (Fig. 13a) of a working temperature window at around 13 °C is evident (Fig. 13b).

In another example, a different chromophore, a spiropyrane derivative, is attached at the free γ-carboxyl group of an E-containing ELP either chemically synthesized or genetically engineered. Figure 14 represents the photochromic

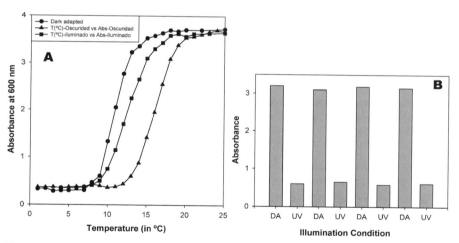

Fig. 13 **A** Temperature profiles of aggregation of 10-mg mL^{-1} water solutions of photoresponsive poly[0.85(VPGVG), 0.15(VPGXG)] (X ≡ L-p-(phenylazo)-phenylalanine) under different illumination regimens. The correspondence between each profile and its illumination condition is indicated in plot. Details on polymer synthesis and illumination conditions can be found in [30]. **B** Photomodulation of phase separation of 10-mg mL^{-1} aqueous samples of poly[0.85(VPGVG), 0.15(VPGXG)] at 13 °C. The prior measurements of the illumination conditions are indicated by the horizontal axis. DA, dark adaptation; UV, UV irradiation. Reproduced with permission from American Chemical Society

reaction for this polymer [122]. As compared to *p*-phenylazobenzenes, spiropyrane compounds show a photoreaction that can be driven by natural cycles of sunlight-darkness without the employment of UV sources, although UV irradiation causes the same effect as darkness but at a higher rate [132].

Again, the difference in polarity between the spiro and merocyanine forms (Fig. 14) is enough to cause a significant T_t shift. Figure 15 shows the turbidity profiles of the polymer in different illumination regimens (Fig. 15a) and the photomodulation of polymer folding and unfolding (Fig. 15b,c).

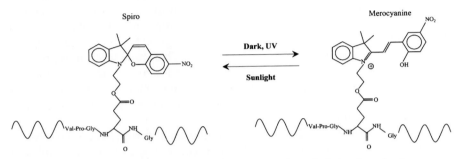

Fig. 14 Photochemical reaction responsible for photochromic behavior of spiropyrane-containing ELP. Reproduced with permission from American Chemical Society

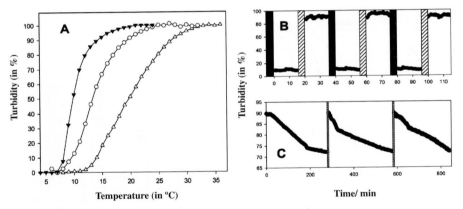

Fig. 15 A Temperature profiles of aggregation of 20-mg mL^{-1} phosphate-buffered (0.01 N, pH 3.5) water solutions of photoresponsive polymer under different illumination regimens. The correspondence between each profile and its illumination condition are indicated in plot. Turbidity was calculated from absorbance values obtained at 600 nm on Cary 50 UV-Vis spectrophotometer equipped with a thermostatized sample chamber. **B** and **C** Photomodulation of phase separation of 5-mg mL^{-1} aqueous samples of photochromic polymer ($T = 14\,°C$, 0.01 N phosphate buffer at pH = 3.5). **A** UV-sunlight *cycles*. *Boxes* in subplot: periods of irradiation: UV, *black boxes*; sunlight, *white boxes*. **B** Darkness-sunlight *cycles*. *Boxes* in subplot: periods of sunlight irradiation. Reproduced with permission from American Chemical Society

The efficiency of the polymer is again outstanding, since just 2.3 spiropyran chromophores per 100 amino acid residues in the polymer backbone were sufficient to render the clear photomodulation shown in Fig. 15.

Different ELP versions responding to pH, light, and other stimuli, such as electrochemical potential or analyte concentrations, can be found in the literature. Most of them were produced by the exclusive use of chemical synthesis in a huge effort, lasting more than a decade, by Prof. Urry's group in a time when the use of genetic engineering to produce PBPs was not sufficiently developed (see, for example, [105]). In some cases, this smart response of the ELPs has already found applications in different fields. For example, Chilkoti et al. have designed thermally and pH-responsive ELPs for targeted drug delivery [77, 80, 133–136], and Kostal et al. have designed tunable ELPs for heavy metal removal [137].

In a different approach in the design of more efficient stimulus-responding ELPs, it is possible to increase and further control the smart behavior of ELPs without increasing the number of sensitive moieties. This is possible if one of the states of that moiety is able to interact with a different compound, while the other state is not, and this interaction causes additional increases in the difference in polarity between both states. This is the basis of the so-called "amplified ΔT_t mechanism", and this has been proved for a p-phenylazobenzene-containing polymer of the kind shown above, poly[0.8(VPGVG), 0.2(VPGXG)], in the presence of α-cyclodextrin (αCD) [121]. The αCD is able to form inclusion compounds with the trans isomer of the p-phenylazobenzene group and not with the cis isomer due to a strong steric hindrance [121] (Fig. 16).

The αCD outer shell has a relatively high polarity, which, of course, is much more polar than the p-phenylazobenzene moiety both in the trans or cis states. The change in polarity between the dark adapted sample (trans isomer buried inside the αCD) and the UV irradiated one (cis isomer unable to form inclusion compounds) led to an enhanced ΔT_t (Fig. 17). Of course, the magnitude of this effect is [αCD] dependent, so it is possible to tune the width and position of the working temperature window just by changing the [αCD].

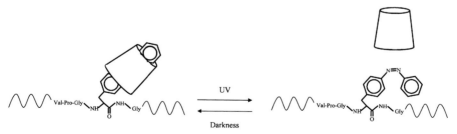

Fig. 16 Schematic diagram of proposed molecular mechanism on interaction between p-phenylazobenzene pendant group and αCD. Reproduced with permission from Wiley

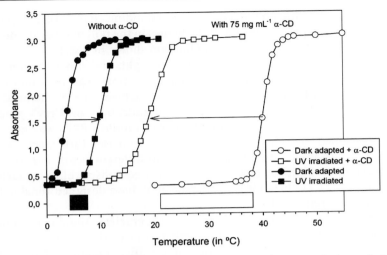

Fig. 17 Temperature profiles of aggregation of 10-mg mL^{-1} water solutions of photoresponsive ELP in absence and presence (75 mg mL^{-1}) of αCD under both illumination regimens. *Circles*: dark-adapted samples; *squares*: UV-irradiated samples. *hollow symbols*: presence of αCD; *filled symbols*, absence of αCD. *Arrows*: sense of displacement of turbidity profile caused by UV irradiation of corresponding dark-adapted sample. *Boxes at bottom*: window of working temperatures open when system is in absence (*filled box*) and presence (*hollow box*) of αCD. Reproduced with permission from Wiley

Table 3 Values of T_t, ΔT_t, offset and gain for a 10-mg mL^{-1} poly[0.8(VPGVG), 0.2 (VPGXG)] water solution in presence of different concentrations of α-CD. DA, dark-adapted samples*; UV, UV-irradiated samples**. Offset and gain as defined in text. Reproduced with permission from Wiley

[α-CD]/ mg mL^{-1}	DA T_t (in °C)	UV T_t (in °C)	ΔT_t (in °C)	Offset (in °C)	Gain
0	3.9	10.0	6.1	–	–
10	20.2	13.7	–6.5	16.3	–1.07
25	26.5	14.7	–11.8	22.6	–1.93
50	33.4	16.2	–17.2	29.5	–2.82
75	40.5	19.5	–21.0	36.6	–3.44

* DA samples were samples kept as the final water solution in the dark for 24–48 h at 5 °C until a stationary transformation of the azo group to the trans isomer was obtained (assessed by UV-Vis spectroscopy).   **UV samples were DA samples further irradiated with UV light. That was made in a standard spectrophotometer quartz cuvette with light from a 500-W Hg arc lamp (model 6285, Oriel) mounted on a lamp housing with an F/1.5-UV-grade fused silica condenser and rear reflector (model 66041, Oriel). UV irradiation was achieved by the use of a band interference filter (340 < λ < 360 nm) from CVI Laser (F10-350.0-4-1.00). The irradiation time needed to obtain a photostationary state was 30 s. The exposure energy irradiation was ca. 4 mW cm^{-2}. Additional information on the irradiation setup can be found elsewhere [123].

As a result, in the αCD/poly[0.8(VPGVG), 0.2(VPGXG)] coupled photoresponsive system, αCD acts much like an amplifier acts on an electronic circuit. αCD promoted a tunable offset, gain, and inversion of the photoresponse of the polymer (Fig. 17 and Table 3). In this way, the polymer photoresponsiveness could be shifted to room or body temperature and with a wider range of working temperatures. Therefore, the use of precise temperature control can be avoided in most conceivable applications, as these applications have a wide range of uses, from photo-operated molecular machines to macroscopic devices (photoresponsive hydrogels, membranes, etc.) and nano- and microdevices (phototransducer particles, photo-operated pumps, etc.). Furthermore, the amplified ΔT_t mechanism is not restricted to photoresponsive ELPs and could be exploited in some other smart ELPs responding to stimuli of a different nature. It also adds a further possibility of control, since the ability of CDs to form inclusion compounds can be controlled by different stimuli in some modified CDs [138–140].

5.7
The Outstanding Biocompatibility of Elastinlike Polymers: The Third Pillar for Extraordinary Biomaterial Designs

The existence of an ITT for ELPs is the base of their remarkable smart and self-assembling properties. A second pillar in the development of extraordinary materials is, evidently, the power of genetic engineering in promoting the easy obtaining of complex and well-defined polymers with controlled and multiple (bio)functionality. Additionally, ELPs show a third property, which is highly relevant when planning the use of these polymers in the most advanced biomedical applications, such us tissue engineering and controlled drug release. This third pillar is the tremendous biocompatibility shown by ELPs.

The complete series of the ASTM-recommended generic biological tests for materials and devices in contact with tissues and tissue fluids and blood demonstrate an unmatched biocompatibility [141]. In spite of the polypeptide nature of these polymers, it has not been possible to obtain monoclonal antibodies against most of them. Apparently, the immune system just ignores these polymers because it cannot distinguish them from natural elastin. Incidentally, it is now believed that the high segmental mobility shown by the β-spiral, the common structural feature of ELPs, greatly helps in preventing the identification of these foreign proteins by the immune system [30, 31]. In addition, the secondary products of their bioabsorption are just simple and natural amino acids.

With this nice set of properties, it is not surprising that the biomedical uses of ELPs seem to be the first area where ELPs will disembark in the market. This is especially true considering that the biomedical (and cosmetic) market shows a clear disposition to quickly adopt those new developments that show superior performance. Additionally, this sector is not so conditioned by the cost associated with the materials used in their devices and developments, as happens in commodity manufacturing and other applications, so the companies producing ELPs will find the biomedical sector a good option for amortizing the cost previously used in the development of all know-how and technology around the production of ELPs.

5.7.1
ELPs for Drug Delivery Purposes: Different Strategies for Molecular Designs

Different versions of ELPs designed for drug delivery purposes can be found in the literature. However, they do not share a common basic strategy on design. On the contrary, ELPs display many different properties that can be useful for drug delivery purposes, i.e., smart behavior (sensitivity to certain stimuli), self-assembly, biocompatibility, etc., so design strategies can be diverse. In fact, the different ELP-based drug delivery systems described to date emphasize exploiting a particular one of those properties.

The first ELP-based drug delivery systems were reported by Urry. They were quite simple devices in which γ-radiated cross-linked poly(VPGVG) hydrogels of different shapes were loaded with a model water-soluble drug (Biebrich Scarlet) [142]. This drug was then released by diffusion. In this simple design, just the extraordinary biocompatibility and the lack of pernicious compounds during the bioresorption of the device were exploited. The designs then became slightly more complicated. The basic VPGVG pentapeptide was functionalized by including some glutamic acids whose free carboxyl groups were used for cross-linking purposes. The cross-linker was of the type that forms caboxyamides, which were selected because of their ability to hydrolyze at a given and controlled rate releasing the polymer chains and, concurrently, any drug entrapped within the cross-linked slabs [143]. This was an apparently simple and conventional degradation-based drug delivery system. However, due to the use of ELPs, the displayed behavior was slightly more complex and efficient. While the cross-link was intact, the carboxyl groups were amidated and, consequently, uncharged. This state of lower polarity yielded a cross-linked ELP material showing a T_t below body temperature. Therefore, the chains were folded at that temperature, the material contracted and deswelled and the polymer chains essentially became insoluble, entrapping the loaded drug quite efficiently in the model drugs studied. When hydrolysis took place on the outer surface of the slab, charged carboxylates appeared, which strongly increased the T_t in this zone (well above body temperature). The skin of the slab became swelled and the loaded drug readily escaped from the outer layer of the device. Additionally, the fully released chains were completely soluble, so they soon diffused and were reabsorbed. This caused the presence of an always fresh surface on the slab and the readiness in the release of the loaded drug within the hydrolyzed surface [143].

For this reason, the kinetics of drug release were almost of the zero-order type and, accordingly, the performance of the system was superior to those made of other equivalent polymers but without showing the ΔT_t mechanism. In general, this statement is more reliable as the size of the loaded drug is higher, since, as no other particular functionalities were added to the polymer chain, in practice, there is no substantial interaction between the drug and the polymer other than the movement constraint of the loaded drug within the polymer matrix. Therefore, a certain degree of uncontrolled diffusion can take place perturbing the kinetics of drug release.

In a different example, and as mentioned in Sect. 5.5, the tendency to form stable, drug-loaded, and nano- and microparticles by some ELPs, especially those based on the (VPAVG) pentapeptide, has facilitated the development of injectable systems for controlled release [127].

Nonetheless, those examples are based on simple polymer formulations that are still far from reaching the full potential of ELPs in developing drug delivery systems. Their smart and self-assembling properties, as well as the deeper knowledge on the molecular basis of the ITT, are only marginally exploited. However, new systems are starting to be published in the literature that already show a more decided bet on exploiting the very special characteristics of ELPs and the powerful way they can be produced, that is, through genetic engineering. For example, Chilkoti's group has produced nice examples of ELPs specially designed for targeting and intracellular drug delivery. They exploited the soluble-insoluble transition of the ELPs to target a solid tumor by local hyperthermia, and then, in the most sophisticated versions of these ELPs, an additional pH responsiveness of these ELPs was used to mimic the membrane disruptive properties of viruses and toxins to cause effective intracellular drug delivery. Among the most evident uses of this kind of advanced drug delivery systems is the more efficient dosage of antitumoral drugs, but these polymers could serve also as alternatives to fusogenic peptides in gene therapy formulations and to enhance the intracellular delivery of protein therapeutics that function in the cytoplasm [40, 80, 133, 135, 136, 144].

On the other hand, the recent deepening knowledge about the molecular characteristics of the ITT has allowed the development of advanced systems for more general drug release that have achieved a practically ideal zero-order drug release kinetics without the concerns caused by previous designs. The first examples are based on Glu-containing ELPs, in which the close vicinities of the γ-carboxyl groups are maintained in a highly hydrophobic environment by positioning phe residues by a precise nanometric design of the polymer sequence in accordance with the β-spiral structure of the folded state [30, 31] in the sense that, once the polymer folds into the β-spiral structure, those phe residues completely surround the free carboxyl group, creating a well-defined battleground where there is a strong competition between the two mutually exclusive forms of hydration, i.e., hydrophobic hydration of the phe residues and hydrophilic hydration of the carboxylate.

The overwhelming presence of phe residues causes extraordinary pK_a shifts of these γ-carboxyl groups toward higher values (the carboxyl group becomes less and less acidic as the number of surrounding phes increases).

Therefore, in neutral or basic pH (including physiological pH), those carboxylate moieties show a strong propensity to neutralize their charge by ion-coupling, i.e., by establishing contacts with positively charged drugs, if this coupling causes an effective decrease in the polarity of the carboxyl vicinities. As a result these polymers, at this neutral or basic pH and in the presence of an adequate oppositely signed drug, form strong insoluble aggregates, which are characterized by a high rate of drug loading and, as implanted, release the drug slowly as it is leached from its coupling on the outer surface of the aggregate. The release rate can be tuned by modifying the hydrophobic environment of the carboxyl by properly choosing the amino acid sequence in the polymer [30, 31]. Once the drug is released and the polymer-drug interaction is lost, and as a consequence of the charged state of the carboxyl group (carboxylate), the polymer unfolds and finally dissolves. At that moment, the interface between the remaining insoluble, still loaded, aggregate and the body fluids are continuously renewing without changing their physicochemical properties for practically the entire functional period of the system. This behavior causes a practically near ideal zero-order release [30, 31].

In the present situation, as demonstrated by the various examples shown above, the different alternatives presented by the extraordinary set of ELP properties, as well as the power of genetic engineering, have shown a remarkable potential for future drug delivery developments. What is more, those independent approaches, exploiting different ELP properties, are not mutually exclusive, so the development of new ELPs combining various strategies of the kind depicted above is foreseeable. As can be easily understood, this could set basis for the development of drug delivery systems with unprecedented efficiency.

5.7.2
ELPs for Tissue Engineering: Introducing Tailored Biofunctionality

Designing a biomedical device is always a tremendous challenge for the material developer. This has been shown above for drug delivery systems, but the most demanding application is likely that of tissue engineering (or as is now preferred, regenerative medicine). When a mature or stem cell divides and spreads in a growing tissue, that cell is passing through the most vulnerable and difficult stage of its life cycle. This is the reason why materials that efficiently work in different biomedical applications can fail when used in for tissue engineering purposes (the failure can be caused both by the material itself and by its biodegradation products).

Additionally, we have to keep in mind that when designing a matrix for tissue engineering, we are trying to substitute the natural extracellular matrix

(ECM), at least transiently. Therefore, many aspects have to be taken into consideration upon designing an adequate artificial ECM. Initially, the material developer must have a decided concern regarding the mechanical properties of the artificial scaffold. It is well known that, when properly attached to the ECM, cells sense the forces to which they are subjected via integrins. Integrins are ubiquitous transmembrane adhesion molecules that mediate the interaction of cells with the ECM. Integrins link cells to the ECM by interacting with the cell cytoskeleton. By this means they couple the deformation of the ECM, as a consequence of the applied forces, with the deformation of the cytoskeleton. The deformed cytoskeleton triggers an intracellular signal transduction cascade that finally causes the expression of those genes related to the rebuilding of the ECM [145]. In this way, the cells are continuously sensing their mechanical environment and responding by producing an ECM that withstands those forces in an adequate manner. In this sense, cells are very efficient force transducers. Therefore, all artificial ECMs have to properly transmit forces from the environment to the growing tissue. Only in this way will the new tissue build the adequate natural ECM that eventually will replace the artificial ECM. In contrast, a stronger or too weak artificial ECM will cause its substitution by a too weak or too dense natural ECM, respectively, which can really compromise the success of the regenerated tissue.

Additionally, we know that the ECM is not just a scaffold showing certain mechanical properties in which the cells attach simply to achieve the necessary tissue consistency and shape. Far from that, the proteins of the natural ECM (fibronectin, collagen, elastin, etc.) contain in their sequence a huge number of bioactive peptides that are of crucial importance in the natural processes of wound healing. Those sequences include, of course, the well-known cell attachment sequences. In natural ECMs we find target domains for specific protease activity. Those proteases, such as the metalloproteinases of the ECM, are only expressed and secreted to the extracellular medium when the tissue wants to remodel its ECM [146]. They act on specific sequences that are present only in the proteins of the ECM, so they cannot cause damage to other proteins in their vicinities. It is also known that some fragments of these hydrolyzed ECM proteins are not just mere debris. Once released they show strong bioactivity, which includes the promotion of cell differentiation, spreading, and angiogenesis, among other activities. Finally, a growing tissue is delicately controlled by a well orchestrated symphony of growth factors and other bioactive substances segregated by the cells. Incidentally, these factors are mainly of a peptide nature.

This is the scenario that a growing tissue expects to find when passing through the difficult circumstances of growth and regeneration. Therefore, this is the situation that we have to (or try to) mimic with our artificial scaffolds designed for tissue engineering. This picture looks quite disheartening and, in fact, one would be hard-pressed to think of a petroleum-based polymer that fulfills a minimum requirement of being reabsorbable, sufficiently biocompatible, and nontoxic (the polymer itself and/or its biodegradation products), having adequate mechanical properties and being able to display or induce a minimum number of needed biofunctionalities. One must not be surprised by the fact that, in spite of the expectations caused by tissue engineering, it has achieved a quite moderate success to date. Among the first properties that seem to be unachievable by conventional polymers is complexity. The set of minimum requirements listed above clearly points to the need for a very complex material that could be well beyond the practical reach of our synthesis technology. This must not surprise us. We are trying to mimic an intrinsically complex natural ECM to a level that, in fact, we have not yet fully uncovered. It is hardly imaginable that such a variety of specific properties and biofunctionalities can be achieved by one of our petroleum-based polymers and in spite of the fact that we really can choose functionality from among an impressive set of different monomers developed by organic chemistry in recent decades.

In spite of the discouraging scenario depicted above, we could be in a position where different options could come to our aid; GEPBPs could represent one of these clear breakthrough alternatives.

Soon after the finding of the extraordinary biocompatibility of the (VPGVG)-based ELPs [141], the capabilities of ELPs for tissue engineering were tested. The first candidates were the simple polymers like poly(VPGVG)s and their cross-linked matrices. Surprisingly, the cross-linked matrices of poly(VPGVG)s when tested for cell adhesion showed that cells do not adhere at all to this matrix and no fibrous capsule forms around it when implanted [147]. Of course, this matrix and other states of the material have a potential use in the prevention of postoperative, posttrauma adhesions [147], but in principle they do not seem to be realistic candidates for tissue engineering.

Nonetheless, this absolute lack of cell adherence is not a drawback; on the contrary, it is highly desirable since it provides us with a starting material with the adequate mechanical properties and biocompatibility and lacks unspecific bioactivities. Very soon those simple molecules were enriched with short peptides having specific bioactivities. Due to the polypeptide nature of the ELPs, those active short sequences were easily inserted in the polymer sequence even though, at that time, chemical synthesis was still the only option for obtaining these polymers. The first active peptides inserted in the polymer chain were the well-known general-purpose cell adhesion peptide RGD (R = L-arginine, G = glycine, and D = L-aspartic acid) and REDV (E = L-glutamic acid and V = L-valine), which is specific to endothelial cells. The results were clear: the bioacivated (VPGVG) derivatives showed a high capacity to promote cell attachment, especially those based on RGD, which showed a cell attachment capability almost equivalent to that of the human fibronectin [148]. Once genetic engineering was finally adopted as the production method, the molecular designs started to increase in complexity.

Different ELP compositions were tested as base polymers. Additionally, the cell attachment domains were not restricted to the exclusive short peptide active domain and were increased in size as more amino acids were placed in such a way as to surround the central active REDV or RGD domains as a way of obtaining a more active cell-binding site [73]. For example, Panitch et al. have shown that by using the longer CS5 region of the human fibronectin, which is an eicosapeptide having the REDV sequence in its central part, the achieved cell adhesion was more effective than the short REDV inserts [84].

However, those still simple GEPBPs were made more complex by the addition of different functionalities such as cross-linking domains [85, 149, 150].

There are now examples based on more complex designs that include various bioactivities and other functionalities in an effort to mimic the complex composition and function of the natural ECM extracellular matrix. Girotti et al. have bioproduced the ELP polymer depicted in Fig. 18 [88].

This last ELP is made from a monomer 87 amino acids in length and has been produced with $n = 10$ (MW = 80 695 Da). The monomer contains four different functional domains in order to achieve an adequate balance of mechanical and bioactive responses. First, the final matrix is designed to show a mechanical response comparable to the natural ECM, so that the matrix is produced over a base of an ELP of the type $(VPGIG)_n$. This basic sequence assures the desired mechanical behavior and outstanding biocompatibility, as discussed above. In addition, this basic composition endows the final polymer with smart and self-assembling capabilities, which are of high interest in the most advanced tissue engineering developments. The second building block is a variation of the first. It has a lysine substituting the isoleucine so the lysine γ-amino group can be used for cross-linking purposes while retaining the properties of elastinlike polymers. The third group is the CS5 human fibronectin domain. This contains the well-known endothelial cell attachment sequence, REDV, immersed in its natural sequence to retain its efficiency. Finally, the polymer also contains elastase target sequences to favor its bioprocessability by natural routes. The chosen elastase target sequence is the hexapeptide VGVAPG, which is found in natural elastin. This sequence is a target for specific proteases of the natural ECM. The leitmotif is that those proteases are only produced and excreted to the extracellular medium once

(VPGIG)$_2$ **(VPGKG) (VPGIG)**$_2$ **(EEIQIGHIP**REDV**DYHLYP) (VPGIG)**$_2$ **(VPGKG) (VPGIG)**$_2$ **(VGVAPG)**$_3$

Fig. 18 Schematic composition of monomer used in ELP design described in text. The scheme shows the different functional domains of monomer, which can be easily identified with their corresponding peptide sequences

the tissue decides that the natural ECM must be remodeled. In this sense, the presence of this specific sequence in the artificial polymer guarantees that the polymer will be bioprocessed only when the growing tissue decides that it is time to substitute it by a natural ECM, while in practice it remains fully functional until that time. In addition, the activity of this domain is not restricted to being an inert target of protease activity. It is well known that these hexapeptides, as they are released by the protease action, have strong cell proliferation activity and other bioactivities related to tissue repairing and healing [151].

Although we are sill far from exploiting the full potential of genetic engineering, this last example impressively shows that we are now able to create materials for tissue engineering purposes whose composition and (bio)functionality are unprecedently closer to the rich complexity in functionality and bioactivity of the natural ECM. This polymer also shows the potential of genetic engineering in producing complex polymers in general, since one can hardly imagine obtaining polymers of the complex composition displayed by this last example by chemical methodologies that, in addition, will likely never be so robust, clean, cheap, and easily scalable.

6
Conclusions

Although the creation by genetic engineering of protein-based materials is still in its infancy, it has already shown extraordinary potential. Very complex, well-defined, and tailored polymers can be obtained by this technique, with a wide range of properties. Examples can be found in bulk materials and fibers with extraordinary mechanical performance as well as the most advanced, functional, self-assembling, and smart materials for biomedical uses and nano(bio)technology. The achievable degree of complexity and the concurrent development of function are unparalleled by other techniques. Complexity can be carried to a limit where the concept of the polymer itself vanishes, with the design and bioproduction of materials in which the monomer is getting bigger and more complex from design to design. We are approaching a concept of the protein in our GEPBPs where, rather than having a polymer made by the repetition of a relatively short monomer or a combination of them, a macromolecule without excessive or no repetition is obtained. In that molecule, the single amino acids are grouped within functional domains. In their turn, those domains are arranged along the polymer chain in a well-defined molecular architecture in which there is no space for randomness. All of this to obtain a material in which an unprecedented given set of structural, physicochemical, and biological functionalities are required and must be fulfilled. In addition, the flexibility of bioproduction is so high that we can surely say that the achievable complexity of the GEPBPs, in terms of macromolecular sequence, is, for the first time, not limited in practice by technological constraints but only by our imagination.

All the above examples were accomplished by a robust and relatively easy technology that, in the near future, could be a serious alternative to conventional polymer chemistry, especially if we take into consideration environmental concerns. By this clean procedure, we can produce economical and complex materials that would outperform the efficiency of the existing petroleum-based polymers. GEPBPs are expanding the limits of macromolecular functionality to territories never before imagined.

Acknowledgements This work was supported by the "Junta de Castilla y León" (VA002/02), by the MCYT (MAT2000-1764-C02, MAT2001-1853-C02-01, MAT2003-01205, and MAT2004-03484-C02-01), and by the European Commission (Marie Curie Research Training Network BioPolySurf MRTN-CT-2004-005516).

References

1. Pintauer T, Matyjaszewski K (2005) Coordin Chem Rev 249:1155
2. Pyun J, Tang CB, Kowalewski T, Frechet JMJ, Hawker CJ (2005) Macromolecules 38:2674
3. Nyce GW, Glauser T, Connor EF, Mock A, Waymouth RM, Hedrick JL (2003) J Am Chem Soc 125:3046
4. Coates GW, Moore DR (2004) Angew Chem Int Edit 43:6618
5. Deming TJ (2002) Adv Drug Deliv Rev 54:1145
6. United Nations Framework Convention on Climate Change (http://unfccc.int/2860.php)
7. Massimini K (ed) (2001) Genetic disorders sourcebook. Omnigraphics, Detroit, MI
8. Cappello J (1992) MRS Bull 17:48
9. McGrath K, Kaplan D (eds) (1997) Protein-based materials. Birkhäuser, Boston
10. Krejchi MT, Atkins EDT, Waddon AJ, Fournier MJ, Mason TL, Tirrell DA (1994) Science 265:1427
11. Capello J, Ferrari F (1994) In: Mobley DP (ed) Plastics from microbes. Hanser/Gardner, Cincinnati, OH, p 35
12. McMillan RA, Lee TAT, Conticello VP (1999) Macromolecules 32:3643
13. Meyer DE, Chilkoti A (2002) Biomacromolecules 3:357
14. McPherson DT, Morrow C, Minehan DS, Wu JG, Hunter E, Urry DW (1992) Biotechnol Prog 8:347
15. Prince JT, McGrath KP, Digirolamo CM, Kaplan DL (1995) Biochemistry 34:10879
16. Guda C, Zhang X, McPherson DT, Xu J, Cherry JH, Urry DW, Daniell H (1995) Biotechnol Lett 17:745
17. Whaley SR, English DS, Hu EL, Barbara PF, Belcher AM (2000) Nature 405:665
18. Lee J, Macosko CW, Urry DW (2001) Biomacromolecules 2:170
19. Girotti A, Reguera J, Arias FJ, Alonso M, Testera AM, Rodríguez-Cabello JC (2004) Macromolecules 37:3396
20. Meyer DE, Chilkoti A (2004) Biomacromolecules 5:846
21. Park C, Yoon J, Thomas EL (2003) Polymer 44:6725

22. Elices M (2000) Structural biological materials: design and structure-property pelationships. Elsevier, London
23. Hinman MB, Jones JA, Lewis RV (2000) Tibtech 18:374
24. Vollrath F, Knight DP (2001) Nature 410:541
25. Shao Z, Vollrath F (2002) Nature 418:741
26. Tatham AS, Shewry PR (2000) Tibs 25:567
27. Gosline J, Lillie M, Carrington E, Guerette P, Ortlepp C, Savage K (2002) Philos Trans R Soc Lond B 357:121
28. Lombardi EC, Kaplan DL (1993) Mater Res Soc Symp Proc 292:3
29. Urry DW, Hugel T, Seitz M, Gaub HE, Sheiba L, Dea J, Xu J, Parker T (2002) Phil Trans R Soc Lond B 357:169
30. Urry DW (2005) What sustains life? Consilient mechanisms for protein-based machines and materials. Springer, Berlin Heidelberg New York
31. Urry DW (2005) Deciphering engineering principles for the design of protein-based nanomachines. In: Renugopalakrishnan V, Lewis R (eds) Protein-based nanotechnology. Kluwer, Dordrecht (in press)
32. Kwon I, Kirshenbaum K, Tirrell DA (2003) J Am Chem Soc 125:7512
33. Yu BY (2002) Adv Drug Deliv Rev 54:1113
34. Bilgiçer B, Fichera A, Kumar K (2001) J Am Chem Soc 123:4393
35. Tang Y, Ghirlanda G, Vaidehi N, Kua J, Mainz DT, Goddard WA, DeGrado WF, Tirrell DA (2001) Biochemistry 40:2790
36. Potekhin SA, Medvedkin VN, Kashparov IA, Venyaminov S (1994) Protein Eng 7:1097
37. Goeden-Wood NL, Keasling JD, Muller SJ (2003) Macromolecules 36:2932
38. Panitch A, Matsuki K, Cantor EJ, Cooper SJ, Atkins EDT, Fournier MJ, Mason TL, Tirrell DA (1997) Macromolecules 30:42
39. Zhang G, Fournier MJ, Mason TL, Tirrell DA (1992) Macromolecules 25:3601
40. Chilkoti A, Dreher MR, Meyer DE (2002) Adv Drug Deliv Rev 54:1093
41. Haider M, Megeed Z, Ghandehari H (2004) J Control Release 95:1
42. Gosline JM, DeMont ME, Denny MW (1986) Endeavour 10:31
43. Gosline JM, Guerette PA, Ortlepp CS, Savage KN (1999) J Exp Biol 202:3295
44. Vollrath F (1992) Sci Am 266:70
45. Vollrath F (2000) J Biotechnol 74:67
46. Gatesy J, Hayashi C, Motriuk D, Woods J, Lewis R (2001) Science 291:2603
47. Scheibel T (2004) Microb Cell Fact 3:14
48. Colgin MA, Lewis RV (1998) Protein Sci 7:667
49. Hayashi CY, Blackledge TA, Lewis RV (2004) Mol Biol Evol 21:1950
50. Guerette PA, Ginzinger DG, Weber BH, Gosline JM (1996) Science 272:112
51. Hinman MB, Lewis RV (1992) J Biol Chem 267:19320
52. Craig CL, Riekel C (2002) Comp Biochem Physiol B Biochem Mol Biol 133:493
53. Sponner A, Unger E, Grosse F, Weisshart K (2004) Biomacromolecules 5:840
54. Xu M, Lewis RV (1990) Proc Natl Acad Sci USA 87:7120
55. Hayashi CY, Lewis RV (1998) J Mol Biol 275:773
56. Beckwitt R, Arcidiacono S (1994) J Biol Chem 269:6661
57. Simmons AH, Ray E, Jelinski LW (1994) Macromolecules 27:5235
58. Parkhe AD, Seeley SK, Gardner K, Thompson L, Lewis RV (1997) J Mol Recog 10:1
59. Bram A, Branden CI, Craig C, Snigireva I, Riekel C (1997) J Appl Cryst 30:390
60. Simmons AH, Michal CA, Jelinski LW (1996) Science 271:84
61. Kummerlen J, Vanbeek J, Vollrath F, Meier B (1996) Macromolecules 29:2920
62. Lewis RV, Hinman M, Kothakota S, Fournier MJ (1996) Protein Expr Purif 7:400

63. Hutchinson E, Thornton J (1994) Protein Sci 3:2207
64. Urry DW, Luan CH, Peng SQ (1995) Ciba Found Symp 192:4
65. Van Dijk AA, Van Wijk LL, Van Vliet A, Haris P, Van Swieten E, Tesser GI, Robillard GT (1997) Protein Sci 6:637
66. Van Beek JD, Hess S, Vollrath F, Meier BH (2002) Proc Natl Acad Sci USA 99:10266
67. Dong Z, Lewis RV, Middaugh CR (1991) Arch Biochem Biophys 284:53
68. Fahnestock SR, Yao Z, Bedzyk LA (2000) J Biotechnol 74:105
69. Huemmerich D, Helsen CW, Quedzuweit S, Oschmann J, Rudolph R, Scheibel T (2004) Biochemistry 43:13604
70. Cappello J, Ferrari F (1994) In: Mobley DP (ed) Plastics from microbes. Hanser, New York, p 35
71. Nagarsekar A, Crissman J, Crissman M, Ferrari F, Cappello J, Ghandehari H (2002) J Biomed Mater Res 62:195
72. Ferrari F, Richardson C, Chambers J, Causey SC, Pollock TJ, Cappello J, Crissman JW (1987) US Patent 5,243,038
73. Urry DW, Pattanaik A, Xu J, Woods TC, McPherson DT, Parker TM (1998) J Biomater Sci Polym Edn 9:1015
74. Lee J, Macosko CW, Urry DW (2001) J Biomater Sci Polym Edn 12:229
75. Nicol A, Gowda DC, Parker TM, Urry DW (1993) J Biomed Mater Res 27:801
76. Urry DW (2003) Elastic protein-based biomaterials: elements of basic science, controlled release and biocompatibility. In: Wise DL, Hasirci V, Yaszemski MJ, Altobelli DE, Lewandrowski KU, Trantolo DJ (eds) Biomaterials handbook–advanced applications of basic sciences and bioengineering. Marcel Dekker, New York
77. Meyer DE, Kong GA, Dewhirst MW, Zalutsky MR, Chilkoti A (2001) Cancer Res 61:1548
78. Trabbic-Carlson K, Setton LA, Chilkoti A (2003) Biomacromolecules 4:572
79. Knight MK, Setton LA, Chilkoti A (2003) Summer Bioengineering Conference, Key Biscayne, FL, 25–29 June 2003
80. Dreher MR, Raucher D, Balu N, Colvin OM, Ludeman SM, Chilkoti A (2003) J Control Release 91:31
81. Hyun J, Lee WK, Nath N, Chilkoti A, Zauscher S (2004) J Am Chem Soc 126:7330
82. Trabbic-Carlson K, Liu L, Kim B, Chilkoti A (2004) Protein Sci 13:3274
83. Deguchi Y, Fournier MJ, Mason TL, Tirrell DA (1994) JMS-Pure Appl Chem A 31:1691
84. Panitch A, Yamaoka T, Fournier MJ, Mason TL, Tirrell DA (1999) Macromolecules 32:1701
85. Di Zio K, Tirrell DA (2003) Macromolecules 36:1553
86. McGrath KP, Fournier MJ, Mason TL, Tirrell DA (1992) J Am Chem Soc 114:727
87. Lumb KJ, Kim PS (1995) Biochemist 34:8642
88. Girotti A, Reguera J, Rodríguez-Cabello JC, Arias FJ, Alonso M, Testera AM (2004) J Mater Sci Mater M 15:479
89. Dinerman AA, Cappello J, Ghandehari H, Hoag SW (2002) Biomaterials 23:4203
90. Nagarsekar A, Crissman J, Crissman M, Ferrari F, Cappello J, Ghandehari H (2003) Biomacromolecules 4:602
91. Cappello J, Crissman JW, Crissman M, Ferrari FA, Textor G, Wallis O, Whitledge JR, Zhou X, Burman D, Aukerman L, Stedronsky ER (1998) J Control Release 53:105
92. Zhou Y, Wu S, Conticello VP (2001) Biomacromolecules 2:111
93. Wright ER, Conticello VP (2002) Adv Drug Deliv Rev 54:1057
94. Nagapudi K, Brinkman WT, Leisen J, Thomas BS, Wright ER, Haller C, Wu X, Apkarian RP, Conticello VP, Chaikof EL (2005) Macromolecules 38:345

95. Ohgo K, Kurano TL, Kumashiro KK, Asakura T (2004) Biomacromolecules 5:744
96. Asakura T, Nitta K, Yang M, Yao J, Nakazawa Y, Kaplan DL (2003) Biomacromolecules 4:815
97. O'Brien JP, Fahnestock SR, Termonia Y, Gardner KH (1998) Adv Mater 10:1185
98. Goldberg I, Salerno AJ, Patterson T, Williams JI (1989) Gene 80:305
99. Shimazu M, Mulchandani A, Chen W (2003) Inc Biotechnol Bioeng 81:74
100. Ayad S, Humphries M, Boot-Handford R, Kadler K, Shuttleworth A (1994) The extracellular matrix facts book. Facts Book Series. Academic, San Diego
101. Urry DW, Luan CH, Harris CM, Parker T (1997) Protein-based materials with a profound range of properties and applications: the elastin ΔT_t hydrophobic paradigm. In: McGrath K, Kaplan D (eds) Proteins and modified proteins as polymeric materials. Birkhäuser, Boston, p 133–177
102. Urry DW (1998) Biopolymers 47:167
103. Gowda DC, Parker TM, Harris RD, Urry DW (1994) Synthesis, characterization and medical applications of bioelastic materials. In: Basava C, Anantharamaiah GM (eds) Peptides: design, synthesis and biological activity. Birkhäuser, Boston, p 81
104. Martino M, Perri T, Tamburro AM (2002) Macromol Biosci 2:319
105. Urry DW (1993) Angew Chem Int Edit Engl 32:819
106. San Biagio PL, Madonia F, Trapane TL, Urry DW (1988) Chem Phys Lett 145:571
107. Urry DW (1997) J Phys Chem B 101:11007
108. Rodríguez-Cabello JC, Alonso M, Pérez T, Herguedas MM (2000) Biopolymers 54:282
109. Tanford C (1973) The hydrophobic effect: formation of micelles and biological membranes. Wiley, New York
110. Pauling L, Marsh E (1952) Proc Natl Acad Sci USA 38:112
111. Urry DW, Trapane TL, Prasad KU (1985) Biopolymers 24:2345
112. Manno M, Emanuele A, Martorana V, San Biagio PL, Bulone D, Palma-Vittorelli MB, McPherson DT, Xu J, Parker TM, Urry DW (2001) Biopolymers 59:51
113. Alonso M, Arranz D, Reboto V, Rodríguez-Cabello JC (2001) Macromol Chem Phys 202:3027
114. Reguera J, Lagaron JM, Alonso M, Reboto V, Calvo B, Rodríguez-Cabello JC (2003) Macromolecules 36:8470
115. Rodríguez-Cabello JC, Reguera J, Alonso M, Parker TM, McPherson DT, Urry DW (2004) Chem Phys Lett 388:127
116. Reading M (1993) Trends Polym Sci 1:248
117. Wunderlich B, Androsch R, Pyda M, Kwon YK (2000) Thermochim Acta 348:181
118. Gill PS, Sauerbrunn SR, Reading M (1993) J Therm Anal 40:931
119. Jorimann U, Widmann G, Riesen R (1999) J Therm Anal Calor 56:639
120. Menczel JD, Judovist L (1998) J Therm Anal 54:419
121. Rodríguez-Cabello JC, Alonso M, Guiscardo L, Reboto V, Girotti A (2002) Adv Mater 14:1151
122. Alonso M, Reboto V, Guiscardo L, San Martín A, Rodríguez-Cabello JC (2000) Macromolecules 33:9480
123. Alonso M, Reboto V, Guiscardo L, Mate V, Rodríguez-Cabello JC (2001) Macromolecules 34:8072
124. Yang G, Woodhouse KA, Yip CM (2002) J Am Chem Soc 124:10648
125. Lee TAT, Cooper A, Apkarian RP, Conticello VP (2000) Adv Mater 12:1105
126. Reguera J, Fahmi A, Moriarty P, Girotti A, Rodríguez-Cabello JC (2004) J Am Chem Soc 126:13212
127. Herrero-Vanrell R, Rincón A, Alonso M, Reboto V, Molina-Martinez I, Rodríguez-Cabello JC (2005) J Control Release 102:113

128. Nath N, Chilkoti A (2003) Anal Chem 75:709
129. Nath N, Chilkoti A (2001) J Am Chem Soc 123:8197
130. Nath N, Chilkoti A (2002) Adv Mater 14:1243
131. Urry DW (2004) Chem Phys Let 399:177
132. Ciardelli F, Fabbri D, Pieroni O, Fissi A (1989) J Am Chem Soc 111:3470
133. Stayton PS, Hoffman AS, Murthy N, Lackey C, Cheung C, Tan P, Klumb LA, Chilkoti A, Wilbur FS, Press OW (2000) J Control Release 65:203
134. Waite JH, Sun C, Lucas JM (2002) Philos Trans R Soc B 357:143
135. Chilkoti A, Dreher MR, Meyer DE, Raucher D (2002) Adv Drug Deliv Rev 54:613
136. Meyer DE, Shin BC, Kong GA, Dewhirst MW, Chilkoti A (2001) J Control Release 74:213
137. Kostal J, Mulchandani A, Chen W (2001) Macromolecules 34:2257
138. Kuwabara T, Nakamura A, Ueno A, Toda F (1994) J Phys Chem 98:6297
139. Chokchainarong S, Fennema OR, Connors KA (1992) Carbohyd Res 232:161
140. Reguera J, Alonso M, Testera AM, López IM, Martín S, Rodríguez-Cabello JC (2004) Carbohyd Polym 57:293
141. Urry DW, Parker TM, Reid MC, Gowda DC (1991) J Bioactive Comp Polym 6:263
142. Urry DW, Gowda DC, Harris CM, Harris RD (1994) Bioelastic materials and the ΔT_t-Mechanism in drug delivery. In: Ottenbrite RM (ed) Polymeric Drugs and Drug Administration. ACS, Washintong DC, chap 2, p 15
143. Urry DW (1990) Polym Mater Sci Eng 63:329
144. Andersson L, Davies J, Duncan R, Ferruti P, Ford J, Kneller S, Mendichi R, Pasut G, Schiavon O, Summerford C, Tirk A, Veronese FM, Vincenzi V, Wu G (2005) Biomacromolecules 6:914
145. Jiuliano RL (2002) Annu Rev Pharmacol Toxicol 42:283
146. Sternlicht MD, Werb Z (2001) Annu Rev Cell Dev Biol 17:463
147. Urry DW, Nicol A, Gowda DC, Hoban LD, McKee A, Williams T, Olsen DB, Cox BA (1993) Medical applications of bioelastic materials. In: Gebelein CG (ed) Biotechnological polymers: medical, pharmaceutical and industrial applications. Technomic, Atlanta, p 82–103
148. Nicol A, Gowda DC, Parker M, Urry DW (1994) Cell adhesive properties of bioelastic materials containing cell attachment sequences. In: Gebelein C, Carraher C (eds) Biotechnology and bioactive polymers. Plenum, New York
149. Welsh ER, Tirrell DA (2000) Biomacromolecules 1:23
150. Nowatzki PJ, Tirrell DA (2004) Biomaterials 25:1261
151. Alix AJ (2001) J Soc Biol 195:181

Organic and Macromolecular Films and Assemblies as (Bio)reactive Platforms: From Model Studies on Structure–Reactivity Relationships to Submicrometer Patterning

Holger Schönherr (✉) · Geerten H. Degenhart · Barbara Dordi ·
Chuan Liang Feng · Dorota I. Rozkiewicz · Alexander Shovsky ·
G. Julius Vancso

MESA⁺ Institute for Nanotechnology and Faculty of Science and Technology, Department of Materials Science and Technology of Polymers, University of Twente, Postbus 217, 7500 AE Enschede, The Netherlands
h.schonherr@tnw.utwente.nl, g.j.vancso@tnw.utwente.nl

1	Introduction: Bioreactive Thin Film Architectures and Patterning Methods	171
1.1	Platforms	173
1.2	Patterning	174
1.3	Surface Chemistry in Ordered Systems	174
1.4	Challenges in Surface Characterization and Analysis	176
2	Ultrathin Organic and Macromolecular Films and Assemblies as (Bio)reactive Platforms	178
2.1	Structure-Reactivity Relationships: Model Studies	178
2.1.1	Hydrolysis of NHS Ester SAMs	178
2.1.2	Aminolysis of NHS Ester SAMs	182
2.1.3	Analysis of Reaction Kinetics on the Nanometer Scale: iCFM on NHS – C10 SAMs	183
2.1.4	Determination of Activation Energies for NHS – C10 Ester SAMs versus NHS Homopolymer Thin Films	188
2.2	Micro- and Nanofabrication of High Loading (Bio)reactive Surfaces	190
2.2.1	Covalent Coupling of Dendrimers to NHS Ester SAMs	192
2.2.2	Micropatterning of Dendrimers by Microcontact Printing	193
2.2.3	Nanopatterning of Dendrimers by Scanning Probe Lithography	195
2.3	Nanofabrication of Patterned Biocompatible Bilayer-Vesicle Architectures	196
2.3.1	Bilayer Formation via Vesicle Fusion	197
2.3.2	Bilayer Architectures on Patterned Supports for Biosensing	197
2.3.3	Directing Vesicle Adsorption to Bilayers by SPL	200
3	Outlook	202
	References	203

Abstract In this contribution we review our recent progress in studies that aim at the understanding of the relationship between structure and surface reactivity of organic thin films on the one hand, and at the micro- and nanofabrication of bioreactive or biocompatible platforms on the other hand. Self-assembled monolayers (SAMs) of n,n'-

dithiobis(N-hydroxysuccinimidyl-n-alkanoate) exposing NHS reactive ester groups were studied as model systems for immobilization reactions of DNA, proteins, and receptors. Reaction kinetics and activation energies were determined quantitatively at length scales ranging from millimeters down to nanometers using, for example, surface infrared spectroscopy and in situ inverted chemical force microscopy (iCFM), respectively. The increase in conformational order with increasing alkane segment length was found to result in reduced reactivity due to steric crowding. This drawback of highly organized monolayer architectures and the inherently limited loading can be circumvented by utilizing well-defined macromolecular thin films. Using amine-terminated polyamidoamine (PAMAM) dendrimers immobilized via soft lithography, as well as scanning probe lithography (SPL) approaches (dip-pen nanolithography, DPN) on NHS ester surfaces, robust micrometer and submicrometer patterned (bio)reactive surfaces, which allow one to achieve high molecular loading in coupling reactions for chip-based assays and sensor surfaces, were fabricated. Covalent coupling afforded the required robustness of the patterned assemblies. Finally, we address micro- and nanopatterned bilayer-based systems. SPL was applied in order to fabricate nanoscale biocompatible supramolecular architectures on solid supports. The adsorption of vesicles onto lipid bilayers was spatially controlled and directed in situ with nanometer-scale precision using SPL. This methodology, which provides a platform for research on proteins incorporated in the lipid bilayers comprising the vesicles, does not require that the vesicles are chemically labeled in order to guide their deposition.

Keywords Biointerfacing · Micropatterning · Nanopatterning · Polymer thin films · Surface reactivity

Abbreviations

SAM	Self-assembled monolayer
NSA	Nonspecific adsorption
NHS	N-Hydroxysuccinimide
PNHSMA	Poly(N-hydroxysuccinimidyl) methacrylate
CFM	Chemical force microscopy
iCFM	Inverted chemical force microscopy
PAMAM	Polyamidoamine
Gn	Generation n
SPL	Scanning probe lithography
AFM	Atomic force microscopy
DPN	Dip-pen nanolithography
µCP	Microcontact printing
2D	Two-dimensional
3D	Three-dimensional
PEG	Poly(ethylene glycol)
DMPC	1,2-Dimyristoyl-sn-glycero-3-phosphatidylcholine
SIMS	Secondary ion mass spectrometry
MALDI-MS	Matrix-assisted laser desorption/ionization mass spectrometry
XPS	X-ray photoelectron spectroscopy
ESCA	Electron spectroscopy for chemical analysis
GIR	Grazing incidence reflection
FTIR	Fourier transform infrared
CA	Contact angle

ODT	Octadecanethiol
JKR	Johnson-Kendall-Roberts
x	Extent of reaction
A_t	Integrated absorbance of IR active band at time t
t	Time
θ_i	Contact angle of SAM exposing group i
χ_i	Surface coverage of component in SAM
$\tau_{1/2}$	Half-life of reaction
F	Force
$F_{\text{pull-off}}$	Pull-off force
W_{12}	Work of adhesion
A	Pre-exponential (Arrhenius) factor
E_a	Activation energy
R	Gas constant
T	Absolute temperature
k_b	Boltzmann constant
h	Planck's constant
$\Delta S^{\ddagger}$	Activation entropy
γ_i	Surface free energy of component i
k	Rate constant
k'	First order or pseudo first order rate constant
k''	Second order rate constant
θ	Normalized surface coverage

1
Introduction: Bioreactive Thin Film Architectures and Patterning Methods

The ability to control and modify the chemical and structural properties of surfaces is crucial to advancements in many areas, including selective and environmentally friendly catalysis [1], electronics [2], (bio)chemical sensing [3–8], and biochemistry [9]. Studies of chemical reactions of surfaces may also provide new routes to tailored surface properties. Such surface reactions allow, for example, the tethering of biologically or biomedically important molecules to surfaces, which has significant importance in chemical biology and microarray technology [10–13]. Many approaches rely on monomolecular or thin organic films to covalently couple active species, such as receptors or protein-repellent polymers, to solid supports [14–18]. In addition, owing to increasing surface-to-volume ratios, chemical reactions that occur on organic or polymeric surfaces play a crucial role in many applications, ranging from the previously mentioned array technologies to nanoclusters [19], nanoreactors [20] and drug delivery [21].

Self-assembled monolayers (SAMs) [22–25] are perhaps the most popular model systems for studying chemistry at interfaces under controlled conditions. In the last decade, countless studies have been performed that involve the chemical modification of monolayers [26–35]. However, systematic studies that aim at unraveling important parameters, such as activation energies, have been scarce. In general, it has been noted that the reactivity of functional groups placed in an ordered monolayer environment will be influenced by many factors, such as solvent, steric and electronic effects [36, 37]. Consequently, chemical reactivity can be affected by confinement in highly ordered architectures, which leads (except in rare cases [38]) to reduced reactivity and incomplete conversions [39–42]. For typical applications in, say, the field of sensing, however, rapid reactions and full conversion are desirable to optimize throughput and to minimize reaction times. Similarly, optimal adsorbate orientation and (bio)availability of the active components must be ensured [43].

Since they are intrinsically two-dimensional (2D) systems, SAMs are limited in terms of the surface density of coupled (bio)molecules. The area requirement for SAMs of alkanethiols on gold is $\sim 20-25$ Å^2/molecule, which corresponds to coverages of $5-4 \times 10^{14}$ molecules/cm^2 [44–46]. Hence approaches that extend the dimensionality have received attention. In addition, applications involving biomolecules, such as proteins, may possess stringent constraints to prevent nonspecific adsorption (NSA). Among the various successful approaches to suppressing NSA, the use of or the modification of surfaces with poly(ethylene glycol) (PEG) have received most attention [47–50]. Alternative approaches include surface modification with hyperbranched polyglycerines [51] and other SAM termini [52, 53].

The surface coverage achieved in PEG immobilization determines the NSA of proteins as well as cell adhesion [54–57]. Thus, precise control of the modification reactions is also desirable also in this context. This control is directly linked to the detailed study of the relevant surface reactions, and in particular to a fundamental understanding of the relation of structure, local order, local surface properties on the one hand to the reaction kinetics, the activation energies and transition state parameters on the other hand. As previously mentioned, systematic studies of such confined reactions on solid supports have been scarce to date [36, 37, 58]. In particular, the direct assessment of the relation of local, nanometer-scale structure and surface properties to chemical reactivity in wet chemical surface reactions has been hampered by instrumental and analytical limitations so far.

Our target is to ultimately fabricate reactive micro- and nanopatterns for the area-selective immobilization of biologically relevant molecules via covalent coupling. In addition to full control of reactivity and pattern sizes, biocompatibility and minimized NSA are important for rendering these systems useful as generic platforms. In this context we review in this contribution our recent efforts in this area. We focus in particular on (1) the elucidation of structure–reactivity relationships, (2) the in situ compositional analysis of wet chemical reactions in monolayer-based systems down to nanometer length scales, and on (3) the application and refinement of various micro- and nanofabrication methods to obtain patterns where we have control over the surface chemical composition over a broad range of length scales.

1.1
Platforms

In principle, several different systems can serve as a basis for the mentioned platforms. As shown in Fig. 1, we will discuss in this contribution a variety of complementary 2D and quasi-3D architectures. SAMs of organothiols, disulfides or sulfides on gold [29, 59] and SAMs of organosilicon compounds on hydroxylated silicon surfaces [60–62] are probably among the best known model systems due to their ease of preparation and the high level of structural and chemical control. Vesicles (liposomes) [63] and substrate-supported lipid bilayers [64–67] are related well-established model systems for biological membranes that allow one to study membrane constituents in a controlled environment, and they can serve as a platform for biosensors based on naturally existing biomolecules present in a milieu that approximates a cell mem-

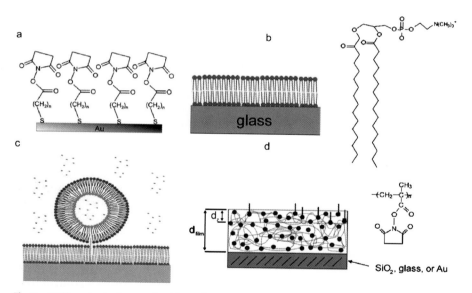

Fig. 1 Different platforms for biomolecule immobilization or biosensor surface modifications: **a** reactive ester-terminated SAM on gold; **b** substrate-supported lipid bilayer on glass (structure of 1.2-dimyristoyl-*sn*-glycero-3-phosphatidylcholine, DMPC); **c** substrate-immobilized lipid vesicle; **d** spin-coated thin film of a reactive homopolymer, such as poly(*N*)-hydroxysuccinimidyl methacrylate (PNHSMA; with tunable thickness d_{film}; the reactive groups are located in a region near the surface with depth d_z; the reactant molecules and reactive moieties in the film are schematically depicted as *bars* and *dots*, respectively)

brane [63]. Finally, substrate-supported (ultra)thin polymer films comprise an alternative platform for interfacing artificial (such as sensor) surfaces with biologically relevant media and systems [18, 68]. Even though structural control on a molecular level is less defined compared to SAMs, the tunable compositions of these systems, their unique polymer properties, such as swelling or presence of entropic forces under certain conditions, their robustness, as well as the facile control of layer thickness over a broad range make these systems attractive for certain applications. Polymers are also promising materials for overcoming the intrinsic limitations of 2D platforms. Such systems and approaches comprise hydrogels [16, 69–72] dendrimers [73–77], hyperbranched polymers [78], polymers prepared by chemical vapor deposition approaches [79], plasma polymers [80, 81], self-assembled polyelectrolyte mutilayers [82] and polymer brushes obtained by grafting-from approaches [83]. From this list we will only treat dendrimer systems in this review.

1.2
Patterning

Patterned surfaces are required for many application platforms [84]. As illustrated with examples from our and our collaborators' work (Fig. 2), SAMs on gold, lipid bilayers, and thin polymer films can be patterned using conventional photolithographic approaches [85], or unconventional approaches, such as soft lithography [86–89] and direct-write scanning probe lithography [90, 91]. Depending on the method utilized, pattern sizes of hundreds of micrometers to sub-100 nm are accessible in principle. The underlying principles of these approaches have been reviewed recently [86–91] and will be discussed, where necessary, in the corresponding sections of the review. Considering the broad range of length scales involved, it is clear that there is a need for a number of complementary approaches to patterned surface functionalization. In order to realize the stated objectives, knowledge of reactivity and its relation to structure of the assembly on the one hand and the analysis of local chemical composition on the other are also required.

1.3
Surface Chemistry in Ordered Systems

Besides the spatial control of surface modification (patterning), control over surface coverage (functional group densities) is a centrally important point. As in any organic chemical reaction, the functional groups involved, the medium and the reaction conditions (such as temperature) influence the reactivity. However, for surface-based reactions, additional factors must be taken into account [36, 37]. These include, among others, steric and anchimeric effects of the reactants, prevented or hindered access of the reactive species from the solution to the reaction centers, or interactions of neighbor-

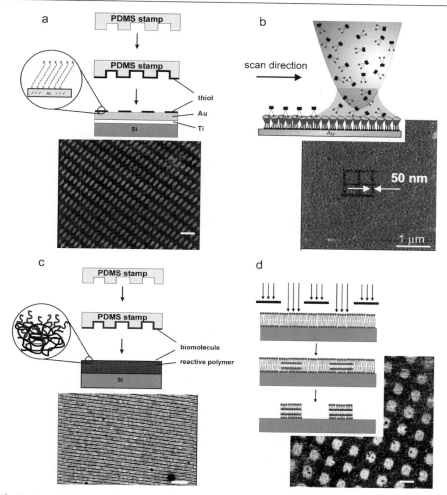

Fig. 2 Atomic force microscopy (AFM) friction images and schematic illustrations of the patterning processes of: **a** microcontact printed SAMs (mercaptoethanol dots in octadecanethiol matrix, scale bar 10 μm); **b** patterned molecular printboards fabricated by supramolecular dip-pen nanolithography (DPN) (reprinted with permission from [92]; Copyright 2004. Wiley VCH); **c** locally hydrolyzed *tert*-butyl acrylate-terminated polymer film on oxidized silicon (soft lithography; scale bar 3 μm) (Feng CL, Vancso GJ, Schönherr H, manuscript submitted to Langmuir); **d** photopatterned bilayer of diacetylene lipid (scale bar 10 μm). Reprinted in part with permission from [93], copyright (1999), American Chemical Society

ing functional groups with the reaction center or enrichment of reactants in disordered layers [94]. For monolayers, additional effects include interactions with the substrate, resulting in altered nucleophilicity and restricted reorientations of functional groups at the monolayer surface [95].

The local environments of the functional groups immobilized in densely packed SAMs, for instance, can also be significantly different from the typical situation in solution. Consequently, the reactivities of these groups may change, as judged from changes in local pK_A [96–100] for example. Similar pK_A changes observed on surface-treated polymers suggest that these phenomena are not limited to perfectly ordered assemblies, but may also be significant in more disordered systems [101, 102]. For optimized surface and interface chemistry in organized molecular assemblies and thin polymer films, it is therefore imperative to understand the underlying *structure–reactivity relationship*. This may include the effect of local order versus disorder and changes in reactivity that may accompany the transition from 2D to 3D architectures.

1.4
Challenges in Surface Characterization and Analysis

Since the effect of functional groups on the reactivities of neighboring functional groups may be highly localized (due to the range of the interaction forces) [103], and since heterogeneities of, say, polymer surfaces also span an enormously wide range, the necessary laterally resolved compositional analysis from micrometer to nanometer length scales is a second point of interest in this review. As reviewed recently [104, 105], there are different approaches that can be used to perform the compositional analysis of organic and polymeric surfaces; however, it was noted that both the experimental procedures and the theoretical background are still far from being fully developed. Laterally averaged chemical composition data, on the other hand, is readily available [106].

The images of the patterned systems shown in Fig. 2 are atomic force microscopy (AFM) friction force images, which display pronounced contrast between areas with different tribological properties [107]. The contrast is related to different surface properties, including surface free energy, and different mechanical responses, for example those arising from differences in molecular packing [108]. While the contrast appears to be sufficient for qualitative analysis, it is difficult to assess the surface coverage of a particular functional group or a particular molecular adsorbate in a quantitative manner based on the friction force contrast. Particularly in systems that are oriented anisotropically in-plane, friction forces on chemically homogeneous surfaces may depend on the relative orientation [109–113]. Complementary approaches comprise AFM force mapping [114–119], as well as various spectroscopies (infrared and Raman) [120], secondary ion mass spectrometry (SIMS) [121, 122], matrix-assisted laser desorption/ionization mass spectrometry (MALDI-MS) [123], X-ray photoelectron spectroscopy (XPS or ESCA) [124, 125], and near-field optical techniques [126] used for imaging.

This review will treat organic and macromolecular films and assemblies as (bio)reactive platforms starting from the analysis of structure–property and consequently structure–reactivity relationships in well-defined model systems (SAMs on Au, Fig. 3). The role of conformational order in determining the reactivity of NHS active esters in hydrolysis and aminolysis reactions will be discussed, as well as the analysis of reaction kinetics on the nanometer scale using inverted chemical force microscopy (iCFM). Then we will extend the SAM-based systems to quasi-3D systems using two complementary ap-

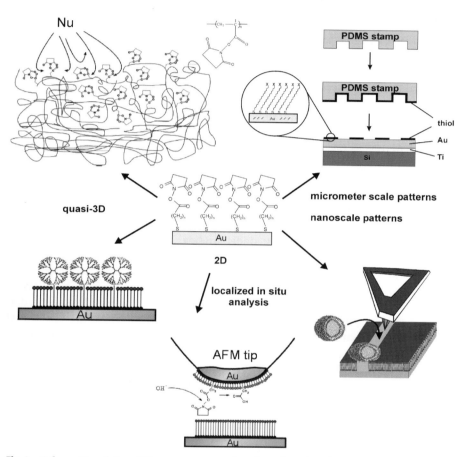

Fig. 3 Schematic of the different aspects of surface functionalization, patterning and analysis treated in this review. The topic is introduced and developed starting from the discussion of well-defined model systems (SAMs on Au). The determination of structure–reactivity relationships, and in particular the way conformational order affects the reactivity of NHS active esters will be discussed. Using iCFM, very localized information on surface reactions can be quantitatively measured in situ for SAM-based systems. The extension of the dimensionality to quasi-3D systems via the immobilization of dendrimers and the fabrication of thin reactive homopolymer films will be addressed, as well as micro- and nanopatterning approaches via soft and scanning probe lithography. Here we discuss SAM-based, as well as bilayer/vesicle-based systems

proaches, namely the fabrication of thin reactive homopolymer films and the immobilization of dendrimers. Finally, micro and nanopatterning via soft and scanning probe lithography will be discussed for SAM-based, bilayer and vesicle-based systems.

2
Ultrathin Organic and Macromolecular Films and Assemblies as (Bio)reactive Platforms

The surface reactivities of ultrathin organic and macromolecular films and assemblies are of central importance to the targeted immobilization reactions of biomolecules. Compared to reactions that occur rapidly in solution, steric effects and locally altered environments may adversely affect reactivity in substrate-supported architectures [36, 37]. Hence the relationship of layer structure to reactivity, highly localized in situ analysis of surface chemical reaction kinetics, and the maximization of surface coverage (molecular loading) by extending the dimensionality of the reactive platform from 2D to quasi-3D will be elaborated on in the following sections.

2.1
Structure-Reactivity Relationships: Model Studies

As mentioned in the *Introduction*, it has been shown that chemical reactivity in ordered ultrathin organic films, such as Langmuir monolayers at the air–water interface [58], or SAMs on solid supports [36, 37], can be distinctly different from reactions carried out in solution. Since the functional groups or molecules involved in these reactions are immobilized at interfaces or on surfaces, these differences can be attributed to "confinement effects" [127]. As shown below, this reduction of reactivity is also present in substrate-supported thin polymer films, albeit to a different extent [128]. The discussion is structured by increasing the level of complexity, starting out with very well defined SAMs on gold.

2.1.1
Hydrolysis of NHS Ester SAMs

We focus initially on the relationship of SAM structure to reactivity for SAMs of activated NHS esters, which are versatile reactive functional groups utilized for the covalent coupling of biologically relevant molecules to surfaces [129–133]. It is well established that the conformational order of SAMs is a function of adsorbate chain length [134]. Since structure (as a result of confinement for example), local order and packing of functional groups appear to be related (see above), differences in conformational order likely result in different reactivities. To this end the conformational order and the kinetics of the base-catalyzed hydrolysis of SAMs of n,n'-dithiobis(N-hydroxysuccinimidyl n-alkanoate) (NHS – Cn, n = 2, 10, 15) were elucidated by grazing angle reflection (GIR) FTIR spectroscopy (Fig. 4) [127].

The FTIR spectra of SAMs of NHS – C2, NHS – C10, and NHS – C15 are shown in Fig. 5. The most prominent bands are the asymmetric C – H stretching vibration, v_{as} (CH$_2$), at $\sim$2920 cm^{-1}, the symmetric C – H stretching vibration, v_s (CH$_2$), at $\sim$2850 cm^{-1}, and the C = O stretching vibration, v (C = O), at $\sim$1748 cm^{-1}. For the complete band assignments and listing of peak positions [131, 132, 135], as well as other complementary characterization data, we refer to [127].

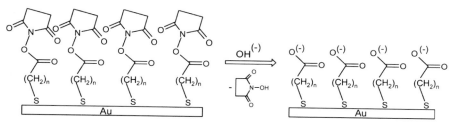

Fig. 4 Structure of NHS ester-functionalized SAM on gold (n = 2, 10, 15) and hydrolysis reaction in aqueous NaOH

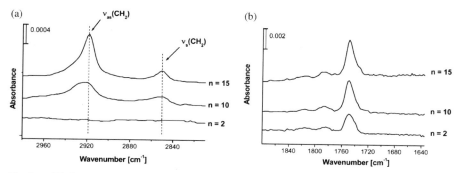

Fig. 5 a High-energy region of GIR-FTIR spectra of SAMs of NHS – C$_n$ with n = 2, 10, and 15 on gold showing the C – H stretching vibrations. b Low-energy region of GIR-FTIR spectra of SAMs of NHS – C$_n$ with n = 2, 10, and 15 on gold showing the succinimidyl and ester carbonyl C = O stretching vibrations. The spectra have been normalized to the absorbance of the C – D stretching vibrations of d_{33}-hexadecanethiol on gold used to record the background spectra. The integrated absorbance of the succinimidyl C = O stretching vibrations in the normalized spectra shown in Fig. 5b suggests a lower coverage for decreasing chain length of the disulfide, provided that the mean orientation of the transition dipole moments is similar. Furthermore, the peak width at half-maximum increases monotonically for decreasing chain length, which is indicative of a more disordered arrangement of the NHS ester end groups in the short chain disulfides. (Reprinted with permission from [127], copyright (2003), American Chemical Society)

The peak positions of the ν_s (CH$_2$) and ν_{as} (CH$_2$) modes for the NHS – C15 monolayers (2850 cm^{-1}, 2918 cm^{-1}) are shifted to lower frequencies compared the NHS – C10 monolayers (2852 cm^{-1}, 2922 cm^{-1}). These modes are unrecognizable in SAMs of NHS – C2, probably due to a broadening of the bands (the broadening of bands attributed to C – H stretching vibrations is obvious even for SAMs of NHS – C10). The band positions are consistent with *near-crystalline packing* of NHS – C15 in SAMs, while SAMs of NHS – C10 and NHS – C2 resemble *more disordered, liquid-like* SAMs [134]. Contact angle (CA) measurements with water as a probe liquid are fully consistent with this interpretation. The hysteresis decreases from 21° and 16° for SAMs of NHS – C2 and NHS – C10, respectively, to 10° for NHS – C15 [127].

The impact of the pronounced conformational differences of these SAMs on their reactivities was assessed by GIR-FTIR and CA measurements for the well known ester hydrolysis in alkaline medium. These measurements were performed in an ex situ mode for samples immersed in the appropriate solutions for variable periods of time followed by extensive rinsing. The kinetics was determined by measuring the decrease in the integrated intensity of the succinimidyl carbonyl band, as shown in Fig. 6a for a NHS – C10 SAM hydrolyzed in 1.00×10^{-2} M NaOH. The strong band at 1748 cm^{-1} decreases in absorbance as the reaction progresses. The extent of the base-catalyzed reaction x can be expressed as a function of hydrolysis time

$$x = \frac{A_0 - A_t}{A_0 - A_\infty}, \tag{1}$$

where A_0 is the integrated absorbance of the succinimide ester carbonyl band at time zero, at time t, and A_∞ is at infinitive time, respectively.

$$\cos\theta_{\exp} = \chi_{NHS} \cos\theta_{NHS} + \chi_{COOH} \cos\theta_{COOH}, \tag{2}$$

where χ_{NHS} and χ_{COOH} are the surface coverages of the two components and $\theta_{NHS} = 59 \pm 2°$ and $\theta_{COOH} = 0°$ are the contact angles of the two pure SAMs.

Similarly, CA measurements were used (in conjunction with the Cassie equation) [136] to estimate the corresponding surface coverages. For conversions of < 50%, FTIR and CA data are in quantitative agreement [128].

As shown in Fig. 6b, the reaction kinetics for identical reaction conditions (1.00×10^{-2} M NaOH at 30 °C) differ significantly for NHS ester SAMs with different chain lengths. While NHS – C2 and NHS – C10 display pseudo first order kinetics with different rate constants (Table 1), NHS – C15 shows the presence of an induction period (see inset in Fig. 6b).

Compared to the hydrolysis reactions of NHS ester model compounds in solution, [136] we observe a decrease in the apparent rate constants by 2-3 orders of magnitude for NHS – C2 and NHS – C10. More strikingly, the reaction of the NHS esters in the highly ordered SAM NHS – C15 shows a different overall kinetic profile. Instead of the expected pseudo first order (exponential) kinetics, sigmoid kinetics with a pronounced induction period are found.

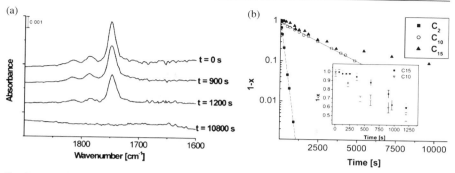

Fig. 6 a GIR-FTIR spectra for NHS – C10 hydrolyzed for different times in 1.00×10^{-2} M NaOH at 30 °C; (Reprinted with permission from [127]; Copyright (2003) American Chemical Society). **b** linearized kinetics plot of hydrolysis for NHS – Cn for n = 2, 10, and 15 (1.00×10^{-2} M NaOH at 30 °C), *inset*: comparison of early stages of hydrolysis of C10 and C15 systems

Table 1 Rate constants and half-lives of the reactions obtained for the NHS – C$_n$ esters

	k''_{FTIR} [M^{-1} s^{-1}]	k''_{CA} [M^{-1} s^{-1}]	$\tau_{1/2}$ (FTIR) [s]	$\tau_{1/2}$ (CA) [s]
n = 2	$(61 \pm 11) \times 10^{-2}$	$(56 \pm 23) \times 10^{-2}$	117 ± 5 [a]	124 ± 5 [a]
n = 10	$(4.5 \pm 0.4) \times 10^{-2}$	$(4.5 \pm 2.3) \times 10^{-2}$	1540 ± 10 [a]	1500 ± 10 [a]
n = 15	–	–	1700 ± 20 [b]	1700 ± 20 [b]
bulk [c]	8700×10^{-2} [c]	8700×10^{-2} [c]	0.8 [c]	0.8 [c]

[a] Calculated as $\tau_{1/2} = \ln 2/k'$ for a base concentration of 1.00×10^{-2} M
[b] Measured for a base concentration of 1.00×10^{-2} M
[c] Data obtained/recalculated from [137]

This change in the rate law with increasing chain length can be attributed to tighter packing of the ester groups as a result of the increasing conformational order. Consequently, access to the hydroxide ions is much more hindered compared to reactions of short chain SAMs [29]. The observed behavior is consistent with a reaction that starts at defect sites and accelerates as more reactive site become accessible as a consequence of the initially reacted ester groups. However, the nature of the induction period is difficult to unravel by FTIR and CA measurements, owing to the lack of spatially resolved information. The surface chemical composition and wettability are assessed as a mean value over almost macroscopic distances (on the order of 10^{12}–10^{14} molecules are probed). Before we elucidate how AFM approaches can be utilized to analyze surface reactions at the relevant length scale in order to help unravel the nature of, say, the previously mentioned induction periods, a second class of reactions of NHS esters are discussed.

2.1.2
Aminolysis of NHS Ester SAMs

The relevance of NHS esters stems from their role as reactive groups that are susceptible to nucleophilic attack, for example from primary amino group-containing molecules (also in aqueous medium). NHS esters are hence frequently utilized to immobilize biomolecules on surfaces via covalent attachment reactions of primary amino groups. Examples include amino end-functionalized DNA, proteins or antibodies [129–133].

As a simple model reaction for such immobilizations, we investigated the reaction of NHS–C10 SAMs and n-butyl amine in aqueous medium [138, 139]. The coupling reaction was followed analogously to the hydrolysis discussed above by ex situ GIR-FTIR and CA measurements. The corresponding FTIR spectra, as well as the reaction kinetics assessed by both methods, are shown in Fig. 7.

During the reaction of SAMs of NHS–C10 with n-butylamine, the appearance of the CH_3 asymmetric in-plane CH stretching mode ($\nu_a(CH_3, ip)$, 2966 cm^{-1}), the CH_3 symmetric CH stretching mode ($\nu_a(CH_3, FR)$, 2879 cm^{-1}), and the amide I (1650 cm^{-1}) and amide II (1550 cm^{-1}) bands are diagnostic of the amide groups formed during the reaction [140]. The kinetics can be determined in a similar way to the hydrolysis by analyzing the integrated absorbance of pronounced bands in the FTIR spectra ($\nu_a(CH_3, ip)$ and succinimide C = O) and by analyzing the CA data using the Cassie equation. The half-lives of the aminolysis reaction, as determined using both methods, are in good agreement ($\tau_{1/2}(FTIR) = 2685 \pm 40$; $\tau_{1/2}(CA) = 2800 \pm 40$).

The deviation of the kinetics from the simple pseudo first order kinetics observed for the hydrolysis is certainly related to the differences in size and nucleophilicity of the attacking nucleophile. Similar to the induction period observed for the hydrolysis of NHS esters in SAMs of NHS–C15 on gold,

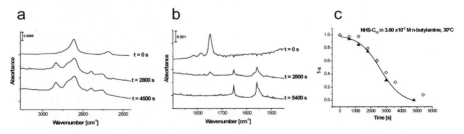

Fig. 7 **a** High-energy region and **b** low-energy region of GIR-FTIR spectra of SAMs of NHS–C10 on gold after different reaction times with 3.00×10^{-2} M aqueous n-butylamine at 30 °C (Reprinted from [139], copyright (2004), with permission from Elsevier). **c** Reaction kinetics obtained from the analysis of the GIR-FTIR and CA data (*right*). (Reprinted from [138], copyright (2004), with permission from Elsevier)

a laterally inhomogeneous reaction, which starts at initiation sites, would offer a plausible explanation. Based on the mean domain size of ~ 5 nm reported for SAMs on gold [141], one may expect that the relevant length scale is < 50–100 nm. However, without high-resolution compositional information acquired with this level of lateral resolution, such an interpretation remains speculative. Hence, the development of new approaches to characterizing local chemical surface compositions is needed.

2.1.3
Analysis of Reaction Kinetics on the Nanometer Scale: iCFM on NHS – C10 SAMs

The apparent limitations of spatially resolved ex situ analysis and interpretation of the reaction kinetics of surface reactions on soft (organic and polymeric) surfaces, using methods such as GIR-FTIR, CA and other established methods (including XPS and SIMS), was highlighted in the previous sections. Methods for performing in situ analysis of the reaction kinetics of wet chemical surface reactions with sufficiently high resolution are largely unknown [104, 105]. One exception is the family of scanning probe microscopies. So-called chemical force microscopy (CFM) [142] has demonstrated its potential for discriminating areas with different chemical compositions down to sub-50 nm length scales [143]. Using chemically functionalized tips, pull-off forces measured in force–displacement measurements contain information about the surface and interfacial free energies of the contacting surfaces and hence constitute a way to estimate surface coverages in simple reactions with high lateral resolution.

To circumvent the problems of instrumental drift during intrinsically slow in situ force mapping of wet chemical reactions on surfaces (a 64×64 pixel^2 map is typically acquired in several minutes), we introduced an AFM-based technique called inverted CFM [29]. In this approach, the reactants are immobilized on the AFM tip and *not* on the flat sample surface (Fig. 8). The flat surface consists of an inert SAM on Au(111). To follow the kinetics of the reactions of the tip-immobilized functional groups, the variation in pull-off forces between the tip coated with the reactant and the inert surface is monitored as a function of time in situ in the reaction medium. The contact area of the tip at the pull-off in such experiments using nonreactive SAMs deposited on atomically flat Au(111) (as inert samples) varies (depending on the surface free energies) between approximately 10 and 100 effectively interacting molecular pairs [138, 139]. However, as the surface characteristics of the inert substrate do *not* vary as a function of position, the pull-off force values only contain compositional information about those reactant groups on the tip that reside inside the tip–sample contact area. As shown below, this approach can provide some information that is lacking about surface reactions that display an induction period.

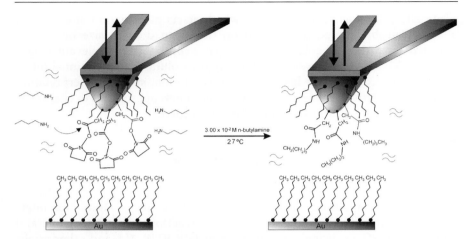

Fig. 8 Schematic drawing of "inverted" chemical force microscopy for the reaction between NHS-esters and *n*-butylamine in aqueous medium. In iCFM the pull-off forces between an AFM tip covered with a SAM of NHS – C10 and an inert octadecanethiol SAM are measured in situ during the conversion of the reactive groups attached to the tip. The interaction between tip and inert surface varies systematically with the extent of the reaction and hence it allows one to quantitatively investigate the reaction kinetics. (Reprinted from reference [139], copyright (2004), with permission from Elsevier)

To obtain a better understanding of the sigmoid kinetics observed for hydrolysis and aminolysis reactions, as discussed in the previous sections, these reactions were investigated on the nanometer scale by iCFM. In these experiments the force required to pull gold-coated AFM tips functionalized with SAMs of NHS – C10 away from contact with an inert octadecanethiol (ODT) SAM on flat Au(111) was monitored in real-time during reaction in aqueous NaOH and *n*-butylamine for hydrolysis and aminolysis, respectively.

As shown in Fig. 9, the pull-off forces (each data point represents the mean value of 200 individual pull-off events) were found to *decrease* for the hydrolysis, while the forces *increased* for the aminolysis. The changes in pull-off force were directly related to changes in surface composition of the contact area at pull-off.

The pull-off force $F_{\text{pull-off}}$ can be expressed as function of tip radius R and work of adhesion (surface energy per unit area) W_{12} as

$$F_{\text{pull-off}} = -3/2\pi R W_{12} \tag{3}$$

W_{12} is a function of the surface free energies of the tip (γ_1), the sample γ_2, and the corresponding interfacial free energy γ_{12} (Eq. 4). If the experiment is carried out in a medium, the γ_i refer to the surface free energy for the surface i

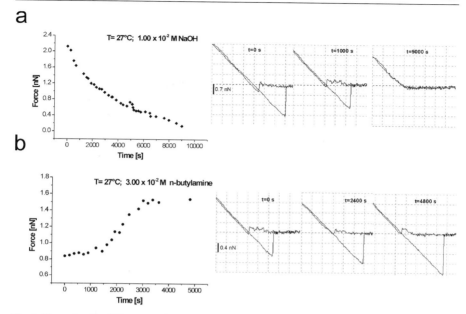

Fig. 9 Plot of pull-off forces as function of reaction time during hydrolysis (**a**) and aminolysis (**b**) of NHS – C10 determined by iCFM. Each data point corresponds to the mean pull-off value of 200 individual pull-off events. Representative force–displacement curves are shown as insets. ($F \propto$ extent of reaction) (Reprinted from [138], copyright (2004), with permission from Elsevier)

in contact with the corresponding medium.

$$W_{12} = \gamma_1 + \gamma_2 - \gamma_{12} \quad (4)$$

The conversion x of ester groups to carboxyl and amide groups was calculated from:

$$x = \frac{F_0 - F_t}{F_0 - F_\infty} \quad (5)$$

where F_0, F_t and F_∞ denote the measured average pull-off forces at $t = 0$, $t = t$, and $t = \infty$, respectively. This equation assumes that the forces change linearly with the work of adhesion.

Each curve represents the kinetics of the reaction occurring *exclusively* in the contact area of an *individual* AFM tip modified with a SAM of NHS – C10. The observed trends are fully consistent with solvent exclusion effects [144]. Increasing conversion leads to a progressively more solvated carboxylate surface in the case of the hydrolysis, while for the aminolysis an increasingly less solvated, hydrophobic surface is obtained.

For the hydrolysis each experiment displayed an exponential decrease in the pull-off force, which can be linearly transformed to surface coverage via

Eq. 3. Thus, the data is consistent with pseudo first order kinetics for the hydrolysis; furthermore, the absence of any induction period points to a spatially homogeneous hydrolysis reaction on the mentioned length scale. The corresponding rate constants are summarized in Table 2.

By contrast, widely different individual force (reaction) profiles were observed for the aminolysis reaction. Most of the profiles showed an induction period, after which the pull-off forces increased and finally leveled off. A number of representative individual traces are shown in Fig. 10a. Figure 10b shows a histogram of the induction periods observed for the aminolysis, as well as a plot of the experimentally determined induction period vs the number of effectively interacting molecular pairs (evaluated by the JKR [145] and the Poisson [146] approaches, respectively).

The experimental data indicate that the aminolysis reaction may spread from initiation or defect sites that are initially accessible for nucleophilic attack. At a very early stage, the reaction proceeds very slowly, as generally seen by FTIR spectroscopy (Fig. 7), because larger numbers of accessible reactive groups in the monolayer must first be generated as a consequence of the initial hydrolysis reaction. As more accessible reactive groups form, the reaction accelerates. The observation of a broad range of induction periods is fully consistent with this model. The reaction can be detected at or just after the start of the experiment (Fig. 10: $t_{\text{ind}} \leq 200$ s), if initiation or defect sites are present in or are close to the small tip-sample contact area. For initiation or defect sites outside of this contact area there are initially no changes in pull-off force. The highly localized observation of the reaction only starts after the reaction has proceeded to the tip–substrate contact area (Fig. 10: $t_{\text{ind}} > 800$ s).

Consistent with this interpretation, the averaged force versus time data reproduces the sigmoid conversion observed on the macroscale (Fig. 7c), as seen by the excellent agreement of the mean half-life of the reaction in 3.00×10^{-2} M aqueous n-butylamine of 2600 ± 240 s and the half-life of

Table 2 Kinetic parameters for reactions of SAMs of NHS – C10

reaction	k'' [M^{-1} s^{-1}]	$\tau_{1/2}$ [s]
hydrolysis (FTIR) at $T = 30$ °C	$(4.5 \pm 0.4) \times 10^{-2}$	1540 ± 10 [a]
hydrolysis (CA) at $T = 30$ °C	$(4.5 \pm 2.3) \times 10^{-2}$	1500 ± 10 [a]
hydrolysis (iCFM) at $T = 27$ °C	$(3.0 \pm 0.2) \times 10^{-2}$	2310 ± 20 [a]
aminolysis (FTIR) at $T = 30$ °C	–	2685 ± 40 [b]
aminolysis (CA) at $T = 30$ °C	–	2800 ± 40 [b]
aminolysis (iCFM) at $T = 27$ °C	–	2600 ± 240 [b]

[a] Calculated as $\tau_{1/2} = ln2/k'$ for a base concentration of 1.00×10^{-2} M.
[b] Measured for 3.00×10^{-2} M aqueous n-butylamine.

2685 ± 40 s estimated from the FTIR data (Table 2) for the aminolysis. From the regression analysis in Fig. 10b we can estimate that the average number of effectively interacting molecular pairs, for which the induction period vanishes, corresponds to 85 ± 4 pairs (Poisson) and 77 ± 3 pairs (JKR) (area ~ 20 nm^2). Based on the interpretation that the reaction starts at defect or initiation sites, this value corresponds to 5×10^{12} defects/cm^2 and an approximate mean distance between neighboring defects of ≥ 5 nm. As there are several thousand pinholes/cm^2 in etch-resistant SAMs of, say, ODT on gold [147], the initiation sites are unlikely to be pinholes, but may be defects in optimal head group packing.

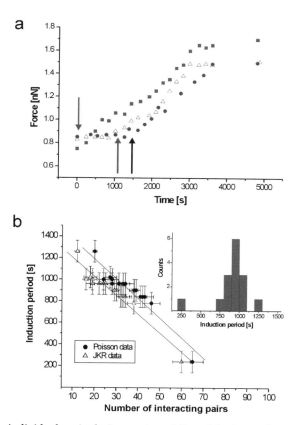

Fig. 10 a Three individual aminolysis reactions followed by iCFM force measurements. The *arrows* indicate three widely different induction periods of ~ 0 s, ~ 1000 s, and ~ 1450 s; **b** Plot of induction period vs number of interacting pairs estimated using both the JKR theory and the Poisson method (*inset*: histogram of induction periods as measured by iCFM during aminolysis of NHS – C10). (Reprinted from [138], copyright (2004), with permission from Elsevier)

In conjunction with the results of the previous sections, it appears that a high degree of conformational order and tight packing in, say, monolayer systems, is detrimental for realizing highly reactive platforms for the immobilization of (bio)molecules with high molecular loading. Direct molecular level evidence by iCFM points to the presence of laterally heterogeneous reactions for highly ordered systems. The difference between the two types of model reactions also underlines the importance of the size and character of the nucleophile for obtaining reactive systems with simple and predictable kinetics. Further insight into the relationship of structural order to reactivity was sought in comparative studies of the temperature dependence of model reactions in SAMs and related spin-coated polymer thin films.

2.1.4
Determination of Activation Energies for NHS – C10 Ester SAMs versus NHS Homopolymer Thin Films

Analysis of the temperature dependence of surface chemical reactions will provide a more detailed insight into the underlying factors that may hamper the corresponding surface reactions. Using the CA approaches introduced above, the surface compositions of SAMs of NHS – C10 and thin films of poly(N-hydroxysuccinimidyl methacrylate) (PNHSMA) on oxidized silicon (Fig. 11) were determined after reaction in alkaline media at different temperatures [128]. FTIR spectroscopy provides complementary information, but owing to their limited surface sensitivity, spectroscopic methods are inferior to CA measurements [148].

The kinetic data show that the NHS ester groups in PNHSMA are hydrolyzed in a reaction that can be described as a pseudo first order reaction for all temperatures with an apparent (second order) rate constant that is ~ 5 times faster than for the SAM of NHS – C10 on gold. In Fig. 11, the corresponding data has been plotted according to the linearized form of the Arrhenius equation (Eq. 6).

$$\ln k'' = \ln A - \frac{E_a}{RT}, \qquad (6)$$

where k'' is the second order rate constant, A the pre-exponential (Arrhenius) factor, E_a the activation energy, R is the gas constant, and T is the absolute temperature.

The activation energy and the Arrhenius pre-factor can be obtained from the slope and the intercept. The latter factor can yield the parameters of the transition state, such as the entropy of the transition state (Eq. 7).

$$\Delta S^{\ddagger} = R(\ln \frac{A}{T} - \ln \frac{k_b}{h} - 1), \qquad (7)$$

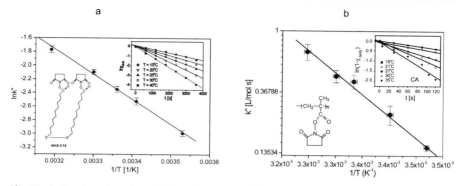

Fig. 11 Arrhenius plots for **a** NHS – C10 SAM and **b** PNHSMA. The *solid lines* correspond to linear least squares fits of the data (*insets*: linearized kinetics for different temperatures for SAM and PNHSMA evaluated based on CA measurements; linearization according to pseudo first order kinetics of the NHS ester surface coverage data calculated from the corresponding CA data using the Cassie equation). (Adapted with permission from [128], copyright (2003), American Chemical Society)

Table 3 Activation energies and estimated parameters characterizing the transition state of the aqueous NaOH

Sample	E_a [kJ/mol]	$\Delta S^{\ddagger}$ (298 K) [J/mol K]
NHS – C10	30 ± 1	– 170
PNHSMA	61 ± 2	– 60

where $\Delta S^{\ddagger}$ is the activation entropy, k_b is the Boltzmann constant, and h is Planck's constant.

The activation energies (Table 3) show a different trend than the rate constants. Compared to SAMs of NHS – C10, the activation energies are significantly higher for the surface reaction of PNHSMA. These observations can be attributed to an increase in mobility and flexibility in the polymer films compared to the SAMs. For the surface region of PNHSMA, the activation entropy is far less negative than for the SAMs, which means that the hydrolysis of NHS – C10 is characterized by a tighter and sterically more demanding transition state.

These data are most consistent with differences in structure between NHS – C10 and PNHSMA – a tightly packed SAM with slight conformational disorder versus an amorphous polymer film in which the NHS ester groups cannot be tightly packed (Fig. 12). In the former case fewer degrees of freedom are available compared to the polymer films. These differences in structure appear to be intimately linked to the kinetic and thermodynamic

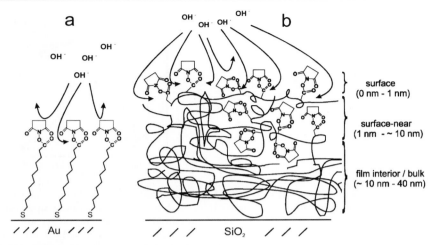

Fig. 12 Schematic of base-catalyzed hydrolysis reaction in **a** SAMs of NHS – C10 and **b** ultrathin films of PNHSMA on oxidized silicon together with the definitions of surface and surface-near regions of the polymer film. The approximate depths in this tentative model were assigned based on the information depths of the techniques (CA: 1 nm [148], IR: the entire film, in other words 40 nm), the fact that only 25% of the NHS ester groups can be hydrolyzed, and that the reaction can be expected to start at the film-solution interface and to proceed homogeneously into the amorphous film. (Reprinted with permission from [128], copyright (2003), American Chemical Society)

parameters of the reactions and the corresponding transition states, respectively. Hence we can conclude that careful design of the organic thin film structure will allow one to control the reactivity in wet chemical reactions, including the immobilization of, say, DNA.

Together with very recent results that show an increase in surface coverage of, for instance, immobilized amino-group-terminated poly(ethylene glycol) (Feng CL, Vancso GJ, Schönherr H, unpublished work) by a factor of 3–4, the much less restricted reactivity of simple reactive homopolymer films is an attractive feature for applications that require robust reactive coatings with high molecular loading. These systems are amenable to the patterning procedures that will be discussed in the following sections. However, the organized assemblies discussed offer the advantage of a higher degree of definition, which facilitates their quantitative characterization and thus the derivation of general guidelines based on these model studies.

2.2
Micro- and Nanofabrication of High Loading (Bio)reactive Surfaces

The drawback of reduced reactivity due to steric crowding found in highly organized monolayer architectures and the inherently limited loading can be

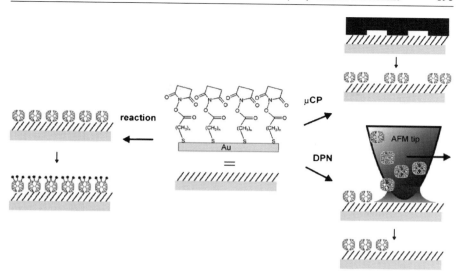

Fig. 13 Schematic of immobilization of amino-terminated PAMAM dendrimers to NHS reactive ester SAMs on gold via covalent bond formation; reaction from solution provides homogeneously covered layers that can be labeled in order to determine the number of retained primary amino groups of the dendrimers. Micro and nanometer-scale patterning is possible via μCP and DPN

circumvented by utilizing well-defined macromolecular thin films. The extension of 2D architectures to the third dimension is an attractive way to increase the loading of (bio)molecules on reactive surfaces and to reduce the effects of steric crowding at the same time. The latter effects have only be considered in the context of actual immobilization chemistry so far. However, it is clear that any biosensor or biochip must present the immobilized species in its active form, such that the interactions to be studied (DNA hybridization, antibody–antigen interactions, and so on) are not hindered by spatial constraints due to tight packing on the sensor or chip surface. For example, the optimized surface coverage for 2D architectures (SAMs) for biotin–streptavidin interactions has been reported to be as low as $\chi = 0.1$ [149].

Recently, reactive platforms based on well-defined macromolecules, such as dendrimers [75, 150], have been introduced as reactive layers that expose chemically accessible functional groups in high densities. These approaches can be extended to micro- and nanoscale patterns by means of microcontact printing (μCP) [86–89] and scanning probe lithography (AFM tip-assisted deposition, also called "dip-pen nanolithography", DPN) [90], as reviewed below (Fig. 13).

2.2.1
Covalent Coupling of Dendrimers to NHS Ester SAMs

To obtain robust reactive ultrathin films with high molecular loadings, in which steric interactions are minimized, covalent attachment of dendrimers to reactive SAMs has been investigated [77]. As shown schematically in Fig. 13, amino group-terminated PAMAM dendrimers can be immobilized on reactive NHS – C10 SAMs by coupling from methanolic solution. The process can be conveniently followed by ex situ FTIR, among other techniques [77]. Upon immobilization, the typical C = O vibration of the NHS ester SAM gradually disappears at the expense of the pronounced amide I and amide II vibrations of the PAMAM dendrimers (Fig. 14a). Since the dendrimers contain a significant number of internal amide bonds that contribute to these latter peaks, complementary experiments with polypropylene imine dendrimers with amine termination (DAB) dendrimers have been carried out.

The dendrimer immobilization can be described by a Langmuir isotherm (Fig. 14b). Complementary XPS analyses in conjunction with labeling of the primary amino groups with trifluoroacetic acid anhydride showed that 28% of all the peripheral primary amino groups are chemically accessible (corresponding to an area requirement for each accessible amino group of $\sim 8.9 \times 10^{-19}$ m^2).

The immobilized G4 PAMAM dendrimers can be directly visualized using intermittent contact (tapping) mode AFM. As shown in Fig. 15, the dendrimers attached to NHS – C10 SAMs appear as globular features with heights

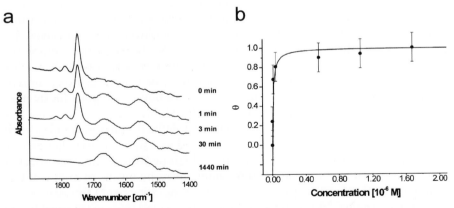

Fig. 14 **a** GIR-FTIR spectra of NHS – C10 SAM after reaction with G4 PAMAM dendrimers for various times (4.5×10^{-6} M methanolic solution of PAMAM G4). **b** Adsorption isotherm of PAMAM G4 on NHS – C10. The *solid line* corresponds to the fit of the Langmuir isotherm (reprinted with permission from [77], copyright (2004), American Chemical Society)

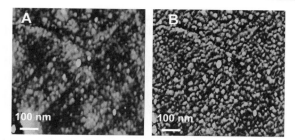

Fig. 15 a Tapping mode AFM height image (acquired in air, z-scale = 5.0 nm) and **b** phase image of NHS–C10 SAM fully covered with PAMAM G4. The triangular terraces of Au(111) are clearly recognizable in the height image (*left*), indicating that a layer of homogeneous thickness has been deposited. (Reprinted with permission from [77], copyright (2004). American Chemical Society)

of ∼ 2 nm and (convoluted) widths of between 10 and 15 nm. These values are consistent with the interpretation that the features are indeed individual dendrimers, considering the theoretical diameter (4.5 nm) [151] and tip convolution effects. The AFM data shows that highly defined layers are formed, because the triangular terraces of the underlying Au(111) substrate can be still recognized. While these layers comply with the requirements identified in the previous sections, patterning appears to be a necessary condition for many biosensor and related applications.

2.2.2
Micropatterning of Dendrimers by Microcontact Printing

Patterning of SAMs can be performed by a multitude of techniques, as reviewed recently [152, 153]. Apart from photolithography using UV light [154, 155] or e-beam lithography [156, 157], microcontact printing has received a lot of attention [86–89]. In this process, an elastomeric stamp is soaked with a solution containing the reactive molecules that should be transferred. Upon establishing conformal contact between the dried stamp and a reactive substrate, transfer of molecules takes place in those regions where contact is established. If diffusion of the ink molecules via the surface or the gas phase can be excluded, patterns of one type of molecule can be prepared. For μCP of thiols on gold, it has been shown that high-quality SAMs are formed [158] and refilling of the uncontacted (unfunctionalized) areas leads to SAMs exposing two functionalities.

The micropatterning of PAMAM dendrimers relies on μCP with a hydrophilized stamp. An UV/ozone treatment increases the surface energy of the PDMS [159, 160] and thus provides the necessary homogeneous loading of the stamp (Fig. 16). Contact mode AFM height and friction images recorded on micropatterned dendrimer surfaces show that elevated areas with high friction can be observed after transfer of G4 PAMAM dendrimers to NHS – C10 SAMs. The friction contrast can be understood via surface energy arguments, as the dendrimers are more hydrophilic than the unmodified NHS ester surface. Preferential adsorption of water to the dendrimer regions will result in considerable capillary forces, that lead to higher friction forces.

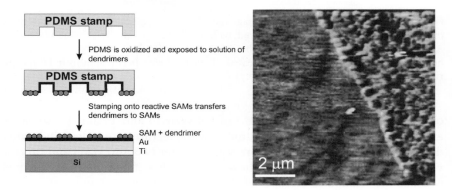

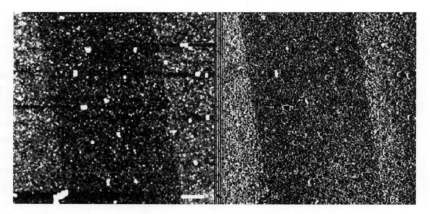

Fig. 16 a *Left*: Schematic of μCP process; *right*: Top view of AFM height image acquired at the border between NHS SAM and dendrimer-modified NHS SAM on an atomically flat Au(111) sample. **b** CM-AFM height and friction images of NHS – C10 SAM patterned with PAMAM G4 by microcontact printing (acquired in air; scale bar 1 μm, the height scale covers 14 nm from dark to bright; the friction forces (a.u.) increase from dark to bright contrast). In these lower resolution images obtained on granular gold, the low friction areas correspond to bare NHS – C10 SAM and the high friction areas to the immobilized dendrimers. (Reprinted in part with permission from [77], copyright (2004), American Chemical Society)

In the higher resolution image (Fig. 16a) one can discern many densely packed globular particles and only a few irregular clusters. Each of the bright spots may represent a single dendrimer molecule. The edge of the stamped area is remarkably sharp (edge roughness $\leq$ apparent width of a single dendrimer). Hence substantial surface diffusion of the dendrimers during or after printing can be ruled out. The diffusion of the "ink" is probably strongly minimized compared to low molar mass inks due to the molecular mass and the covalent attachment. Thus, in principle, higher resolution can be achieved. Our data agree with results reported by the group of Reinhoudt, who used "heavyweight" molecules [161], and by Huck and coworkers, who studied μCP with dendrimers on silicon substrates [162, 163].

The ease of μCP with dendrimers and the high level of definition of the transferred pattern indicate that μCP with this high molecular weight "ink" provides a interesting method of patterning with possibly sub-μm features. Hence, the simple and cost-effective fabrication of functionalized high definition arrays appears to be possible using microcontact printing. In the data presented, the limiting factor for the smallest attainable feature size is represented by the dimensions of the stamp.

2.2.3
Nanopatterning of Dendrimers by Scanning Probe Lithography

The patterning strategy (high molecular mass adsorbate and robust covalent attachment) can be extended to sub-100 nm sized patterns by exploiting AFM-tip assisted transfer of PAMAM dendrimers. By scanning surfaces with an AFM tip, which has been previously coated with the dendrimers, molecules can be deposited onto, say, silicon, mica or SAMs of NHS – C10. In the case of mica and oxidized silicon substrates, the originally deposited patterns were detectable, but the AFM friction images showed that dendrimer molecules may have diffused over the substrates. Based on the currently available data, spontaneous diffusion or tip-induced effects cannot be differentiated. It is, however, clear that the patterns produced do not possess sufficient stability and definition to be of any use.

By contrast, when a NHS – C10 SAM on gold was used as a substrate in DPN experiments, stable patterns were deposited, as observed in AFM friction images. By scanning an AFM tip inked with G4 PAMAM dendrimers (5.6×10^{-5} M methanolic solution) over a NHS – C10 SAM, patterns with sub-100 nm line widths could be fabricated via DPN. In Fig. 17a, a friction force AFM image of lines 2 μm long and 50 ± 20 nm wide is shown; in Fig. 17b the lines are 1 μm long and 70 ± 10 nm wide.

The observation of stable patterns underlines the importance of covalent attachment to achieve robust patterns, and it confirms the overall strategy employed. Thus, NHS – C10 SAMs can be easily and rapidly functionalized with PAMAM dendrimers via amide linkage formation in a very simple

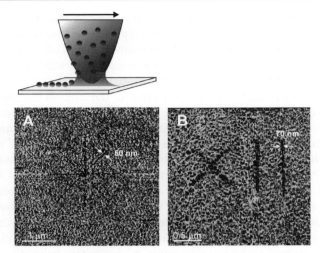

Fig. 17 *Top*: Schematic of DPN process; *bottom*: Sequence of LFM force images (acquired in air; friction forces (a.u.) increase from dark to bright contrast) of arrays of lines with mean widths (± standard deviation) of 50 ± 20 nm and 70 ± 10 nm produced by DPN of G4 PAMAM dendrimers on NHS – C10 SAMs on granular gold. The contrast in the LFM scans is reversed compared to the microcontact-printed patterns, which were scanned with a clean Si$_3$N$_4$ tip. As also observed in an independent study [92], the remaining "ink" on the AFM tip used for DPN alters the relative magnitude of the friction forces in this situation. (Reprinted with permission from [77], copyright (2004), American Chemical Society)

and straightforward process. Using μCP and DPN, micron- and sub-100 nm-scale patterns have been produced. The resolutions of the patterns obtained in our work are probably limited only by the size of the stamps and the scanned areas in DPN. Together with the demonstrated quasi-3D architecture, which allows one to achieve high molecule loading in coupling reactions for chip-based assays and sensor surfaces, these layers constitute an interesting platform for the attachment of biomolecules via exposed primary amino groups.

2.3
Nanofabrication of Patterned Biocompatible Bilayer-Vesicle Architectures

Lipid bilayers and surface immobilized vesicles provide an alternative architecture for the micro- and nanofabrication of bioreactive and biocompatible platforms [63–67]. In recent years, the modification of solid surfaces with biological molecules has been widely studied as a means to obtain biomimetic interfaces for biomedical and environmental applications. Among the various formats of functionalized interfaces investigated, substrate-supported lipid bilayers have received considerable attention. Proteins have been successfully incorporated and shown to be functional in substrate-supported lipid bilayers [163], and the need for a water layer between bilayer and substrate in order to protect sensitive proteins from malfunction or denaturation has been realized which has prompted significant research in, for example, the area of polymer-tethered lipid membranes [63, 165, 166].

As a viable alternative, intact vesicles have been immobilized on solid supports and studied for possible applications in the area of biotechnology [167–170] and to develop chemosensors [171]. There are two main immobilization approaches: (a) immobilization via interaction of complementary DNA fragments that are exposed on the surface and the vesicle membrane, respectively [167, 168]; (b) immobilization mediated by specific streptavidin-biotin interactions [169]. These vesicle systems possess the advantage that the underlying substrate interferes only marginally or not at all with the membrane properties of the immobilized vesicles [172, 173]. Here we discuss a scanning probe lithography-based, label-free method to guide vesicle adsorption to a specific location in the substrate-supported bilayer membrane [174].

2.3.1
Bilayer Formation via Vesicle Fusion

Bilayers on surfaces, including SAM of thiols with hydroxyl end groups, can be formed by vesicle fusion [175, 176]. The process, as depicted schematically in Fig. 18, can be followed conveniently by in situ AFM measurements [177, 178]. As shown by various authors, the vesicle surface coverage, the mechanism of adsorption and bilayer formation, and the vesicle dimensions are directly accessible. Using appropriate models, the adhesion potential and the critical rupture radius of the vesicles can be calculated [179].

2.3.2
Bilayer Architectures on Patterned Supports for Biosensing

As mentioned, substrate-supported lipid bilayers are attractive systems for studying embedded proteins and constructing biosensors. For applications including molecular separation [180], lipid bilayer compartments or patterned bilayers have been utilized [66, 181, 182]. Different approaches to obtaining patterned bilayers have been described recently, including photopolymerization [183], mechanical manipulation [184–186], or the use of prepatterned supports [187–189].

A convenient strategy combines the use of prepatterned SAMs prepared by µCP and bilayer formation by vesicle fusion (Sect. 2.3.1) [164]. As shown schematically in Fig. 19, this approach comprises, in a first step, the patterning of a SAM (a cholesterol-terminated thiol is transferred to the gold substrate, the remaining areas are back-filled with the hydrophilic hydroxy-

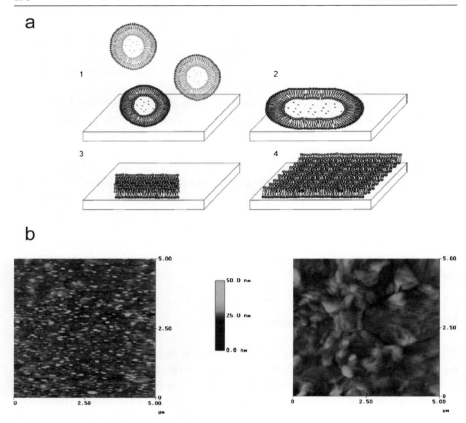

Fig. 18 a Four-step scenario of supported bilayer formation via vesicle fusion comprising 1 vesicle adsorption, 2 fusion of vesicles at the surface to form larger vesicles, 3 rupture of the fused vesicles resulting in bilayer discs, and finally 4 merging of the discs. **b** *Left*: AFM height image of DMPC vesicles (nominal diameter of 50 nm) adsorbed to mercaptoethanol SAM on annealed gold; *right*: AFM height image of DMPC bilayer on mercaptoethanol SAM on annealed gold at increased solution concentrations (images were acquired in buffer at minimized force)

terminated mercaptoethanol). Among the advantages of the subsequent deposition of a lipid bilayer and lipid monolayer on the hydrophilic and hydrophobic areas, respectively, are reduced leakage currents in electrochemical detection, spatial control of the in-plane bilayer architecture (size, shape, and distribution of bilayer areas), and the possibility of incorporating transmembrane proteins localized in bilayer regions in which they are adsorbed (in other words, the possibility of restricting their lateral motion) [164].

Nanometer-scale characterization of the fabricated architectures is again important. Figure 20 shows typical friction force scans of the patterned monolayer samples before and after bilayer formation. In some of the experi-

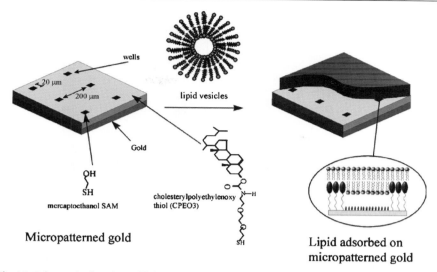

Fig. 19 Schematic drawing of bilayer deposition by vesicle fusion on patterned SAMs prepared by microcontact printing. (Reprinted with permission from [164], copyright (1999), American Chemical Society)

ments, chemical modification of gold-coated AFM tips with octadecanethiol was used to enhance the contrast in the imaging medium [142].

Prior to the unrolling of the vesicles (Fig. 20a), the friction observed in *water* on the mercaptoethanol part is *lower* than on the cholesterol part. The contrast observed is dominated by hydrophobic forces. The mercaptoethanol-functionalized parts of the sample are solvated to a much higher degree than the hydrophobic cholesterol parts [144]. The friction forces show the same trend, which indicates that the adhesion forces dominate the interaction between tip and surface in this case.

After unrolling the vesicles, the friction contrast was reversed in measurements in water (Fig. 20b). In this case the mercaptoethanol areas show *higher* friction. The observed contrast cannot be explained by different forces between tip and surface functional groups because the functional groups exposed at the surface are the *same*. However, the mechanical properties of the lipid monolayer on the more rigid CPEO3 part are different to the more fluid lipid bilayer on top of the mercaptoethanol [190]. At a given imaging force, the AFM tip penetrates more into the lipid bilayer, and so the contact area between tip and sample is increased, resulting in more pronounced energy dissipation and thus higher friction force.

Using electrochemical impedance measurements, it was demonstrated that these micropatterned thiol-terminated lipophilic SAMs can be used to support lipid membranes that meet the key criteria required for use as potential biosensors: they are integral enough (sufficiently blocking) for lipid bilayer

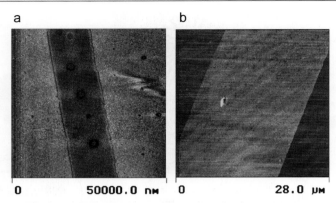

Fig. 20 a AFM friction force micrograph (friction forces increase from low (*dark*) to high (*bright*)) measured on patterned SAM (*stripe*: mercaptoethanol) prior to bilayer deposition. **b** Corresponding friction force micrograph acquired after bilayer formation. (Reprinted in part with permission from [164], copyright (1999), American Chemical Society)

ion channel selectivity to be observed (demonstrated for valinomycin and gramicidin A) [164]; they are formed over hydrophilic SAM regions (mercaptoethanol) and so should have a water layer under the bilayer, which is important for the addition of more complex proteins, especially large ion channels. The bilayers also appear to be relatively fluid (from comparing the frictional forces of the lipid covered cholesterol and lipid covered mercaptoethanol areas).

2.3.3
Directing Vesicle Adsorption to Bilayers by SPL

Instead of utilizing (patterned) substrate-supported membranes for protein studies, and so on, vesicles can be immobilized on suitable substrates. The adsorption of vesicles onto lipid bilayers can be spatially controlled and directed in situ, in principle, with nanometer-scale precision using an AFM-based approach [174]. This strategy enables one to fabricate patterned vesicle arrays without the need to implement molecular recognition units in the vesicles, and hence is applicable to a broad range of systems.

The strategy consists of scanning, say, a previously formed 1.2-dimyristoyl-*sn*-glycero-3-phosphatidylcholine (DMPC) bilayer on a mercaptoethanol SAMs in the presence of DMPC vesicles with an AFM tip, followed by immobilization of vesicles from solution to altered areas of the SAM-supported bilayer (Fig. 21). In the "writing step", patterns are scanned using normal forces of 30–50 nN repeatedly (the bilayer was visibly damaged for forces $> \sim 80\text{--}100$ nN). The interaction of the tip with the bilayer leads to a local modification of the layer. This alteration presumably changes the adhesion

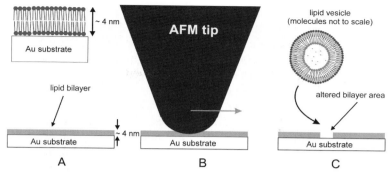

Fig. 21 Schematic of AFM tip-assisted immobilization drawn approximately to scale (bilayer thickness: ∼ 4 nm; tip radius: 20 nm; vesicle diameter ∼ 40 nm): **a** An intact defect-free DMPC bilayer is formed on a mercaptoethanol SAM on gold (SAM is omitted from schematic); **b** subsequent scanning with an AFM tip at high force leads to local damage of the bilayer; **c** in these areas (the schematic drawing does not imply any molecular detail concerning the damage created in step (**b**)) vesicles will adsorb from the solution and stay immobilized. (Reprinted with permission from [174], copyright (2004), American Chemical Society)

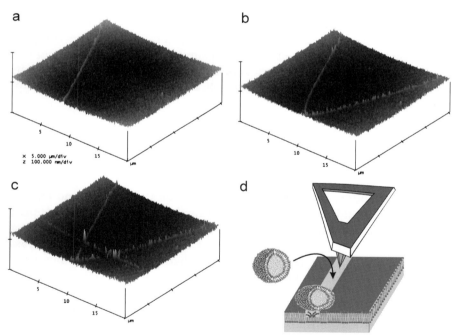

Fig. 22 The stepwise fabrication of vesicle patterns is shown in the sequence of AFM images **a–c** (image size: 20 μm × 20 μm). The scanning probe lithographic modification and the adsorption process of vesicles in solution onto the altered part of the bilayer is depicted schematically in **d** (no molecular-level structural details of the AFM tip-induced line are implied in the schematic)

potential [191, 192] such that vesicle adsorption is possible; in fact the deposited vesicles are much more strongly adsorbed and resist shear forces much better compared to the situation on glass or intact bilayers. As shown in Fig. 22, the resulting assembly can be imaged by contact mode AFM non-invasively using imaging forces of < 1 nN.

The stepwise nanofabrication of a vesicle pattern is shown in Fig. 22. A first line of vesicles is observed after scanning a single scan line under a load of ~ 40 nN for one minute (Fig. 22a). The next AFM images show the result of scanning a second and a third line at angles of 60° relative to the first and second line, respectively (Figs. 22b and 22c). The vesicles were found to possess similar dimensions to those adsorbed from solution onto the bare SAMs [174], so lines of *individual* vesicles were deposited.

The guided vesicle deposition is attributed to a very localized AFM tip-induced alteration of the original DMPC bilayer, which results in the adsorption of vesicles from the supernatant solution. Vesicle patterns with widths equal to the vesicle size can be fabricated over lengths exceeding 25 μm [174]. Based on an estimate of the tip–sample contact area, the line width is given by the width of the vesicle adsorbed on the bilayer.

In conclusion, this novel method of lipid vesicle immobilization on substrate-supported lipid bilayers in a spatially confined manner may serve as a platform for research on proteins incorporated in the lipid bilayers comprising the vesicles. Owing to their structural similarities to the cell membrane, lipid bilayers and substrate-immobilized vesicles provide interesting platforms for studies of incorporated proteins, an area that will see progressive growth in the near future.

3
Outlook

In this contribution, recent advances in our studies on organic and macromolecular films and assemblies for future applications as (bio)reactive platforms were briefly reviewed. Emphasis was placed on the model character of each system investigated. It is clear that these model systems may possess limitations and that system-specific peculiarities can be very important in applications, or where the coupling of specific proteins (for instance) is concerned. However, the acknowledgement of the importance of structural, conformational and compositional characterization on the relevant length scale, the close relationship between structure and reactivity for different architectures, and the possibilities for unconventional micro- and nanofabrication of reactive platforms provide a set of general guidelines that enable one to design reactive platforms in a specific context.

While highly organized monolayer approaches appear to be appealing in many ways, the limited reactivity and limited attainable surface coverages are clear drawbacks. In fact, analysis of the results summarized in this contribution shows that, for a number of scenarios, compositionally and structurally defined yet *disordered* systems possess clear advantages. The extension from 2D to quasi-3D constitutes a generic strategy for increasing the surface coverage in coupling reactions, while stability and diffusion-related problems necessitate the crosslinking of polymeric systems in hydrogel formats [70–72, 193].

Combinations of the very simple spin-coated reactive polymer films discussed in Sect. 2.1.4 with the micro- and nanopatterning approaches studied and refined in model studies on well-defined macromolecular (dendrimer) systems are currently being investigated with substantial success. Thus, the lessons learned in these model studies can be applied to practical formats in order to provide reactive micro- and nanopatterned platforms for the development of biosensors, biochips (DNA, proteins, saccharides, and so on) and studies of cell–cell and cell–substrate interactions.

Acknowledgements The authors would like to thank Dr. Mark A. Hempenius, Dr. Kenichi Morigaki, Dr. Toby Jenkins, Prof. Dr. Stephen D. Evans, Dr. Joseph M. Johnson, Dr. Peter Lenz, Prof. Dr. Curtis W. Frank, Prof. Dr. Steven G. Boxer, Dr. Victor Chechik, and Prof. Dr. Helmut Ringsdorf for very fruitful and stimulating discussions. This work has been financially supported in part by the European Community's Human Potential Programme under contract HPRN-CT-1999-00151 (G.J.V.), the EU Socrates program (D.I.R.), the MESA$^+$ Institute for Nanotechnology of the University of Twente, the Council for Chemical Sciences of the Netherlands Organization for Scientific Research (CW-NWO) (B.D.), also in the framework of the *vernieuwingsimpuls* program (H.S., C.L.F., A.S.).

References

1. Clark JH, Macquarrie DJ (1996) Chem Soc Rev 25:303
2. Mirkin CA, Ratner MA (1992) Annu Rev Phys Chem 43:719
3. Ricco AJ, Crooks RM, Osbourn GC (1998) Acc Chem Res 31:289
4. Paolesse R, Di Natale C, Macagnano A, Davide F, Boschi T, D'Amico A (1998) Sens Actuators B Chem 47:70
5. Everhart DS (1999) Chemtech 29:30
6. Wessa T, Gopel W (1998) Fres J Anal Chem 361:239
7. Storri S, Santoni T, Minunni M, Mascini M (1998) Biosens Bioelectron 13:347
8. Wang J (1992) ACS Symp Ser 487:125
9. Allara DL (1995) Biosens Bioelectron 10:771
10. Fodor SPA, Read JL, Pirrung MC, Stryer L, Lu AT, Solas D (1991) Science 251:767
11. Schena M, Shalon D, Davis RW, Brown PO (1995) Science 270:467
12. Chee M, Yang R, Hubbell E, Berno A, Huang XC, Stern D, Winkler J, Lockhart DJ, Morris MS, Fodor SPA (1996) Science 274:610
13. Pirrung MC (2002) Angew Chem Int Ed 41:1277
14. Case MA, McLendon GL, Hu Y, Vanderlick TK, Scoles G (2003) Nano Lett 3:425
15. Liu GY, Xu S, Qian YL (2000) Acc Chem Res 33:457
16. Sigal GB, Bamdad C, Barberis A, Strominger J, Whitesides GM (1996) Anal Chem 68:490

17. Prime KL, Whitesides GM (1991) Science 252:1164
18. Jerome C, Gabriel S, Voccia S, Detrembleur C, Ignatova M, Gouttebaron R, Jerome R (2003) Chem Commun 2500
19. Tannenbaum R, Hakanson C, Zeno A, Tirrell M (2002) Langmuir 18:5592
20. Han WQ, Fan SS, Li QQ, Hu YD (1997) Science 277:1287
21. Godwin A, Hartenstein M, Muller AHE, Brocchini S (2001) Angew Chem Int Ed 40:594
22. Ulman A (1991) An introduction to ultrathin organic films: from Langmuir-Blodgett to self-assembly. Academic, New York
23. Dubois LH, Nuzzo RG (1992) Annu Rev Phys Chem 43:437
24. Ulman A (1996) Chem Rev 96:1533
25. Schreiber F (2004) J Phys Condens Matter 16:R881
26. Ferretti S, Paynter S, Russell DA, Sapsford KE, Richardson DJ (2000) Trends Anal Chem 19:530
27. Mrksich M (2000) Chem Soc Rev 29:267
28. Ostuni E, Yan L, Whitesides GM (1999) Colloids Surf B Biointerfaces 15:3
29. Schönherr H, Chechik V, Stirling CJM, Vancso GJ (2000) J Am Chem Soc 122:3679
30. Schönherr H, Chechik V, Stirling CJM, Vancso GJ (2001) ACS Symp Ser 781:36
31. Prucker O, Ruhe J (1998) Macromolecules 31:592
32. Prucker O, Ruhe J (1998) Macromolecules 31:602
33. Evans SD, Johnson SR, Ringsdorf H, Williams LM, Wolf H (1998) Langmuir 14:6436
34. Lockhart DJ, Winzeler EA (2000) Nature 405:827
35. Zhu H, Snyder M (2002) Curr Opin Cell Biol 14:173
36. Chechik V, Crooks RM, Stirling CJM (2000) Adv Mater 12:1161
37. Sullivan TP, Huck WTS (2003) Eur J Org Chem 17
38. Tollner K, PopovitzBiro R, Lahav M, Milstein D (1997) Science 278:2100
39. Neogi P, Neogi S, Stirling CJM (1993) J Chem Soc Chem Commun 1134
40. VanRyswyk H, Turtle ED, WatsonClark R, Tanzer TA, Herman TK, Chong PY, Waller PJ, Taurog AL, Wagner CE (1996) Langmuir 12:6143
41. Wang JH, Kenseth JR, Jones VW, Green JBD, McDermott MT, Porter MD (1997) J Am Chem Soc 119:12796
42. Bertilsson L, Liedberg B (1993) Langmuir 9:141
43. Su XD, Wu YJ, Robelek R, Knoll W (2005) Langmuir 21:348
44. Camillone N, Chidsey CED, Liu GY, Scoles G (1993) J Chem Phys 98:3503
45. Poirier GE (1997) Chem Rev 97:1117
46. Jaschke M, Schönherr H, Wolf H, Butt HJ, Bamberg E, Besocke MK, Ringsdorf H (1996) J Phys Chem 100:2290
47. Prime KL, Whitesides GM (1993) J Am Chem Soc 115:10714
48. Harder P, Grunze M, Dahint R, Whitesides GM, Laibinis PE (1998) J Phys Chem B 102:426
49. Lussi JW, Michel R, Reviakine I, Falconnet D, Goessl A, Csucs G, Hubbell JA, Textor M (2004) Prog Surf Sci 76:55
50. Amirgoulova EV, Groll J, Heyes CD, Ameringer T, Rocker C, Moller M, Nienhaus GU (2004) Chem Phys Chem 5:552
51. Siegers C, Biesalski M, Haag R (2004) Chem Eur J 10:2831
52. Holmlin RE, Chen XX, Chapman RG, Takayama S, Whitesides GM (2001) Langmuir 17:2841
53. Ostuni E, Chapman RG, Holmlin RE, Takayama S, Whitesides GM (2001) Langmuir 17:5605
54. Kane RS, Deschatelets P, Whitesides GM (2003) Langmuir 19:2388

55. Ostuni E, Chapman RG, Liang MN, Meluleni G, Pier G, Ingber DE, Whitesides GM (2001) Langmuir 17:6336
56. Chapman RG, Ostuni E, Liang MN, Meluleni G, Kim E, Yan L, Pier G, Warren HS, Whitesides GM (2001) Langmuir 17:1225
57. Huang NP, Michel R, Voros J, Textor M, Hofer R, Rossi A, Elbert DL, Hubbell JA, Spencer ND (2001) Langmuir 17:489
58. Ahmad J, Astin KB (1990) Langmuir 6:1797
59. Nuzzo RG, Allara DL (1983) J Am Chem Soc 105:4481
60. Sagiv J (1980) J Am Chem Soc 102:92
61. Finklea HO, Robinson LR, Blackburn A, Richter B, Allara DL, Bright T (1986) Langmuir 2:239
62. Sabatani E, Rubinstein I, Maoz R, Sagiv J (1987) Electroanal Chem 219:365
63. Ringsdorf H, Schlarb B, Venzmer J (1988) Angew Chem Int Ed 27:113
64. Sackmann E (1996) Science 271:43
65. Plant AL (1993) Langmuir 9:2764
66. Boxer SG (2000) Curr Opin Chem Biol 4:704
67. Groves JT, Boxer SG (2002) Acc Chem Res 35:149
68. Khademhosseini A, Jon S, Suh KY, Tran TNT, Eng G, Yeh J, Seong J, Langer R (2003) Adv Mater 15:1995
69. Malmqvist M, Karlsson R (1997) Curr Opin Chem Biol 1:378
70. Gehrke SH, Vaid NR, McBride JF (1998) Biotechnol Bioeng 58:416
71. Putka CS, Gehrke SH, Willis M, Stafford D, Bryant J (2002) Biotechnol Bioeng 80:139
72. Toomey R, Freidank D, Ruhe J (2004) Macromolecules 37:882
73. Niemeyer CM, Blohm D (1999) Angew Chem Int Ed 38:2865
74. Schulze A, Downward J (2001) Nat Cell Biol 3:E190
75. Benters R, Niemeyer CM, Drutschmann D, Blohm D, Wohrle D (2002) Nucleic Acids Res 30:e1
76. Pathak S, Singh AK, McElhanon JR, Dentinger PM (2004) Langmuir 20:6075
77. Degenhart GH, Dordi B, Schönherr H, Vancso GJ (2004) Langmuir 20:6216
78. Rowan B, Wheeler MA, Crooks RM (2002) Langmuir 18:9914
79. Lahann J, Balcells M, Rodon T, Lee J, Choi IS, Jensen KF, Langer R (2002) Langmuir 18:3632
80. Chen Q, Forch R, Knoll W (2004) Chem Mater 16:614
81. Zhang ZH, Chen Q, Knoll W, Foerch R, Holcomb R, Roitman D (2003) Macromolecules 36:7689
82. Zhou X, Wu L, Zhou J (2004) Langmuir 20:8877
83. Ruhe J, Golze S, Freidank D, Mohry S, Klapproth H (2001) Abstr Pap Am Chem Soc 221:U356
84. Schena M (2003) Microarray analysis. Wiley-Liss, Hoboken NJ
85. Falconnet D, Koenig A, Assi T, Textor M (2004) Adv Funct Mater 14:749
86. Kumar A, Biebuyck HA, Whitesides GM (1994) Langmuir 10:1498
87. Wilbur JL, Kumar A, Kim E, Whitesides GM (1994) Adv Mater 6:600
88. Kumar A, Whitesides GM (1994) Science 263:60
89. Xia YN, Whitesides GM (1998) Annu Rev Mater Sci 28:153
90. Ginger DS, Zhang H, Mirkin CA (2004) Angew Chem Int Ed 43:30
91. Wouters D, Schubert US (2004) Angew Chem Int Ed 43:2480
92. Auletta T, Dordi B, Mulder A, Sartori A, Onclin S, Bruinink CM, Peter M, Nijhuis CA, Beijleveld H, Schönherr H, Vancso GJ, Casnati A, Ungaro R, Ravoo BJ, Huskens J, Reinhoudt DN (2004) Angew Chem Int Ed 43:369
93. Morigaki K, Schönherr H, Frank CW, Knoll W (2003) Langmuir 19:6994

94. Chechik V, Stirling CJM (1998) Langmuir 14:99
95. Chechik V, Stirling CJM (1997) Langmuir 13:6354
96. Creager SE, Clarke J (1994) Langmuir 10:3675
97. Chatelier RC, Drummond CJ, Chan DYC, Vasic ZR, Gengenbach TR, Griesser HJ (1995) Langmuir 11:4122
98. Lee TR, Carey RI, Biebuyck HA, Whitesides GM (1994) Langmuir 10:741
99. Bain CD, Whitesides GM (1989) Langmuir 5:1370
100. Hu K, Bard AJ (1997) Langmuir 13:5114
101. Schönherr H, Hruska Z, Vancso GJ (2000) Macromolecules 33:4532
102. Schönherr H, van Os MT, Förch R, Timmons RB, Knoll W, Vancso GJ (2000) Chem Mater 12:3689
103. Israelachvili JN (1991) Intermolecular and surface forces, 2nd edn. Academic, London
104. Wirth MJ (ed) (1999) Chem Rev 99(10):2843–3152 (special issue)
105. Vancso GJ, Hillborg H, Schönherr H (2005) Adv Polym Sci 182:55
106. Ulman A (ed) (1995) Characterization of organic thin films. Butterworth-Heinemann
107. Overney RM, Meyer E, Frommer J, Brodbeck D, Luthi R, Howald L, Guntherodt HJ, Fujihira M, Takano H, Gotoh Y (1992) Nature 359:133
108. Carpick RW, Salmeron M (1997) Chem Rev 97:1163
109. Nisman R, Smith P, Vancso GJ (1994) Langmuir 10:1667
110. Schönherr H, Vancso GJ (1997) Macromolecules 30:6391
111. Vancso GJ, Schönherr H (1999) In: Tsukruk VV, Wahl KJ (eds) Microstructure and microtribology of polymer surfaces (ACS Symp Ser vol 741). American Chemical Society, New York, p 317
112. Vancso GJ, Snétivy D, Schönherr H (1998) In: Ratner BD, Tsukruk VV (eds) Scanning probe microscopy of polymers (ACS Symp Ser vol 694). American Chemical Society, New York, p 67
113. Schönherr H, Kenis PJA, Engbersen JFJ, Harkema S, Hulst R, Reinhoudt DN, Vancso GJ (1998) Langmuir 14:2801
114. Mizes HA, Loh KG, Miller RJD, Ahuja SK, Grabowski EF (1991) Appl Phys Lett 59:2901
115. Joyce SA, Houston JE, Michalske TA (1992) Appl Phys Lett 60:1175
116. Van der Werf KO, Putman CAJ, De Grooth BG, Greve J (1994) Appl Phys Lett 65:1195
117. Baselt DR, Baldeschwieler JD (1994) J Appl Phys 76:33
118. Berger CEH, Van der Werf KO, Kooyman RPH, De Grooth BG, Greve J (1995) Langmuir 11:4188
119. Radmacher M, Cleveland JP, Fritz M, Hansma HG, Hansma PK (1994) Appl Surf Sci 66:2159
120. Schmidt P, Kolarik J, Lednicky F, Dybal J, Lagaron JM, Pastor JM (2000) Polymer 41:4267
121. Benninghoven A (1994) Angew Chem Int Ed 33:1023
122. Yang ZP, Belu AM, Liebmann-Vinson A, Sugg H, Chilkoti A (2000) Langmuir 16:7482
123. Stoeckli M, Farmer TB, Caprioli RM (1999) J Am Soc Mass Spectrom 10:67
124. Fulghum JE (1999) J Electron Spectrosc Relat Phenom 100:331
125. Turner NH, Schreifels JA (2000) Anal Chem 72:99R
126. DeAro JA, Weston KD, Buratto SK, Lemmer U (1997) Chem Phys Lett 277:532
127. Dordi B, Schönherr H, Vancso GJ (2003) Langmuir 19:5780
128. Schönherr H, Feng CL, Shovsky A (2003) Langmuir 19:10843
129. Frey BL, Corn RM (1996) Anal Chem 68:3187

130. Lahiri J, Isaacs L, Tien J, Whitesides GM (1999) Anal Chem 71:777 (and references therein)
131. Wagner P, Kernen P, Hegner M, Ungewickell E, Semenza G (1994) FEBS Lett 356:267
132. Wagner P, Hegner M, Kernen P, Zaugg F, Semenza G (1996) Biophys J 70:135
133. Wojtyk JTC, Morin KA, Boukherroub R, Wayner DDM (2002) Langmuir 18:6081
134. Porter MD, Bright TB, Allara DL, Chidsey CED (1987) J Am Chem Soc 109:3559
135. Duhachek SD, Kenseth JR, Casale GP, Small GJ, Porter MD, Jankowiak R (2000) Anal Chem 72:3709
136. Cassie ABD (1948) Discuss Faraday Soc 3:11
137. Cline GW, Hanna SB (1988) J Org Chem 53:3583
138. Dordi B, Pickering JP, Schönherr H, Vancso GJ (2004) Eur Polym J 40:939
139. Dordi B, Pickering JP, Schönherr H, Vancso GJ (2004) Surf Sci 570:57
140. Frutos AG, Brockman JM, Corn RM (2000) Langmuir 16:2192
141. Delamarche E, Michel B, Gerber C, Anselmetti D, Guntherodt HJ, Wolf H, Ringsdorf H (1994) Langmuir 10:2869
142. Frisbie CD, Rozsnyai LF, Noy A, Wrighton MS, Lieber CM (1994) Science 265:2071
143. Noy A, Vezenov DV, Lieber CM (1997) Annu Rev Mater Sci 27:381
144. Sinniah SK, Steel AB, Miller CJ, ReuttRobey JE (1996) J Am Chem Soc 118:8925
145. Johnson KL, Kendall K, Roberts AD (1971) Proc R Soc Lond A 324:301
146. Stevens F, Lo YS, Harris JM, Beebe TP (1999) Langmuir 15:207
147. Wilbur J, Whitesides GM (1999) In: G Timp (ed) Nanotechnology. Springer, Berlin Heidelberg New York, p 339
148. Laibinis PE, Bain CD, Nuzzo RG, Whitesides GM (1995) J Phys Chem 99:7663
149. Spinke J, Liley M, Schmitt FJ, Guder HJ, Angermaier L, Knoll W (1993) J Chem Phys 99:7012
150. Benters R, Niemeyer CM, Wohrle D (2001) Chembiochem 2:686
151. Lackowski WM, Campbell JK, Edwards G, Chechik V, Crooks RM (1999) Langmuir 15:7632
152. Smith RK, Lewis PA, Weiss PS (2004) Prog Surf Sci 75:1
153. Yan L, Huck WTS, Whitesides GM (2004) J Macromol Sci Polym Rev C44:175
154. Tarlov MJ, Burgess DRF, Gillen G (1993) J Am Chem Soc 115:5305
155. Huang JY, Dahlgren DA, Hemminger JC (1994) Langmuir 10:626
156. David C, Muller HU, Volkel B, Grunze M (1996) Microelectron Eng 30:57
157. Mendes PM, Preece JA (2004) Curr Opin Colloid Interf Sci 9:236
158. Larsen NB, Biebuyck H, Delamarche E, Michel B (1997) J Am Chem Soc 119:3017
159. Hillborg H, Gedde UW (1998) Polymer 39:1991
160. Hillborg H, Tomczak N, Olah A, Schönherr H, Vancso GJ (2004) Langmuir 20:785
161. Liebau M, Huskens J, Reinhoudt DN (2001) Adv Funct Mater 11:147
162. Li HW, Kang DJ, Blamire MG, Huck WTS (2002) Nano Lett 2:347
163. Li HW, Muir BVO, Fichet G, Huck WTS (2003) Langmuir 19:1963
164. Jenkins ATA, Boden N, Bushby RJ, Evans SD, Knowles PF, Miles RE, Ogier SD, Schönherr H, Vancso GJ (1999) J Am Chem Soc 121:5274
165. Beyer D, Elender G, Knoll W, Kuhner M, Maus S, Ringsdorf H, Sackmann E (1996) Angew Chem Int Ed 35:1682
166. Heibel C, Maus S, Knoll W, Rühe J (1998) ACS Symp Ser 695:104
167. Yoshina-Ishii C, Boxer SG (2003) J Am Chem Soc 125:3696
168. Svedhem S, Pfeiffer I, Larsson C, Wingren C, Borrebaeck C, Hook F (2003) Chembiochem 4:339
169. Jung LS, Shumaker-Parry JS, Campbell CT, Yee SS, Gelb MH (2000) J Am Chem Soc 122:4177

170. Stamou D, Duschl C, Delamarche E, Vogel H (2003) Angew Chem Int Ed 42:5580
171. Kim JM, Ji EK, Woo SM, Lee HW, Ahn DJ (2003) Adv Mater 15:1118
172. Boukobza E, Sonnenfeld A, Haran G (2001) J Phys Chem B 105:12165
173. Stanish I, Santos JP, Singh A (2001) J Am Chem Soc 123:1008
174. Schönherr H, Rozkiewicz DI, Vancso GJ (2004) Langmuir 20:7308
175. Watts TH, Brian AA, Kappler JW, Marrack P, McConnell HM (1984) Proc Natl Acad Sci USA 81:7564
176. Watts TH, Gaub HE, McConnell HM (1986) Nature 320:179
177. Reviakine I, Brisson A (2000) Langmuir 16:1806
178. Kumar S, Hoh JH (2000) Langmuir 16:9936
179. Schönherr H, Johnson JM, Lenz P, Frank CW, Boxer SG (2004) Langmuir 20:11600
180. van Oudenaarden A, Boxer SG (1999) Science 285:1046
181. Kung LA, Kam L, Hovis JS, Boxer SG (2000) Langmuir 16:6773
182. Groves JT, Boxer SG (2002) Acc Chem Res 35:149
183. Morigaki K, Baumgart T, Offenhausser A, Knoll W (2001) Angew Chem Int Ed 40:172
184. Cremer PS, Groves JT, Kung LA, Boxer SG (1999) Langmuir 15:3893
185. Hovis JS, Boxer SG (2000) Langmuir 16:894
186. Hovis JS, Boxer SG (2001) Langmuir 17:3400
187. Groves JT, Ulman N, Boxer SG (1997) Science 275:651
188. Jenkins ATA, Bushby RJ, Evans SD, Knoll W, Offenhausser A, Ogier SD (2002) Langmuir 18:3176
189. Cremer PS, Yang TL (1999) J Am Chem Soc 121:8130
190. Dufrene YF, Barger WR, Green JBD, Lee GU (1997) Langmuir 13:4779
191. Lipowsky R, Seifert U (1991) Mol Cryst Liq Cryst 202:17
192. Seifert U (1997) Adv Phys 46:13
193. Gong P, Grainger DW (2004) Surf Sci 570:67 (and references therein)

Development of a Biodegradable Composite Scaffold for Bone Tissue Engineering: Physicochemical, Topographical, Mechanical, Degradation, and Biological Properties

M. Navarro · C. Aparicio · M. Charles-Harris · M. P. Ginebra · E. Engel · J. A. Planell (✉)

Center of Reference for Bioengineering of Catalonia (CREBEC),
Dept. Materials Science and Metallurgy, Technical University of Catalonia (UPC),
Avda. Diagonal 647, 08028 Barcelona, Spain
josep.a.planell@upc.es

1	Introduction	210
2	Development of the Composite Material	211
2.1	Calcium Phosphate Soluble Glasses	212
2.2	PLA/Calcium Phosphate Glass Composite Material	215
3	Surface Characterization	216
3.1	Roughness	217
3.2	Wettability	218
4	Protein Adsorption	220
5	Biological Behavior	221
6	Development of a Bioabsorbable Composite Scaffold	223
6.1	Solvent Casting	223
6.1.1	Macroporosity	224
6.1.2	Mechanical Properties	226
6.2	Phase Separation	227
7	Conclusion	229
	References	230

Abstract The development of synthetic materials and their use in tissue engineering applications has attracted much attention in recent years as an option for trabecular bone grafting. Bioabsorbable polyesters of the poly(α-hydroxy acids) family, and specifically polylactic acid (PLA), are well known bioabsorbable materials and are currently used for numerous biomedical applications. The incorporation of an inorganic phase, such as a soluble calcium phosphate glass in the $P_2O_5 - CaO - Na_2O - TiO_2$ system, into the polymeric matrix enhances the mechanical integrity of the material. In fact, the flexural elastic modulus increases from 3.2 to 10 GPa with 50 wt/wt % of glass particles. It also improves the biological behavior and modifies the degradation pattern of the polymer. The presence of glass particles accelerates the material degradation and induces the formation of calcium phosphate precipitates in the surface of the composite. Therefore, the combination of a bioabsorbable polymer such as

PLA with a soluble calcium phosphate glass leads to a fully degradable composite material with a high bone regenerative potential. The success of a 3D scaffold depends on several parameters that go from the macro- to the nanoscale. The solvent and casting technique, together with particulate leaching, allows the elaboration of 95%-porosity scaffolds with a well interconnected macro- and microporosity. Factors such as surface chemistry, surface energy, and topography can highly affect the cell-material response. Indeed, the addition of glass particles in the PLA matrix modifies the material surface properties such as wettability AI (Area index or real-surface-area/nominal-area ratio) and roughness, improving the cell response and inducing morphological changes in the cytoskeleton of the osteoblasts. This study offers valuable insight into the parameters affecting cell-scaffold behavior, and discusses the special relevance that a comprehensive characterization and manufacturing control of the composite surface can have for monitoring the biological–synthetic interactions.

Keywords Bioabsorbable composite scaffold · Bone tissue engineering · Osteoblast cell culture · Protein adsorption · Wettability

Abbreviations
AI	Area index or real-surface-area/nominal area ratio
CaP	Calcium phosphate
E	Young's modulus
ECM	Extracellular matrix
FCS	Fetal calf serum
G5	$44{,}5P_2O_5 - 44{,}5CaO - 6Na_2O - 5TiO_2$ glass (molar composition)
HV	Vickers microhardness
ICP-MS	Inductively coupled plasma-mass spectroscopy
MTT	Tetrazolium-salt assay
Mw	Molecular weight
PLA	Polylactic acid
SBF	Simulated body fluid
S_a	Average 3D roughness
Sku	Kurtosis of the 3D surface texture
Ssk	Skewness of the 3D surface texture
T_g	Glass transition temperature
Wa	Work of adhesion

1
Introduction

Nowadays, autografts, allografts, and xenografts are used for the restoration of bone injuries. Although the use of these grafts has presented satisfactory results under certain conditions, there are some restrictions associated with donor site scarcity, rejection, diseases transfer, and elevated harvesting costs. Due to the numerous drawbacks these grafts present, research has focused on the development of alternative synthetic materials.

Bioabsorbable polymers such as aliphatic polyesters from the poly (α-hydroxy acids) family, especially polylactic acid (PLA), are well known bioabsorbable materials and are widely used for biomedical applications such as sutures, pins, screws and drug delivery systems [1–4]. Given the biocompatibility and biodegradability features PLA presents, its use in tissue engineering applications has attracted much attention in recent years. Thereby, the development of PLA biodegradable porous scaffolds represents a promising alternative for trabecular bone grafting.

The incorporation of an inorganic phase into the polymeric matrix may enhance the mechanical integrity of the material, as well as its biological behavior, and can also modify the degradation mechanism of the polymer. Some calcium phosphate ceramics and biological glasses have been used with this aim [5–7]. Specifically, calcium phosphate (CaP) glasses are well suited for bone remodeling given that they possess a chemical composition close to that of the mineral phase of bone and that their solubility rate can be adjusted by controlling their chemical composition.

Therefore, the combination of a bioabsorbable polymer such as PLA with a soluble CaP glass leads to a fully degradable composite material with a high bone regenerative potential.

The success of a 3D scaffold depends on several parameters that range from the macro- to the nanoscale. Macro- and microporosity, as well as interconnectivity, are of great importance in promoting tissue ingrowth, vascularization, and the delivery of nutrients throughout the newly formed tissue. The attachment and adhesion of the cells on the material surface is also of paramount importance. These are protein-mediated processes, where factors such as surface chemistry, surface energy, and topography can affect the cell-material response [8]. Indeed, surface characteristics at all dimensional scales affect the adsorption of proteins. Differences in protein adsorption (type of adsorbed proteins, orientation and conformation, and the kinetics of adsorption) lead to variations in the number of cells and their force of adhesion to the substrate [9, 10]. This is a process mediated by the interactions between the adsorbed proteins and integrins, which are cell membrane proteins [11]. The cell adhesion process triggers different chemical and mechanical signals, thus influencing the regulation of cell survival, proliferation and differentiation, which in turn determine cell function within a defined tissue.

This review offers some insight into the parameters affecting the cell-scaffold behavior from the macro- to the microscale, from the bulk to the surface, and discusses the special relevance that a comprehensive characterization and manufacturing control of the composite surface might have in monitoring the biological–synthetic interactions.

2
Development of the Composite Material

The resorption rate of a biomaterial in vivo involves a very complex mechanism that depends on numerous variables and involves both the material physicochemical features and biological events, including protein- and cell-mediated processes. Among the physicochemical properties, the solubility of the material plays an important role and significantly affects the biomaterial's stability in vivo. Thus, if the material's solubility rate is too high, it will be resorbed by passive dissolution due to the physiological fluids without stimulating tissue turnover, i.e., the resorption/regeneration process mediated by bone cells during bone remodeling. In contrast, if the solubility of the material is too low, it will remain in the body for a long period of time, and bone remodeling will not take place adequately. The use of materials with a moderate solubility rate induces an active resorption process, which is lead by cells and resembles the biological bone remodeling process. Hence, the control of degradation kinetics is a key point in the design of bioabsorbable materials for regenerative purposes.

2.1
Calcium Phosphate Soluble Glasses

Calcium phosphate glasses represent an interesting alternative, since the solubility of these glasses can be adjusted depending on their chemical composition. This fact presents an important advantage over crystalline calcium phosphates.

The structural unit of phosphate glasses is the PO_4 tetrahedron. The basic phosphate tetrahedra form long chains and rings that give rise to the 3D vitreous network [12]. These phosphate chains and rings may be interrupted by the incorporation of certain ions, generating nonbridging oxygens in the glass structure. The incorporation of other modifying ions can lead to the creation of ionic cross-links between nonbridging oxygens of two different chains, thus reinforcing the glass network. Therefore, depending on the modifiers present in the vitreous structure, long-term or short-term soluble phosphate glasses can be obtained [6, 13–15].

Previous studies show that the addition of CaO, Na_2O and TiO_2 into the phosphate network allows control of the solubility and mechanical properties of these glasses within certain ranges [15–17]. Both CaO and TiO_2 enhance glass stability, particularly TiO_2 given its small ionic radius and the large charge on the Ti^{4+} ion [18, 19]. The characteristics of the Ti^{4+} ion allow it to penetrate into the vitreous arrangement, inducing a higher degree of reticulation in the glass network.

The CaP glass of the $44,5P_2O_5 - 44,5CaO - 6Na_2O - 5TiO_2$ (molar composition) system, coded G5, is a good candidate since it presents a good chemical stability (see Fig. 1) as well as good mechanical properties.

In vitro degradation studies on the G5-glass were performed with SBF [20] (an acellular and aproteic fluid that has an ionic concentration similar to that of human blood plasma, see Table 1) at physiological temperature. ICP-MS analyses showed that G5-glass dissolution occurs uniformly, which means

Development of a Biodegradable Composite Scaffold

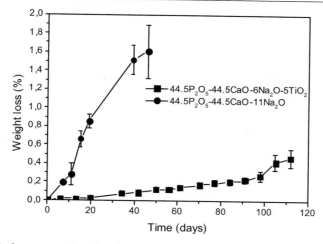

Fig. 1 Weight loss versus dissolution time for two different CaP glasses during degradation in SBF at 37 °C. Error bars not shown if smaller than symbols

Table 1 SBF and human blood plasma ionic composition and concentration (mM)

Ion	SBF	Human plasma
Na^+	142.0	142.0
K^+	5.0	5.0
Mg^{2+}	1.5	1.5
Ca^{2+}	2.5	2.5
Cl^-	147.8	103.0
HCO_3^-	4.2	27.0
HPO_4^{2-}	1.0	1.0
SO_4^{2-}	0.5	0.5

that none of the ions conforming the glass network is released preferentially. In addition, in vitro analysis revealed that during dissolution water diffuses into the glass surface and surrounds the external PO_4 chains, creating a hydrated layer. When the phosphate polymeric chains have been completely surrounded by the aqueous medium, the hydrated chains separate from the bulk of the material and leach into the solution. Due to the homogeneous superficial dissolution process, the mechanical properties of the glass are maintained throughout the degradation period [21].

Biocompatibility studies of the G5-glass, performed with human skin fibroblasts and osteoblast-like human cells from a cell line coded MG63, have shown that this material as well as its degradation products are noncytotoxic [22].

Cell differentiation studies are used to follow the development of cell phenotype by analyzing the concentration of two proteins directly related to bone extracellular matrix mineralization: alkaline phosphatase and the osteocalcin. Cell differentiation studies performed on the G5 glass have shown that it induces an earlier differentiation of the osteoblastic cells than the polysterene plate controls (unpublished data) (see Fig. 2). Consequently, a faster bone formation could be obtained.

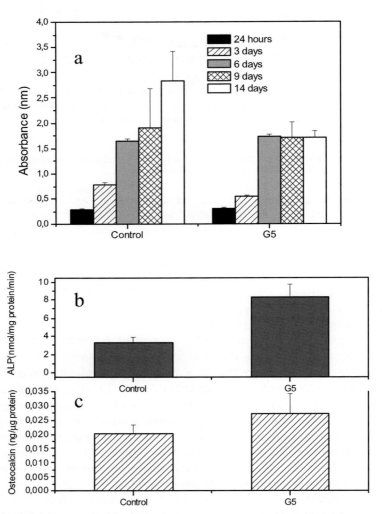

Fig. 2 a MTT results of the effect of the G5 glass on MG63 cells, showing cell proliferation. **b** Alkaline phoshatase activity (ALP) values of MG63 cells after 11 days of culture. **c** Osteocalcin concentration values of the MG63 cells after 11 days of culture

Table 2 Properties of the G5 glass (mean values ± standard deviation)

Properties	G5 glass
T_g [°C]	532.9
HV [HV0.2]	431.1 ± 7.8
E [GPa]	71.1 ± 1.7
Solubility rate in distilled water [g cm^{-2} h^{-1}]	$3.13 \cdot 10^{-6} \pm 1.38 \cdot 10^{-7}$
Solubility rate in SBF [g cm^{-2} h^{-1}]	$3.2 \cdot 10^{-7} \pm 1.03 \cdot 10^{-7}$

Recently, in vivo studies have also revealed a good biocompatibility and guidance of the newly formed tissue to the G5-glass surface, which confirms its osteoconductive potential. In an in vivo study using a rabbit model, the percentage of new bone formation with implanted glass particles was comparable to that obtained for the autologous bone (control), after 12 weeks of implantation [23]. The properties of the G5-glass are summarized in Table 2.

2.2
PLA/Calcium Phosphate Glass Composite Material

Given the advantages of incorporation of an inorganic phase into a polymeric matrix, the G5-glass has been combined with a 95L/5DL-PLA in order to develop a nonporous 2D fully resorbable composite material that could be used in different load-bearing bone-repairing situations.

In general, the incorporation of the G5-glass particles in the polymer improves the flexural mechanical properties of PLA, modifies its degradation behavior, and induces interesting changes in the material surface morphology.

PLA flexural mechanical properties are very low in comparison with cortical bone properties. Therefore, PLA properties are insufficient for high load-bearing applications. Addition of the inorganic phase into the PLA matrix leads to a rise in the mechanical properties of the material, to more nearly approach the mechanical properties of bone and, thus, allowing a better load-transfer to the newly formed tissue [24]. Former studies have shown that the mechanical properties of nonporous materials, especially the Young's modulus (E), undergo a significant increase (from 3.2 to 10 GPa) with the incorporation of 50% by weight of glass particles. However, the PLA matrix has a saturation limit for enveloping the particles, and the efficiency of the G5 particles seems to decrease as the percentage of particles exceeds this limit of approximately 60%.

The presence of glass particles modifies the in vitro degradation pattern of the polymer. In general, the degradation of PLA depends on several factors, which include its crystallinity, molecular weight, dimensions, composition,

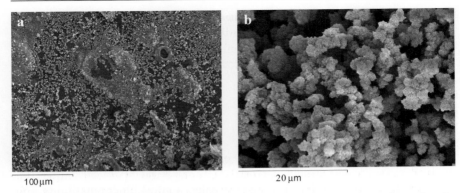

Fig. 3 Composite surface (**a**) and microstructure of the CaP precipitate (**b**) formed at the material surface after 6 weeks of immersion in SBF at 37 °C

and the pH of the surrounding medium. Nevertheless, in spite of the influence these factors may have on the degradation of PLA, it is well known that the degradation mechanism of this polymer is a bulk mechanism autocatalyzed by carboxyl end groups formed by chain cleavage [25, 26].

The addition of G5 particles into the polymer matrix implies the presence of PLA/G5 interfaces at the surface, which allows the penetration of the aqueous fluid into the interior of the composite. This fact, combined with the glass reactivity in aqueous media, induces the formation of surface microcracks. These facilitate both fluid penetration, which accelerates degradation of the polymer chains, and the release of the degradation by-products. At the same time, the degradation products of the glass act as buffering agents that interfere with the autocatalytic process. All these events lead to a higher mass loss and a higher crystallinity, and to a lower Mw loss of the PLA/G5 composite in comparison to the PLA polymer.

On the other hand, the G5 particles react with SBF, giving rise to a globular CaP amorphous structure (see Fig. 3) that emerges in the composite material (manuscript submitted), with a Ca/P ratio close to 1.5. This CaP precipitate could enhance the interaction between the bone cells and the material during bone regeneration since this amorphous CaP is a transient structure to hydroxyapatite, which is the mineral phase of bone with a higher Ca/P ratio.

3
Surface Characterization

The cell adhesion process is critical to most bone regeneration applications [27, 28]. In general, cell adhesion to synthetic substrates is a protein-mediated process. Thus, the amount, type, and activity of the adsorbed proteins on the material surface are a key issue, though the individual role of each parameter is not clear. Numerous studies have shown that the characteristics of the adsorbed proteins and the cell behavior depend strongly on surface properties such as hydrophilicity, surface energy, and the topography of the substrate surface [8].

3.1
Roughness

Roughness and texture are two of the properties that most influence the biological behavior of synthetic materials. On one hand, it is well known that when the topographical features of the surface roughness follow a regular disposition (columns, grooves, etc.), cells are oriented by the pattern and have limited motility [29, 30]. This behavior, which is a consequence of the micro- and/or nanometer texture, is called cell guiding [31].

On the other hand, higher roughness in anisotropic-topographical surfaces is related to better attachment, adhesion, and differentiation of the osteoblast cells onto synthetic materials [8, 9]. This is because osteoblasts can extend from peak to peak and take optimal shapes for their "accommodation". These optimal shapes lead to changes in the cytoskeleton that also favor, via biochemical signals, osteoblast behavior. However, roughness must be of the order of cell dimensions for osteoblasts to "feel" the topographical features [32]. This means that roughness must be in the micrometer range with a maximum and minimum value for the height of and the distance between the peaks/valleys. Consequently, the calculation not only of amplitude roughness parameters but also of spatial and/or hybrid parameters is of paramount relevance.

The influence of roughness in the nanometer scale on cell behavior is controversial, and can be due to the changes that it induces in other physicochemical properties, such as wettability and Z-potential [33]. This will mainly influence the layer of proteins that are adsorbed on those surfaces. This

Table 3 Roughness parameters values (white light optical interferometry) of PLA and PLA/G5 before and after being polished (mean values ± standard deviation)

Material	S_a [μm]	Sku	Ssk	AI
PLA/G5				
Polished	0.238 ± 0.11	53.1 ± 27	−5.25 ± 2.7	1.09 ± 0.02
Unpolished	0.411 ± 0.04	4.8 ± 2	−0.40 ± 0.43	1.15 ± 0.10
PLA				
Polished	0.054 ± 0.01	11.2 ± 8	−0.99 ± 0.7	1.01 ± 0.00
Unpolished	0.372 ± 0.12	8.3 ± 7	0.97 ± 1.1	1.19 ± 0.06

knowledge, and the comments of the previous paragraph, suggest the use of several roughness characterization techniques in order to cover all dimensional scales, from micro to nano.

Polished and unpolished PLA and PLA/G5 have nontextured surfaces with nanometer roughness (Table 3). Consequently, as explained above, different surfaces will influence their biological response by the changes that roughness provokes in properties such as wettability.

3.2
Wettability

According to some authors, contact angle and work of adhesion (Wa) are the best wettability properties to predict the material–cell interactions at the initial stages of contact [34, 35]. Therefore, contact angle measurements have been performed to evaluate the hydrophilicity of the composite material. The G5-glass possesses a hydrophilic surface, so the incorporation of glass particles in the PLA matrix reduces the hydrophobic behavior of the polymer (Table 4). Thus, depending on the quantity of glass incorporated into the PLA/G5 composite material, the biomaterial surface wettability can be adjusted to obtain different degrees of hydrophilicity.

There is some debate about the affinity of proteins to hydrophobic surfaces. Some authors sustain the hydrophobic affinity theory [36], while others prefer the hydrophilic affinity theory [37]. The results obtained from studies carried out with culture medium supplemented with 10% fetal calf serum (FCS) suggested that the complex mixture of proteins present in FCS presented a higher affinity for the hydrophilic surfaces. Furthermore, the Wa values suggested that the mixture of proteins adsorbed better on hydrophilic surfaces, though the type of protein and their adsorption speed onto the surfaces is still unknown. However, the use of dynamic contact angle techniques (as has been confirmed with other materials) could help identify the velocity of adsorption and the number of steps of adsorption, desorption, and/or adsorption/desorption that lead to the final interaction between the substrate and the proteins in the culture medium (Fig. 4). The contact angles obtained for the two different fluids are shown in Table 5.

Table 4 Effect of the weight percent of G5 glass on the polished composite wettability (mean values ± standard deviation)

Composition	Contact Angle with distilled water (°)
0 wt % G5 glass	73.56 ± 1.50
20 wt % G5 glass	72.86 ± 1.60
50 wt % G5 glass	67.56 ± 1.70

Development of a Biodegradable Composite Scaffold

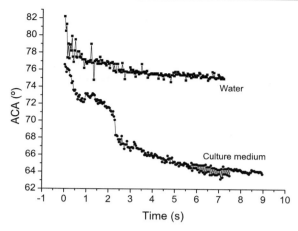

Fig. 4 Dynamic contact angles showing the different advancing contact angle (ACA) evolution during the time of interaction of the distilled water and the culture medium on titanium samples. The abrupt increase at $t \approx 1$ s and the abrupt decrease at $t \approx 2$ s of the ACA values indicate processes of adsorption, desorption or adsorption/desorption of proteins on the surface

Table 5 Contact angle values (°) at $t = 0$ s, with different fluids, on the surface of PLA, G5 glass, and the composite material (mean values ± standard deviation)

Material	Contact angle	
	Distilled water	Culture medium
PLA	73.59 ± 0.98	78.31 ± 0.84
PLA/G5	67.56 ± 1.71	68.82 ± 2.02
G5	29.80 ± 0.97	42.05 ± 1.76

The wettability of a surface is known to be affected by its topography, as discussed in the previous section. This statement has been corroborated in the case of the PLA/glass composite material (Table 6). Indeed, composite material specimens with a rough surface presented contact angle values significantly higher than the values reported for the polished materials. Thus, surface roughness leads to differences in the wettability of the surface. This change in the wettability behavior is dependant on the behavior of the ideally nonrough surface of the material studied [38]. For hydrophilic surfaces (contact angle < 90°), the higher the roughness, the lower the contact angle, which could be related to a more hydrophilic surface. For clearly hydrophobic surfaces (contact angle > 90°), the higher the roughness, the higher the contact angle, which could be related to a more hydrophobic surface. Never-

Table 6 Effect of roughness and sterilization on the composite wettability (mean values ± standard deviation) of PLA/G5

PLA/G5 material	Contact angle with distilled water [°]
Polished	67.56 ± 1.71
Rough	82.88 ± 4.03
Polished and sterilized	64.94 ± 1.79

theless, the limit value of 90° has been discussed and materials with contact angles close to 90° could not follow the general rule [39]. If the roughness is sufficiently high, the peaks of the roughness can retain fluid leading to metastable states of the drop that give a increasing value of contact angle. As a consequence, further studies must be made on the influence of roughness on wettability, which is a subject of special interest, as explained in previous sections.

On the other hand, the sterilization processes may somehow affect the surface structure of the material and, therefore, its wettability. In our case, ethylene oxide was chosen as the sterilization technique over autoclaving and gamma-irradiation, since this technique neither modifies the structure nor degrades the component materials of the PLA/G5 composite. The wettability results obtained for the sterilized materials showed a slight increment in the material hydrophilicity (Table 6). The mechanism by which the sterilization process modifies the material surface is still not clear. However, this may be due to a reaction between the sterilization agent and PLA, leading to a hydroxyl or similar group that would increase the hydrophilicity of the composite. For other materials studied in our laboratory, the changes in wetting behavior due to the sterilization treatment can be attributed to the changes that the small amount of contamination remaining on the biomaterial surface induces in the Lewis-basic component of the surface energy. Ethylene oxide sterilization changes the titanium surface from being an electron donor (nontreated) to bipolar (sterilized).

4
Protein Adsorption

Cell adhesion takes place in two different stages. The first stage consists of the adsorption of a layer of proteins that selectively adhere onto the biomaterial surface, and is completed in an interval from seconds up to a few minutes [40]. This is mainly mediated by the surface properties. The second stage involves cell adhesion onto the layer of proteins. This is a more complex process, mediated by extracellular matrix (ECM) proteins, cell membrane proteins, and cytoskeletal proteins [41]. The cell membrane proteins, and in particular the integrins, interact with the layer of adsorbed proteins, the ECM proteins, and cytoskeletal proteins in order to promote the adhesion of cells to the materials. The interactions between ECM proteins and the integrin-receptor binding domains are of great importance since they have crucial effects on cell function. Indeed, the protein–integrin interactions can affect cell adhesion, motility, conformation, and differentiation. Thereby, the interactions between these proteins with the substrate and with the cells are of paramount importance [42].

Fibronectin, vitronectin, and type I collagen are some of the most representative ECM proteins involved in cell adhesion processes, therefore adsorption studies with these proteins and the PLA/G5 composite material have been performed. Preliminary studies have shown that all proteins adhere better to the G5 (the most hydrophilic material) than to the other materials. Vitronectin presented the best adhesion with PLA (the most hydrophobic material), and the PLA/glass composite presented an intermediate behavior. Further experiments are being conducted to evaluate the direct implication of the main proteins present in ECM to regulate cell proliferation and differentiation in the studied materials, and to obtain information on how the quality of the surface (physicochemical and topographical) influences the adsorbed protein layer.

5
Biological Behavior

In vitro models are the first approach used to understand the cell–substrate interaction and biocompatibility of the materials. Cell cultures are ideal systems for the analysis of a specific cell type under certain conditions because they avoid the complexity of the numerous variables involved in in vivo studies. It is not possible, however, to directly extrapolate in vitro results to in vivo results. Indeed, in a previous study performed with two CaP glass formulations with different solubilities, in vitro studies indicated that the differences in solubility affect cell cultures [23, 43]. However, in vivo, the differences in solubility were not evident and the two materials presented good biocompatibility.

Cell cultures performed with MG63 osteoblast-like human cells have indicated that the composite material is noncytotoxic and that the initial attachment of the cells to the PLA, G5, and PLA/G5 substrates, is better for the G5-glass (the most hydrophilic material) than for the other two materials, PLA (the most hydrophobic material) being the substrate with the lowest amount of attached cells. Besides, proliferation and differentiation assays have suggested that the most hydrophilic surface triggered the differentiation process earlier than the hydrophobic surfaces (Fig. 5). Furthermore, SEM

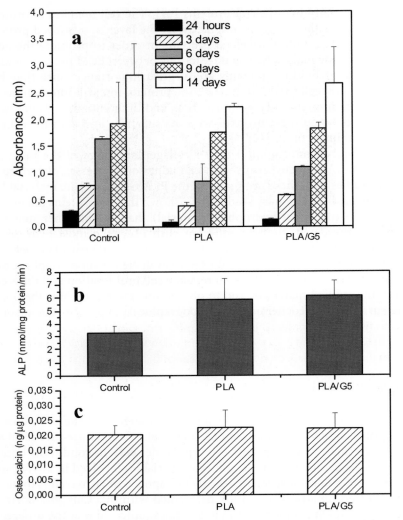

Fig. 5 a MTT results of the effect of the PLA and PLA/G5 composite material on MG63 cell proliferation. **b** Alkaline phoshatase activity (ALP) values of MG63 cells after 11 days of culture. **c** Osteocalcin values of the MG63 cells after 11 days of culture

images have shown significant differences in the morphology of the cells cultured on the substrates with flat or rough surfaces. PLA and G5 flat surfaces presented flat extended cells, while the composite rough material induced conformational changes in the cell cytoskeleton. These changes were mirrored in more rounded cells (Fig. 6).

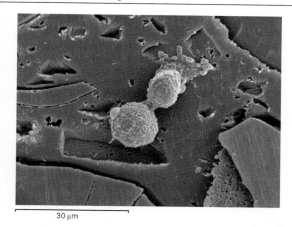

Fig. 6 MG63 osteoblast-like cells on a G5/PLA composite surface showing round shapes

6
Development of a Bioabsorbable Composite Scaffold

The composite material made of PLA and the G5-glass has been used to make scaffolds for tissue engineering. Tissue engineering can be briefly defined as the "... engineering of living tissues ..." [44]. In other words, living cells are grown, either in vitro or in vivo, on degradable scaffolds. The scaffolds should offer:

1. A 3D highly porous interconnected network.
2. Adequate mechanical properties relative to the site of implantation and the cells' requirements.
3. Biocompatibility and bioresorbability.
4. A suitable surface quality – physical, chemical and topographical properties – for cell attachment [45]. Thus, the scaffolds should act as surrogate extracellular matrices until the cells create their own [46].

Various fabrication methods have been developed in order to attain the 3D scaffold characteristics. In the case of synthetic polymer or polymer-matrix composite scaffolds, the methods include [47]: solvent casting and particle leaching, phase separation, extrusion, gas foaming, and free form fabrication. Each method presents certain advantages with respect to others, ranging from ease of manufacture to control of the microstructure/nanostructure. Solvent casting and phase separation methods have been studied at our laboratory.

6.1
Solvent Casting

The solvent casting method was developed by Mikos et al. [48] amongst others for pure PLA, and several authors have used the method to manufac-

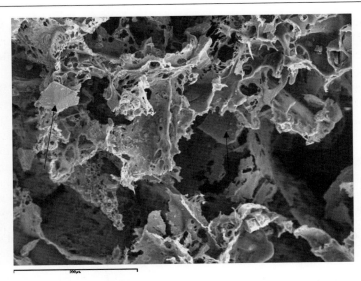

Fig. 7 SEM image of a composite scaffold produced by solvent casting and particle leaching. The black arrows indicate glass particles. The magnification bar corresponds to 200 μm

ture composite scaffolds [49–52]. The method consists of dissolving a polymer in a solvent, and adding particles of a leachable porogen: salt, glucose, etc. The mixture forms a thick paste, which is left to dry in air or under vacuum until the solvent has evaporated completely. The porogen is then dissolved in water by soaking the paste for several days, leaving behind a network of interconnected pores (Fig. 7). In the case of composites, the second phase (i.e., the glass particles) is added with the porogen and remains within the structure after the porogen is leached out. The advantage of the solvent casting method is that it is a simple and fairly reproducible method which does not require sophisticated apparatus. The disadvantages include thickness limitations intrinsic to the particle leaching process and limited mechanical properties. Further, some authors question the homogeneity and interconnection of the pores in the scaffolds, as well as the presence of residual porogen [53]. As with the solid composite material, the addition of glass particles is meant to increase bioactivity and reinforce mechanical properties.

6.1.1
Macroporosity

The morphology, magnitude, and interconnection of the scaffolds' porosity are critical factors in assessing their viability as tissue engineering devices. The structure of the scaffolds and their porosity should transmit the cues for cell adhesion, proliferation, and differentiation, as well as allowing the delivery of nutrients and waste products. It is thus very important to quantify the porosity and to understand which factors play an important role in its tailoring.

A typical solvent cast scaffold is manufactured with approximately 90 wt % of porogen, which produces between 85 and 95% porosities. The influence of scaffold composition on its macroporosity was studied thoroughly at our laboratory using NaCl as a porogen (unpublished data). The magnitude of the porosity is mainly influenced by the wt % of NaCl particles, whereas the pore morphology is chiefly affected by the NaCl particle size (Fig. 8). Neither the wt % nor the size of the G5-glass particles affected the porosity of the composite scaffolds. The interconnection of the pores becomes obvious at high porosities, and is implicit to the particle leaching method if no NaCl remains. The solvent casting method produces a very homogeneous distribution of the glass particles, as can be seen in (Fig. 9).

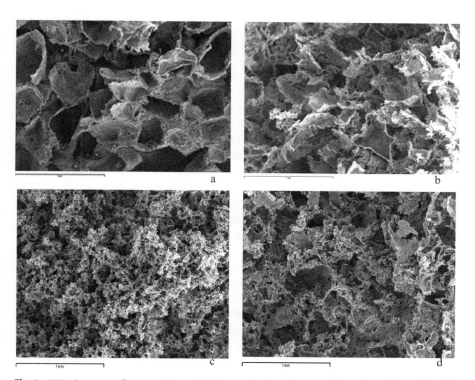

Fig. 8 SEM images of composite scaffolds made by solvent and casting illustrating the effects of changing porogen particle size and weight percent on the porosity. **a** 75 wt % and large particle size, **b** 94 wt % of porogen and large particle size, **c** 94 wt % and small particle size, and **d** 94 wt % of porogen and large particle size. All magnification bars correspond to 1 mm

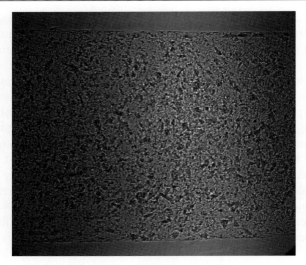

Fig. 9 Synchrotron radiation X-ray projection of a scaffold made by solvent casting. The image reveals the homogeneity of the glass particle distribution. The height of the sample seen in the image is approximately 1 mm and the length seen in the image is approximately 1.5 mm

6.1.2
Mechanical Properties

The mechanical properties of the scaffolds are usually measured by performing compression tests. For scaffolds with 85–95% porosities, stiffness ranged between 100 and 150kPa, yield stresses ranged between 25 and 35kPa, and yield strains ranged between 15 and 60%. Similar values for stiffness are reported in the literature for these porosity levels [54, 55]. Yield properties, however, are often not reported or are poorly defined, and are thus difficult to compare.

The stiffness of the scaffolds decreases as their porosity and wt % of the G5-glass phase increases. The negative effect of the porosity is logical since a higher porosity means less material is supporting the compressive force. The effect of the G5 particles on stiffness may seem surprising, though it is in accordance with composite material mechanics, in which an increase in Young's modulus is mainly attained by introducing a reinforcing phase in the form of fibers, not particles. This effect may be suppressed by improving the adhesion between the PLA matrix and the glass particles.

The glass phase does, however, reinforce the scaffolds' yield properties significantly (Fig. 10). The yield properties are perhaps the most critical mechanical properties because they guarantee the integrity of the macroporous network, which is vital for the cells.

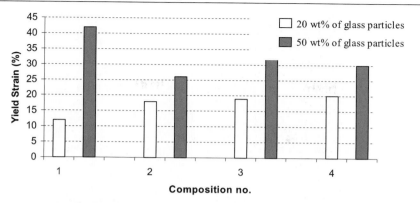

Fig. 10 Differences between yield strain for scaffolds with 20 and 50 wt % of glass particles. The size of the porogen and glass particles varies between compositions 1–4

As a consequence of these results and those of the previous section, the wt % of G5-glass particle can be increased to improve yield properties and potential bioactivity of the scaffolds, without affecting the scaffold's macroporosity.

6.2
Phase Separation

The phase separation technique may prove to be a useful alternative for manufacturing composite biodegradable scaffolds with specific properties. Phase separation of polylactide solutions was first developed by Schugens et al. [56, 57] to produce PLA scaffolds. Later, several authors applied this technique to composite scaffolds [58–61], and have even combined it with solvent casting [62]. The method consists of inducing a solid–liquid or liquid–liquid phase separation of a polylactide solution. The polymer is dissolved in a solvent, often dioxane, and quenched at a certain temperature ranging from 0 °C to −196 °C. The solutions are finally freeze-dried for several days at around 10^{-2} Torr. This method creates a very distinct microstructure (Fig. 11), which can be controlled by varying certain processing parameters such as the quenching temperature, the freeze-drying temperature, and the polymer concentration.

Preliminary studies using phase separation for fabrication of G5/PLA scaffolds show promising results in certain critical aspects. The glass particle in Fig. 12 seems entrapped within the scaffold's microstructure and may contribute to its stiffness. The relative anisotropy of the microstructure may be exploited for specific applications such as nerve regeneration [56]. Furthermore, the phase separation method creates both macro- and microporosity, which would enable cell adhesion in the macropores, and allow the infiltra-

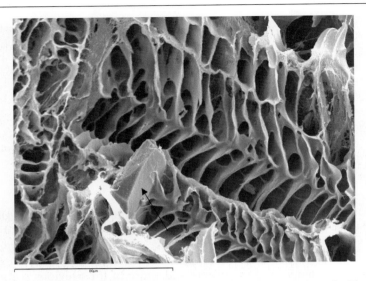

Fig. 11 SEM image of a composite scaffold produced by phase separation. The black arrow indicates a glass particle. The magnification bar corresponds to 80 μm

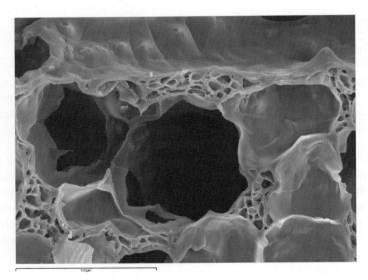

Fig. 12 SEM image of a phase separated scaffold. Two distinct pore sizes can be observed. The magnification bar corresponds to 100 μm

tion of vital blood vessels through the micropores (Fig. 12). From a topographical point of view, the phase separation technique also provides interesting results. Apart from controlling the macrostructure and porosity, the

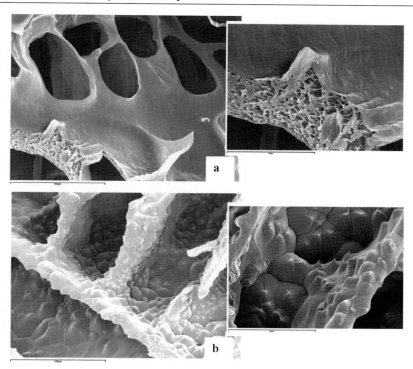

Fig. 13 SEM images of scaffolds prepared by phase separation with different topographical features. **a** Shows a scaffold with a relatively smooth surface even at high magnifications (image to the right). A microstructure within the pore walls is also visible. **b** Shows a scaffold with a distinct microporosity and nanosize wave-like features. The magnification bars of the images to the left correspond to 100 µm, those on the images to the right correspond to 40 µm

processing conditions produce a variety of micro- and nanotopographical features (Fig. 13). As explained before, taking the advantages related to both micro- and nanotopographical roughness, the cell behavior could be adequately affected and even effectively tailored.

7
Conclusion

The development of scaffolds made of a biodegradable composite (PLA/calcium phosphate glass) for bone tissue engineering applications is of major interest as an alternative to existing bone grafts, and is being pursued at our laboratory. The degradability, mechanical properties, and quality of the porosity of these scaffolds have been thoroughly characterized. Moreover, surface properties such as topography, surface energy, and wettability in different dimensional scales were measured in order to correlate them to the biological response of the constructs. The interaction between the synthetic material and biological entities (proteins and cells) is the key issue in determining the success of the potential scaffold. Thus, control of the material's surface quality by means of the fabrication process is our main challenge in this exciting field of research.

References

1. Middleton JC, Tipton AJ (2000) Biomaterials 21:2335
2. Athanasiou KA, Agrawal CM, Barber FA, Burkhart SS (1998) Arthroscopy 14:726
3. An Y, Woolf SFR (2000) Biomaterials 21:2635
4. Rokkanen P (2000) Biomaterials 21:2607
5. Burnie J, Gilchrist T (1981) Biomaterials 2:244
6. Vogel P, Wange P, Hartmann P (1997) Glasstech Ber Glass Sci Tech 70:220
7. Franks K, Abrahams I, Knowles JC (2000) J Mater Sci-Mater Med 11:609
8. Boyan BD, Dean DD, Lohmann CH, Cochran DL, Sylvia VL, Schwartz Z (2001) The titanium bone-cell interface in vitro: the role of the surface in promoting osseointegration. In: Brunette DM, Tengvall P, Textor M, Thomsen P (eds) Titanium in medicine: material science, surface science, engineering, biological responses and medical applications. Springer, Berlin Heidelberg New York, p 561
9. Anselme K (2000) Biomaterials 21:667
10. Iuliano DJ, Saavedra SS, Truskey GA (1993) J Biomed Mater Res 27:1103
11. Albelda SM, Buck CA (1990) FASEB J 4:2068
12. Van Wazer JR (1958) Phosphorous and its compounds. Interscience, New York
13. Lin ST, Krebs SL, Kadiyala S, Leong KW, Lacourse WC, Kumar B (1994) Biomaterials 15:1057
14. Clement J, Torres P, Gil FJ, Planell JA, Terradas R, Martinez S (1999) J Mater Sci-Mater Med 10:437
15. Clement J, Eckeberg L, Martínez S, Ginebra MP, Gil FJ, Planell JA (1998) Key Eng Mater 11:141
16. Clement J, Manero JM, Planell JA, Avila G, Martinez S (1999) J Mater Sci-Mater Med 10:729
17. Navarro M, Ginebra MP, Clement J, Martinez S, Avila G, Planell JA (2003) J Am Ceramic Soc 86:1345
18. Clement J, Avila G, Navarro M, Martínez S, Ginebra MP, Planell JA (2001) Chemical durability and mechanical properties of calcium phosphate glasses with the addition of Fe_2O_3, TiO_2 and ZnO. Key Eng Mater 192–195:621
19. Navarro M, Clement J, Ginebra MP, Martinez S, Avila G, Planell JA (2002) Key Eng Mater 218–220:275
20. Kokubo T, Kushitani H, Sakka S, Kitsugi T, Yamamuro T (1990) J Biomed Mater Res 24:721
21. Navarro M, Ginebra MP, Clement J, Martinez S, Avila G, Planell JA (2003) J Am Ceramic Soc 86:1345
22. Navarro M, Ginebra MP, Planell JA (2003) J Biomed Mater Res Part A 67A:1009
23. Navarro M, Sanzana E, Planell JA, Ginebra MP, Torres P (2005) Key Eng Mater 284–286:893

24. Raiha JE (1992) Clin Mater 10:35
25. Vert M, Li SM, Spenlehauer G, Guerin P (1992) J Mater Sci-Mater Med 3:432
26. Huffman KR, Casey DJ (1985) J Polym Sci Part A-Polym Chem 23:1939
27. Garcia AJ, Keselowsky BG (2002) Crit Rev Eukaryotic Gene Expression 12:151
28. Hench LL, Polak JM (2002) Science 295:1014
29. Curtis ASG, Clark P (1992) Crit Rev Biocomp 5:343
30. Curtis ASG, Wilkinson C (1997) Biomaterials 18:1573
31. Zinger O, Anselme K, Denzer A, Habersetzer P, Wieland M, Jeanfils J, Hardouin P, Landolt D (2004) Biomaterials 25:2695
32. Boyan BD, Schwartz Z (1999) Modulation of osteogenesis via implant surface design. In: Davies JE (ed) Bone engineering. Em squared, Toronto, p 232
33. Aparicio C (2004) PhD Thesis, Technical University of Catalonia
34. Vogler EA (1993) Interfacial chemistry in biomaterials science. In: Berg JC (ed) Wettability, surfactant science series. Dekker, New York, chap 4, p 183
35. Lampin M, WarocquierClerout R, Legris C, Degrange M, SigotLuizard MF (1997) J Biomed Mater Res 36:99
36. Keselowsky BG, Collard DM, Garcia AJ (2004) Biomaterials 25:5947
37. Vogler EA (1998) Adv Colloid Interface Sci 74:69
38. Shibuichi S, Onda T, Satoh K, Tsujii J (1996) Phys Chem 100:19512
39. Bico J, Thiuele U, Quéré D (2002) Coll Surf A: Physicochem Eng Aspects 206:41
40. Meyer AE, Beyer RE, Naatiella JR, Meenaghan MA (1988) J Oral Implantol 14:363
41. Lüthen F, Lange R, Becker P, Rychly J, Beck U, Nebe B (2005) Biomaterials 26:2423
42. Horbett TA (1996) Proteins: structure, properties, and adsorption to surfaces. In: Ratner BD, Hoffman AS, Choen FJ, Leomns JE (eds) Biomaterials science. Academic, San Diego, p 133
43. Navarro M, Ginebra MP, Planell JA (2003) J Biomed Mater Res Part A 67A:1009
44. Griffith LG, Naughton G (2002) Science 295:1009
45. Hutmacher DW (2000) Biomaterials 21:2529
46. Sipe JD (2002) Ann NY Acad Sci 961:1
47. Liu X, Ma PX (2004) Ann Biomed Eng 32:477
48. Mikos AG, Thorsen AJ, Czerwonka LA, Bao Y, Langer R, Winslow DN, Vacanti JP (1994) Polymer 35:1068
49. Thomson RC, Yaszemski MJ, Powers JM, Mikos AG (1998) Biomaterials 19:1935
50. Marra KG, Szem JW, Kumta PN, DiMilla PA, Weiss LE (1999) J Biomed Mater Res 47:324
51. Kasuga T, Maeda H, Kato K, Nogami M, Hata K, Ueda M (2003) Biomaterials 24:3247
52. Liu Q, de Wijn JR, van Blitterswijk CA (1998) J Biomed Mater Res 40:490
53. Nam YS, Park TG (1999) J Biomed Mater Res 47:8
54. Ma PX, Choi J (2003) Tissue Eng 7:23
55. Spaans CJ, Belgraver VW, Rienstra O, de Groot JH, Veth RPH, Pennings AJ (2000) Biomaterials 21:2453
56. Schugens C, Maquet V, Grandfils C, Jerome R, Teyssie P (1996) J Biomed Mater Res 30:449
57. Schugens C, Maquet V, Grandfils C, Jerome R, Teyssie P (1996) Polymer 37:1027
58. Roether JA, Boccaccini AR, Hench LL, Maquet V, Gautier S, Jérôme R (2002) Biomaterials 23:3871
59. Zhang Y, Zhang M (2001) J Biomed Mater Res 55:304
60. Ciapetti G, Ambrosio L, Savarino L, Granchi D, Cenni E, Baldini N, Pagani S, Guizzardi S, Causa F, Giunti A (2003) Biomaterials 24:3815
61. Zhang R, Ma PX (1999) J Biomed Mater Res 44:446
62. Cai Q, Yang J, Bei J, Wang S (2002) Biomaterials 23:4483

Author Index Volumes 101–200

Author Index Volumes 1–100 see Volume 100

de Abajo, J. and *de la Campa, J. G.*: Processable Aromatic Polyimides. Vol. 140, pp. 23–60.
Abe, A., Furuya, H., Zhou, Z., Hiejima, T. and *Kobayashi, Y.*: Stepwise Phase Transitions of Chain Molecules: Crystallization/Melting via a Nematic Liquid-Crystalline Phase. Vol. 181, pp. 121–152.
Abetz, V. and *Simon, P. F. W.*: Phase Behaviour and Morphologies of Block Copolymers. Vol. 189, pp. 125–212.
Abetz, V. see Förster, S.: Vol. 166, pp. 173–210.
Adolf, D. B. see Ediger, M. D.: Vol. 116, pp. 73–110.
Advincula R.: Polymer Brushes by Anionic and Cationic Surface-Initiated Polymerization (SIP). Vol. 197, pp. 107–136.
Aharoni, S. M. and *Edwards, S. F.*: Rigid Polymer Networks. Vol. 118, pp. 1–231.
Akgun, B. see Brittain, W. J.: Vol. 198, pp. 125–147.
Alakhov, V. Y. see Kabanov, A. V.: Vol. 193, pp. 173–198.
Albertsson, A.-C. and *Varma, I. K.*: Aliphatic Polyesters: Synthesis, Properties and Applications. Vol. 157, pp. 99–138.
Albertsson, A.-C. see Edlund, U.: Vol. 157, pp. 53–98.
Albertsson, A.-C. see Söderqvist Lindblad, M.: Vol. 157, pp. 139–161.
Albertsson, A.-C. see Stridsberg, K. M.: Vol. 157, pp. 27–51.
Albertsson, A.-C. see Al-Malaika, S.: Vol. 169, pp. 177–199.
Albrecht, K., Mourran, A. and *Moeller, M.*: Surface Micelles and Surface-Induced Nanopatterns Formed by Block Copolymers. Vol. 200, pp. 57–70.
Al-Hussein, M. see de Jeu, W. H.: Vol. 200, pp. 71–90.
Allegra, G. and *Meille, S. V.*: Pre-Crystalline, High-Entropy Aggregates: A Role in Polymer Crystallization? Vol. 191, pp. 87–135.
Allen, S. see Ellis, J. S.: Vol. 193, pp. 123–172.
Al-Malaika, S.: Perspectives in Stabilisation of Polyolefins. Vol. 169, pp. 121–150.
Alonso, M. see Rodríguez-Cabello, J. C.: Vol. 200, pp. 119–168.
Altstädt, V.: The Influence of Molecular Variables on Fatigue Resistance in Stress Cracking Environments. Vol. 188, pp. 105–152.
Améduri, B., Boutevin, B. and *Gramain, P.*: Synthesis of Block Copolymers by Radical Polymerization and Telomerization. Vol. 127, pp. 87–142.
Améduri, B. and *Boutevin, B.*: Synthesis and Properties of Fluorinated Telechelic Monodispersed Compounds. Vol. 102, pp. 133–170.
Ameduri, B. see Taguet, A.: Vol. 184, pp. 127–211.
Amir, R. J. and *Shabat, D.*: Domino Dendrimers. Vol. 192, pp. 59–94.
Amselem, S. see Domb, A. J.: Vol. 107, pp. 93–142.
Anantawaraskul, S., Soares, J. B. P. and *Wood-Adams, P. M.*: Fractionation of Semicrystalline Polymers by Crystallization Analysis Fractionation and Temperature Rising Elution Fractionation. Vol. 182, pp. 1–54.

Andrady, A. L.: Wavelenght Sensitivity in Polymer Photodegradation. Vol. 128, pp. 47–94.
Andreis, M. and *Koenig, J. L.:* Application of Nitrogen-15 NMR to Polymers. Vol. 124, pp. 191–238.
Angiolini, L. see Carlini, C.: Vol. 123, pp. 127–214.
Anjum, N. see Gupta, B.: Vol. 162, pp. 37–63.
Anseth, K. S., Newman, S. M. and *Bowman, C. N.:* Polymeric Dental Composites: Properties and Reaction Behavior of Multimethacrylate Dental Restorations. Vol. 122, pp. 177–218.
Antonietti, M. see Cölfen, H.: Vol. 150, pp. 67–187.
Aoki, H. see Ito, S.: Vol. 182, pp. 131–170.
Aparicio, C. see Navarro, M.: Vol. 200, pp. 209–232.
Arias, F. J. see Rodríguez-Cabello, J. C.: Vol. 200, pp. 119–168.
Armitage, B. A. see O'Brien, D. F.: Vol. 126, pp. 53–58.
Arnal, M. L. see Müller, A. J.: Vol. 190, pp. 1–63.
Arndt, M. see Kaminski, W.: Vol. 127, pp. 143–187.
Arnold, A. and *Holm, C.:* Efficient Methods to Compute Long-Range Interactions for Soft Matter Systems. Vol. 185, pp. 59–109.
Arnold Jr., F. E. and *Arnold, F. E.:* Rigid-Rod Polymers and Molecular Composites. Vol. 117, pp. 257–296.
Arora, M. see Kumar, M. N. V. R.: Vol. 160, pp. 45–118.
Arshady, R.: Polymer Synthesis via Activated Esters: A New Dimension of Creativity in Macromolecular Chemistry. Vol. 111, pp. 1–42.
Aseyev, V. O., Tenhu, H. and *Winnik, F. M.:* Temperature Dependence of the Colloidal Stability of Neutral Amphiphilic Polymers in Water. Vol. 196, pp. 1–86.
Auer, S. and *Frenkel, D.:* Numerical Simulation of Crystal Nucleation in Colloids. Vol. 173, pp. 149–208.
Auriemma, F., de Rosa, C. and *Corradini, P.:* Solid Mesophases in Semicrystalline Polymers: Structural Analysis by Diffraction Techniques. Vol. 181, pp. 1–74.

Bahar, I., Erman, B. and *Monnerie, L.:* Effect of Molecular Structure on Local Chain Dynamics: Analytical Approaches and Computational Methods. Vol. 116, pp. 145–206.
Baietto-Dubourg, M. C. see Chateauminois, A.: Vol. 188, pp. 153–193.
Ballauff, M. see Dingenouts, N.: Vol. 144, pp. 1–48.
Ballauff, M. see Holm, C.: Vol. 166, pp. 1–27.
Ballauff, M. see Rühe, J.: Vol. 165, pp. 79–150.
Balsamo, V. see Müller, A. J.: Vol. 190, pp. 1–63.
Baltá-Calleja, F. J., González Arche, A., Ezquerra, T. A., Santa Cruz, C., Batallón, F., Frick, B. and *López Cabarcos, E.:* Structure and Properties of Ferroelectric Copolymers of Poly(vinylidene) Fluoride. Vol. 108, pp. 1–48.
Baltussen, J. J. M. see Northolt, M. G.: Vol. 178, pp. 1–108.
Barnes, M. D. see Otaigbe, J. U.: Vol. 154, pp. 1–86.
Barnes, C. M. see Satchi-Fainaro, R.: Vol. 193, pp. 1–65.
Barsett, H. see Paulsen, S. B.: Vol. 186, pp. 69–101.
Barshtein, G. R. and *Sabsai, O. Y.:* Compositions with Mineralorganic Fillers. Vol. 101, pp. 1–28.
Barton, J. see Hunkeler, D.: Vol. 112, pp. 115–134.
Baschnagel, J., Binder, K., Doruker, P., Gusev, A. A., Hahn, O., Kremer, K., Mattice, W. L., Müller-Plathe, F., Murat, M., Paul, W., Santos, S., Sutter, U. W. and *Tries, V.:* Bridging the Gap Between Atomistic and Coarse-Grained Models of Polymers: Status and Perspectives. Vol. 152, pp. 41–156.

Bassett, D. C.: On the Role of the Hexagonal Phase in the Crystallization of Polyethylene. Vol. 180, pp. 1–16.
Batallán, F. see Baltá-Calleja, F. J.: Vol. 108, pp. 1–48.
Batog, A. E., Pet'ko, I. P. and *Penczek, P.*: Aliphatic-Cycloaliphatic Epoxy Compounds and Polymers. Vol. 144, pp. 49–114.
Batrakova, E. V. see Kabanov, A. V.: Vol. 193, pp. 173–198.
Baughman, T. W. and *Wagener, K. B.*: Recent Advances in ADMET Polymerization. Vol. 176, pp. 1–42.
Baum, M. see Brittain, W. J.: Vol. 198, pp. 125–147.
Becker, O. and *Simon, G. P.*: Epoxy Layered Silicate Nanocomposites. Vol. 179, pp. 29–82.
Bell, C. L. and *Peppas, N. A.*: Biomedical Membranes from Hydrogels and Interpolymer Complexes. Vol. 122, pp. 125–176.
Bellon-Maurel, A. see Calmon-Decriaud, A.: Vol. 135, pp. 207–226.
Bennett, D. E. see O'Brien, D. F.: Vol. 126, pp. 53–84.
Bergbreiter, D. E. and *Kippenberger, A. M.*: Hyperbranched Surface Graft Polymerizations. Vol. 198, pp. 1–49.
Berry, G. C.: Static and Dynamic Light Scattering on Moderately Concentraded Solutions: Isotropic Solutions of Flexible and Rodlike Chains and Nematic Solutions of Rodlike Chains. Vol. 114, pp. 233–290.
Bershtein, V. A. and *Ryzhov, V. A.*: Far Infrared Spectroscopy of Polymers. Vol. 114, pp. 43–122.
Bhargava, R., Wang, S.-Q. and *Koenig, J. L*: FTIR Microspectroscopy of Polymeric Systems. Vol. 163, pp. 137–191.
Bhat, R. R., Tomlinson, M. R., Wu, T. and *Genzer, J.*: Surface-Grafted Polymer Gradients: Formation, Characterization, and Applications. Vol. 198, pp. 51–124.
Biesalski, M. see Rühe, J.: Vol. 165, pp. 79–150.
Bigg, D. M.: Thermal Conductivity of Heterophase Polymer Compositions. Vol. 119, pp. 1–30.
Binder, K.: Phase Transitions in Polymer Blends and Block Copolymer Melts: Some Recent Developments. Vol. 112, pp. 115–134.
Binder, K.: Phase Transitions of Polymer Blends and Block Copolymer Melts in Thin Films. Vol. 138, pp. 1–90.
Binder, K. see Baschnagel, J.: Vol. 152, pp. 41–156.
Binder, K., Müller, M., Virnau, P. and *González MacDowell, L.*: Polymer+Solvent Systems: Phase Diagrams, Interface Free Energies, and Nucleation. Vol. 173, pp. 1–104.
Bird, R. B. see Curtiss, C. F.: Vol. 125, pp. 1–102.
Biswas, M. and *Mukherjee, A.*: Synthesis and Evaluation of Metal-Containing Polymers. Vol. 115, pp. 89–124.
Biswas, M. and *Sinha Ray, S.*: Recent Progress in Synthesis and Evaluation of Polymer-Montmorillonite Nanocomposites. Vol. 155, pp. 167–221.
Blankenburg, L. see Klemm, E.: Vol. 177, pp. 53–90.
Blickle, C. see Brittain, W. J.: Vol. 198, pp. 125–147.
Blumen, A. see Gurtovenko, A. A.: Vol. 182, pp. 171–282.
Bogdal, D., Penczek, P., Pielichowski, J. and *Prociak, A.*: Microwave Assisted Synthesis, Crosslinking, and Processing of Polymeric Materials. Vol. 163, pp. 193–263.
Bohrisch, J., Eisenbach, C. D., Jaeger, W., Mori, H., Müller, A. H. E., Rehahn, M., Schaller, C., Traser, S. and *Wittmeyer, P.*: New Polyelectrolyte Architectures. Vol. 165, pp. 1–41.
Bolze, J. see Dingenouts, N.: Vol. 144, pp. 1–48.
Bosshard, C.: see Gubler, U.: Vol. 158, pp. 123–190.

Boutevin, B. and *Robin, J. J.*: Synthesis and Properties of Fluorinated Diols. Vol. 102, pp. 105–132.
Boutevin, B. see Améduri, B.: Vol. 102, pp. 133–170.
Boutevin, B. see Améduri, B.: Vol. 127, pp. 87–142.
Boutevin, B. see Guida-Pietrasanta, F.: Vol. 179, pp. 1–27.
Boutevin, B. see Taguet, A.: Vol. 184, pp. 127–211.
Bowman, C. N. see Anseth, K. S.: Vol. 122, pp. 177–218.
Boyd, R. H.: Prediction of Polymer Crystal Structures and Properties. Vol. 116, pp. 1–26.
Boyes, S. G. see Brittain, W. J.: Vol. 198, pp. 125–147.
Bracco, S. see Sozzani, P.: Vol. 181, pp. 153–177.
Briber, R. M. see Hedrick, J. L.: Vol. 141, pp. 1–44.
Brittain, W. J., Boyes, S. G., Granville, A. M., Baum, M., Mirous, B. K., Akgun, B., Zhao, B., Blickle, C. and *Foster, M. D.*: Surface Rearrangement of Diblock Copolymer Brushes—Stimuli Responsive Films. Vol. 198, pp. 125–147.
Bronnikov, S. V., Vettegren, V. I. and *Frenkel, S. Y.*: Kinetics of Deformation and Relaxation in Highly Oriented Polymers. Vol. 125, pp. 103–146.
Brown, H. R. see Creton, C.: Vol. 156, pp. 53–135.
Bruza, K. J. see Kirchhoff, R. A.: Vol. 117, pp. 1–66.
Buchmeiser M. R.: Metathesis Polymerization To and From Surfaces. Vol. 197, pp. 137–171.
Buchmeiser, M. R.: Regioselective Polymerization of 1-Alkynes and Stereoselective Cyclopolymerization of a, w-Heptadiynes. Vol. 176, pp. 89–119.
Budkowski, A.: Interfacial Phenomena in Thin Polymer Films: Phase Coexistence and Segregation. Vol. 148, pp. 1–112.
Bunz, U. H. F.: Synthesis and Structure of PAEs. Vol. 177, pp. 1–52.
Burban, J. H. see Cussler, E. L.: Vol. 110, pp. 67–80.
Burchard, W.: Solution Properties of Branched Macromolecules. Vol. 143, pp. 113–194.
Butté, A. see Schork, F. J.: Vol. 175, pp. 129–255.

Calmon-Decriaud, A., Bellon-Maurel, V., Silvestre, F.: Standard Methods for Testing the Aerobic Biodegradation of Polymeric Materials. Vol. 135, pp. 207–226.
Cameron, N. R. and *Sherrington, D. C.*: High Internal Phase Emulsions (HIPEs)-Structure, Properties and Use in Polymer Preparation. Vol. 126, pp. 163–214.
de la Campa, J. G. see de Abajo, J.: Vol. 140, pp. 23–60.
Candau, F. see Hunkeler, D.: Vol. 112, pp. 115–134.
Canelas, D. A. and *DeSimone, J. M.*: Polymerizations in Liquid and Supercritical Carbon Dioxide. Vol. 133, pp. 103–140.
Canva, M. and *Stegeman, G. I.*: Quadratic Parametric Interactions in Organic Waveguides. Vol. 158, pp. 87–121.
Capek, I.: Kinetics of the Free-Radical Emulsion Polymerization of Vinyl Chloride. Vol. 120, pp. 135–206.
Capek, I.: Radical Polymerization of Polyoxyethylene Macromonomers in Disperse Systems. Vol. 145, pp. 1–56.
Capek, I. and *Chern, C.-S.*: Radical Polymerization in Direct Mini-Emulsion Systems. Vol. 155, pp. 101–166.
Cappella, B. see Munz, M.: Vol. 164, pp. 87–210.
Carlesso, G. see Prokop, A.: Vol. 160, pp. 119–174.
Carlini, C. and *Angiolini, L.*: Polymers as Free Radical Photoinitiators. Vol. 123, pp. 127–214.
Carter, K. R. see Hedrick, J. L.: Vol. 141, pp. 1–44.
Casas-Vazquez, J. see Jou, D.: Vol. 120, pp. 207–266.

Chan, C.-M. and *Li, L.*: Direct Observation of the Growth of Lamellae and Spherulites by AFM. Vol. 188, pp. 1–41.
Chandrasekhar, V.: Polymer Solid Electrolytes: Synthesis and Structure. Vol. 135, pp. 139–206.
Chang, J. Y. see Han, M. J.: Vol. 153, pp. 1–36.
Chang, T.: Recent Advances in Liquid Chromatography Analysis of Synthetic Polymers. Vol. 163, pp. 1–60.
Charles-Harris, M. see Navarro, M.: Vol. 200, pp. 209–232.
Charleux, B. and *Faust, R.*: Synthesis of Branched Polymers by Cationic Polymerization. Vol. 142, pp. 1–70.
Chateauminois, A. and *Baietto-Dubourg, M. C.*: Fracture of Glassy Polymers Within Sliding Contacts. Vol. 188, pp. 153–193.
Chen, P. see Jaffe, M.: Vol. 117, pp. 297–328.
Chern, C.-S. see Capek, I.: Vol. 155, pp. 101–166.
Chevolot, Y. see Mathieu, H. J.: Vol. 162, pp. 1–35.
Chim, Y. T. A. see Ellis, J. S.: Vol. 193, pp. 123–172.
Choe, E.-W. see Jaffe, M.: Vol. 117, pp. 297–328.
Chow, P. Y. and *Gan, L. M.*: Microemulsion Polymerizations and Reactions. Vol. 175, pp. 257–298.
Chow, T. S.: Glassy State Relaxation and Deformation in Polymers. Vol. 103, pp. 149–190.
Chujo, Y. see Uemura, T.: Vol. 167, pp. 81–106.
Chung, S.-J. see Lin, T.-C.: Vol. 161, pp. 157–193.
Chung, T.-S. see Jaffe, M.: Vol. 117, pp. 297–328.
Clarke, N.: Effect of Shear Flow on Polymer Blends. Vol. 183, pp. 127–173.
Coenjarts, C. see Li, M.: Vol. 190, pp. 183–226.
Cölfen, H. and *Antonietti, M.*: Field-Flow Fractionation Techniques for Polymer and Colloid Analysis. Vol. 150, pp. 67–187.
Colmenero, J. see Richter, D.: Vol. 174, pp. 1–221.
Comanita, B. see Roovers, J.: Vol. 142, pp. 179–228.
Comotti, A. see Sozzani, P.: Vol. 181, pp. 153–177.
Connell, J. W. see Hergenrother, P. M.: Vol. 117, pp. 67–110.
Corradini, P. see Auriemma, F.: Vol. 181, pp. 1–74.
Creton, C., Kramer, E. J., Brown, H. R. and *Hui, C.-Y.*: Adhesion and Fracture of Interfaces Between Immiscible Polymers: From the Molecular to the Continuum Scale. Vol. 156, pp. 53–135.
Criado-Sancho, M. see Jou, D.: Vol. 120, pp. 207–266.
Curro, J. G. see Schweizer, K. S.: Vol. 116, pp. 319–378.
Curtiss, C. F. and *Bird, R. B.*: Statistical Mechanics of Transport Phenomena: Polymeric Liquid Mixtures. Vol. 125, pp. 1–102.
Cussler, E. L., Wang, K. L. and *Burban, J. H.*: Hydrogels as Separation Agents. Vol. 110, pp. 67–80.
Czub, P. see Penczek, P.: Vol. 184, pp. 1–95.

Dalton, L.: Nonlinear Optical Polymeric Materials: From Chromophore Design to Commercial Applications. Vol. 158, pp. 1–86.
Dautzenberg, H. see Holm, C.: Vol. 166, pp. 113–171.
Davidson, J. M. see Prokop, A.: Vol. 160, pp. 119–174.
Davies, M. C. see Ellis, J. S.: Vol. 193, pp. 123–172.
Degenhart, G. H. see Schönherr, H.: Vol. 200, pp. 169–208.
Den Decker, M. G. see Northolt, M. G.: Vol. 178, pp. 1–108.

Desai, S. M. and *Singh, R. P.*: Surface Modification of Polyethylene. Vol. 169, pp. 231–293.
DeSimone, J. M. see Canelas, D. A.: Vol. 133, pp. 103–140.
DeSimone, J. M. see Kennedy, K. A.: Vol. 175, pp. 329–346.
Dhal, P. K., Holmes-Farley, S. R., Huval, C. C. and *Jozefiak, T. H.*: Polymers as Drugs. Vol. 192, pp. 9–58.
DiMari, S. see Prokop, A.: Vol. 136, pp. 1–52.
Dimonie, M. V. see Hunkeler, D.: Vol. 112, pp. 115–134.
Dingenouts, N., Bolze, J., Pötschke, D. and *Ballauf, M.*: Analysis of Polymer Latexes by Small-Angle X-Ray Scattering. Vol. 144, pp. 1–48.
Dodd, L. R. and *Theodorou, D. N.*: Atomistic Monte Carlo Simulation and Continuum Mean Field Theory of the Structure and Equation of State Properties of Alkane and Polymer Melts. Vol. 116, pp. 249–282.
Doelker, E.: Cellulose Derivatives. Vol. 107, pp. 199–266.
Dolden, J. G.: Calculation of a Mesogenic Index with Emphasis Upon LC-Polyimides. Vol. 141, pp. 189–245.
Domb, A. J., Amselem, S., Shah, J. and *Maniar, M.*: Polyanhydrides: Synthesis and Characterization. Vol. 107, pp. 93–142.
Domb, A. J. see Kumar, M. N. V. R.: Vol. 160, pp. 45–118.
Dordi, B. see Schönherr, H.: Vol. 200, pp. 169–208.
Doruker, P. see Baschnagel, J.: Vol. 152, pp. 41–156.
Dubois, P. see Mecerreyes, D.: Vol. 147, pp. 1–60.
Dubrovskii, S. A. see Kazanskii, K. S.: Vol. 104, pp. 97–134.
Dudowicz, J. see Freed, K. F.: Vol. 183, pp. 63–126.
Duncan, R., Ringsdorf, H. and *Satchi-Fainaro, R.*: Polymer Therapeutics: Polymers as Drugs, Drug and Protein Conjugates and Gene Delivery Systems: Past, Present and Future Opportunities. Vol. 192, pp. 1–8.
Duncan, R. see Satchi-Fainaro, R.: Vol. 193, pp. 1–65.
Dunkin, I. R. see Steinke, J.: Vol. 123, pp. 81–126.
Dunson, D. L. see McGrath, J. E.: Vol. 140, pp. 61–106.
Dyer D. J.: Photoinitiated Synthesis of Grafted Polymers. Vol. 197, pp. 47–65.
Dziezok, P. see Rühe, J.: Vol. 165, pp. 79–150.

Eastmond, G. C.: Poly(e-caprolactone) Blends. Vol. 149, pp. 59–223.
Ebringerová, A., Hromádková, Z. and *Heinze, T.*: Hemicellulose. Vol. 186, pp. 1–67.
Economy, J. and *Goranov, K.*: Thermotropic Liquid Crystalline Polymers for High Performance Applications. Vol. 117, pp. 221–256.
Ediger, M. D. and *Adolf, D. B.*: Brownian Dynamics Simulations of Local Polymer Dynamics. Vol. 116, pp. 73–110.
Edlund, U. and *Albertsson, A.-C.*: Degradable Polymer Microspheres for Controlled Drug Delivery. Vol. 157, pp. 53–98.
Edwards, S. F. see Aharoni, S. M.: Vol. 118, pp. 1–231.
Eisenbach, C. D. see Bohrisch, J.: Vol. 165, pp. 1–41.
Ellis, J. S., Allen, S., Chim, Y. T. A., Roberts, C. J., Tendler, S. J. B. and *Davies, M. C.*: Molecular-Scale Studies on Biopolymers Using Atomic Force Microscopy. Vol. 193, pp. 123–172.
Endo, T. see Yagci, Y.: Vol. 127, pp. 59–86.
Engel, E. see Navarro, M.: Vol. 200, pp. 209–232.
Engelhardt, H. and *Grosche, O.*: Capillary Electrophoresis in Polymer Analysis. Vol. 150, pp. 189–217.

Engelhardt, H. and *Martin, H.*: Characterization of Synthetic Polyelectrolytes by Capillary Electrophoretic Methods. Vol. 165, pp. 211–247.
Eriksson, P. see Jacobson, K.: Vol. 169, pp. 151–176.
Erman, B. see Bahar, I.: Vol. 116, pp. 145–206.
Eschner, M. see Spange, S.: Vol. 165, pp. 43–78.
Estel, K. see Spange, S.: Vol. 165, pp. 43–78.
Estevez, R. and *Van der Giessen, E.*: Modeling and Computational Analysis of Fracture of Glassy Polymers. Vol. 188, pp. 195–234.
Ewen, B. and *Richter, D.*: Neutron Spin Echo Investigations on the Segmental Dynamics of Polymers in Melts, Networks and Solutions. Vol. 134, pp. 1–130.
Ezquerra, T. A. see Baltá-Calleja, F. J.: Vol. 108, pp. 1–48.

Fatkullin, N. see Kimmich, R.: Vol. 170, pp. 1–113.
Faust, R. see Charleux, B.: Vol. 142, pp. 1–70.
Faust, R. see Kwon, Y.: Vol. 167, pp. 107–135.
Fekete, E. see Pukánszky, B.: Vol. 139, pp. 109–154.
Fendler, J. H.: Membrane-Mimetic Approach to Advanced Materials. Vol. 113, pp. 1–209.
Feng, C. L. see Schönherr, H.: Vol. 200, pp. 169–208.
Fetters, L. J. see Xu, Z.: Vol. 120, pp. 1–50.
Fontenot, K. see Schork, F. J.: Vol. 175, pp. 129–255.
Förster, S., Abetz, V. and *Müller, A. H. E.*: Polyelectrolyte Block Copolymer Micelles. Vol. 166, pp. 173–210.
Förster, S. and *Schmidt, M.*: Polyelectrolytes in Solution. Vol. 120, pp. 51–134.
Foster, M. D. see Brittain, W. J.: Vol. 198, pp. 125–147.
Freed, K. F. and *Dudowicz, J.*: Influence of Monomer Molecular Structure on the Miscibility of Polymer Blends. Vol. 183, pp. 63–126.
Freire, J. J.: Conformational Properties of Branched Polymers: Theory and Simulations. Vol. 143, pp. 35–112.
Frenkel, D. see Hu, W.: Vol. 191, pp. 1–35.
Frenkel, S. Y. see Bronnikov, S. V.: Vol. 125, pp. 103–146.
Frick, B. see Baltá-Calleja, F. J.: Vol. 108, pp. 1–48.
Fridman, M. L.: see Terent'eva, J. P.: Vol. 101, pp. 29–64.
Fuchs, G. see Trimmel, G.: Vol. 176, pp. 43–87.
Fuhrmann-Lieker, T. see Pudzich, R.: Vol. 199, pp. 83–142.
Fukuda, T. see Tsujii, Y.: Vol. 197, pp. 1–47.
Fukui, K. see Otaigbe, J. U.: Vol. 154, pp. 1–86.
Funke, W.: Microgels-Intramolecularly Crosslinked Macromolecules with a Globular Structure. Vol. 136, pp. 137–232.
Furusho, Y. see Takata, T.: Vol. 171, pp. 1–75.
Furuya, H. see Abe, A.: Vol. 181, pp. 121–152.

Galina, H.: Mean-Field Kinetic Modeling of Polymerization: The Smoluchowski Coagulation Equation. Vol. 137, pp. 135–172.
Gan, L. M. see Chow, P. Y.: Vol. 175, pp. 257–298.
Ganesh, K. see Kishore, K.: Vol. 121, pp. 81–122.
Gaw, K. O. and *Kakimoto, M.*: Polyimide-Epoxy Composites. Vol. 140, pp. 107–136.
Geckeler, K. E. see Rivas, B.: Vol. 102, pp. 171–188.
Geckeler, K. E.: Soluble Polymer Supports for Liquid-Phase Synthesis. Vol. 121, pp. 31–80.
Gedde, U. W. and *Mattozzi, A.*: Polyethylene Morphology. Vol. 169, pp. 29–73.

Gehrke, S. H.: Synthesis, Equilibrium Swelling, Kinetics Permeability and Applications of Environmentally Responsive Gels. Vol. 110, pp. 81–144.

Geil, P. H., Yang, J., Williams, R. A., Petersen, K. L., Long, T.-C. and *Xu, P.*: Effect of Molecular Weight and Melt Time and Temperature on the Morphology of Poly(tetrafluorethylene). Vol. 180, pp. 89–159.

de Gennes, P.-G.: Flexible Polymers in Nanopores. Vol. 138, pp. 91–106.

Genzer, J. see Bhat, R. R.: Vol. 198, pp. 51–124.

Georgiou, S.: Laser Cleaning Methodologies of Polymer Substrates. Vol. 168, pp. 1–49.

Geuss, M. see Munz, M.: Vol. 164, pp. 87–210.

Giannelis, E. P., Krishnamoorti, R. and *Manias, E.*: Polymer-Silicate Nanocomposites: Model Systems for Confined Polymers and Polymer Brushes. Vol. 138, pp. 107–148.

Van der Giessen, E. see Estevez, R.: Vol. 188, pp. 195–234.

Ginebra, M. P. see Navarro, M.: Vol. 200, pp. 209–232.

Girotti, A. see Rodríguez-Cabello, J. C.: Vol. 200, pp. 119–168.

Godovsky, D. Y.: Device Applications of Polymer-Nanocomposites. Vol. 153, pp. 163–205.

Godovsky, D. Y.: Electron Behavior and Magnetic Properties Polymer-Nanocomposites. Vol. 119, pp. 79–122.

Gohy, J.-F.: Block Copolymer Micelles. Vol. 190, pp. 65–136.

Golze, S. see Korczagin, I.: Vol. 200, pp. 91–118.

González Arche, A. see Baltá-Calleja, F. J.: Vol. 108, pp. 1–48.

Goranov, K. see Economy, J.: Vol. 117, pp. 221–256.

Goto, A. see Tsujii, Y.: Vol. 197, pp. 1–47.

Gramain, P. see Améduri, B.: Vol. 127, pp. 87–142.

Granville, A. M. see Brittain, W. J.: Vol. 198, pp. 125–147.

Grein, C.: Toughness of Neat, Rubber Modified and Filled β-Nucleated Polypropylene: From Fundamentals to Applications. Vol. 188, pp. 43–104.

Greish, K. see Maeda, H.: Vol. 193, pp. 103–121.

Grest, G. S.: Normal and Shear Forces Between Polymer Brushes. Vol. 138, pp. 149–184.

Grigorescu, G. and *Kulicke, W.-M.*: Prediction of Viscoelastic Properties and Shear Stability of Polymers in Solution. Vol. 152, p. 1–40.

Grimsdale, A. C. and *Müllen, K.*: Polyphenylene-type Emissive Materials: Poly(*para*-phenylene)s, Polyfluorenes, and Ladder Polymers. Vol. 199, pp. 1–82.

Gröhn, F. see Rühe, J.: Vol. 165, pp. 79–150.

Grosberg, A. Y. and *Khokhlov, A. R.*: After-Action of the Ideas of I. M. Lifshitz in Polymer and Biopolymer Physics. Vol. 196, pp. 189–210.

Grosberg, A. and *Nechaev, S.*: Polymer Topology. Vol. 106, pp. 1–30.

Grosche, O. see Engelhardt, H.: Vol. 150, pp. 189–217.

Grubbs, R., Risse, W. and *Novac, B.*: The Development of Well-defined Catalysts for Ring-Opening Olefin Metathesis. Vol. 102, pp. 47–72.

Gubler, U. and *Bosshard, C.*: Molecular Design for Third-Order Nonlinear Optics. Vol. 158, pp. 123–190.

Guida-Pietrasanta, F. and *Boutevin, B.*: Polysilalkylene or Silarylene Siloxanes Said Hybrid Silicones. Vol. 179, pp. 1–27.

van Gunsteren, W. F. see Gusev, A. A.: Vol. 116, pp. 207–248.

Gupta, B. and *Anjum, N.*: Plasma and Radiation-Induced Graft Modification of Polymers for Biomedical Applications. Vol. 162, pp. 37–63.

Gurtovenko, A. A. and *Blumen, A.*: Generalized Gaussian Structures: Models for Polymer Systems with Complex Topologies. Vol. 182, pp. 171–282.

Gusev, A. A., Müller-Plathe, F., van Gunsteren, W. F. and *Suter, U. W.*: Dynamics of Small Molecules in Bulk Polymers. Vol. 116, pp. 207–248.

Gusev, A. A. see Baschnagel, J.: Vol. 152, pp. 41–156.
Guillot, J. see Hunkeler, D.: Vol. 112, pp. 115–134.
Guyot, A. and *Tauer, K.*: Reactive Surfactants in Emulsion Polymerization. Vol. 111, pp. 43–66.

Hadjichristidis, N. and *Pispas, S.*: Designed Block Copolymers for Ordered Polymeric Nanostructures. Vol. 200, pp. 37–56.
Hadjichristidis, N., Pispas, S., Pitsikalis, M., Iatrou, H. and *Vlahos, C.*: Asymmetric Star Polymers Synthesis and Properties. Vol. 142, pp. 71–128.
Hadjichristidis, N., Pitsikalis, M. and *Iatrou, H.*: Synthesis of Block Copolymers. Vol. 189, pp. 1–124.
Hadjichristidis, N. see Xu, Z.: Vol. 120, pp. 1–50.
Hadjichristidis, N. see Pitsikalis, M.: Vol. 135, pp. 1–138.
Hahn, O. see Baschnagel, J.: Vol. 152, pp. 41–156.
Hakkarainen, M.: Aliphatic Polyesters: Abiotic and Biotic Degradation and Degradation Products. Vol. 157, pp. 1–26.
Hakkarainen, M. and *Albertsson, A.-C.*: Environmental Degradation of Polyethylene. Vol. 169, pp. 177–199.
Halary, J. L. see Monnerie, L.: Vol. 187, pp. 35–213.
Halary, J. L. see Monnerie, L.: Vol. 187, pp. 215–364.
Hall, H. K. see Penelle, J.: Vol. 102, pp. 73–104.
Hamley, I. W.: Crystallization in Block Copolymers. Vol. 148, pp. 113–138.
Hammouda, B.: SANS from Homogeneous Polymer Mixtures: A Unified Overview. Vol. 106, pp. 87–134.
Han, M. J. and *Chang, J. Y.*: Polynucleotide Analogues. Vol. 153, pp. 1–36.
Harada, A.: Design and Construction of Supramolecular Architectures Consisting of Cyclodextrins and Polymers. Vol. 133, pp. 141–192.
Haralson, M. A. see Prokop, A.: Vol. 136, pp. 1–52.
Harding, S. E.: Analysis of Polysaccharides by Ultracentrifugation. Size, Conformation and Interactions in Solution. Vol. 186, pp. 211–254.
Hasegawa, N. see Usuki, A.: Vol. 179, pp. 135–195.
Hassan, C. M. and *Peppas, N. A.*: Structure and Applications of Poly(vinyl alcohol) Hydrogels Produced by Conventional Crosslinking or by Freezing/Thawing Methods. Vol. 153, pp. 37–65.
Hawker, C. J.: Dentritic and Hyperbranched Macromolecules Precisely Controlled Macromolecular Architectures. Vol. 147, pp. 113–160.
Hawker, C. J. see Hedrick, J. L.: Vol. 141, pp. 1–44.
He, G. S. see Lin, T.-C.: Vol. 161, pp. 157–193.
Hedrick, J. L., Carter, K. R., Labadie, J. W., Miller, R. D., Volksen, W., Hawker, C. J., Yoon, D. Y., Russell, T. P., McGrath, J. E. and *Briber, R. M.*: Nanoporous Polyimides. Vol. 141, pp. 1–44.
Hedrick, J. L., Labadie, J. W., Volksen, W. and *Hilborn, J. G.*: Nanoscopically Engineered Polyimides. Vol. 147, pp. 61–112.
Hedrick, J. L. see Hergenrother, P. M.: Vol. 117, pp. 67–110.
Hedrick, J. L. see Kiefer, J.: Vol. 147, pp. 161–247.
Hedrick, J. L. see McGrath, J. E.: Vol. 140, pp. 61–106.
Heine, D. R., Grest, G. S. and *Curro, J. G.*: Structure of Polymer Melts and Blends: Comparison of Integral Equation theory and Computer Sumulation. Vol. 173, pp. 209–249.
Heinrich, G. and *Klüppel, M.*: Recent Advances in the Theory of Filler Networking in Elastomers. Vol. 160, pp. 1–44.
Heinze, T. see Ebringerová, A.: Vol. 186, pp. 1–67.
Heinze, T. see El Seoud, O. A.: Vol. 186, pp. 103–149.

Heller, J.: Poly (Ortho Esters). Vol. 107, pp. 41–92.
Helm, C. A. see Möhwald, H.: Vol. 165, pp. 151–175.
Hemielec, A. A. see Hunkeler, D.: Vol. 112, pp. 115–134.
Hempenius, M. A. see Korczagin, I.: Vol. 200, pp. 91–118.
Hergenrother, P. M., Connell, J. W., Labadie, J. W. and *Hedrick, J. L.*: Poly(arylene ether)s Containing Heterocyclic Units. Vol. 117, pp. 67–110.
Hernández-Barajas, J. see Wandrey, C.: Vol. 145, pp. 123–182.
Hervet, H. see Léger, L.: Vol. 138, pp. 185–226.
Hiejima, T. see Abe, A.: Vol. 181, pp. 121–152.
Hikosaka, M., Watanabe, K., Okada, K. and *Yamazaki, S.*: Topological Mechanism of Polymer Nucleation and Growth – The Role of Chain Sliding Diffusion and Entanglement. Vol. 191, pp. 137–186.
Hilborn, J. G. see Hedrick, J. L.: Vol. 147, pp. 61–112.
Hilborn, J. G. see Kiefer, J.: Vol. 147, pp. 161–247.
Hillborg, H. see Vancso, G. J.: Vol. 182, pp. 55–129.
Hillmyer, M. A.: Nanoporous Materials from Block Copolymer Precursors. Vol. 190, pp. 137–181.
Hiramatsu, N. see Matsushige, M.: Vol. 125, pp. 147–186.
Hirasa, O. see Suzuki, M.: Vol. 110, pp. 241–262.
Hirotsu, S.: Coexistence of Phases and the Nature of First-Order Transition in Poly-N-isopropylacrylamide Gels. Vol. 110, pp. 1–26.
Höcker, H. see Klee, D.: Vol. 149, pp. 1–57.
Holm, C. see Arnold, A.: Vol. 185, pp. 59–109.
Holm, C., Hofmann, T., Joanny, J. F., Kremer, K., Netz, R. R., Reineker, P., Seidel, C., Vilgis, T. A. and *Winkler, R. G.*: Polyelectrolyte Theory. Vol. 166, pp. 67–111.
Holm, C., Rehahn, M., Oppermann, W. and *Ballauff, M.*: Stiff-Chain Polyelectrolytes. Vol. 166, pp. 1–27.
Holmes-Farley, S. R. see Dhal, P. K.: Vol. 192, pp. 9–58.
Hornsby, P.: Rheology, Compounding and Processing of Filled Thermoplastics. Vol. 139, pp. 155–216.
Houbenov, N. see Rühe, J.: Vol. 165, pp. 79–150.
Hromádková, Z. see Ebringerová, A.: Vol. 186, pp. 1–67.
Hu, W. and *Frenkel, D.*: Polymer Crystallization Driven by Anisotropic Interactions. Vol. 191, pp. 1–35.
Huber, K. see Volk, N.: Vol. 166, pp. 29–65.
Hugenberg, N. see Rühe, J.: Vol. 165, pp. 79–150.
Hui, C.-Y. see Creton, C.: Vol. 156, pp. 53–135.
Hult, A., Johansson, M. and *Malmström, E.*: Hyperbranched Polymers. Vol. 143, pp. 1–34.
Hünenberger, P. H.: Thermostat Algorithms for Molecular-Dynamics Simulations. Vol. 173, pp. 105–147.
Hunkeler, D., Candau, F., Pichot, C., Hemielec, A. E., Xie, T. Y., Barton, J., Vaskova, V., Guillot, J., Dimonie, M. V. and *Reichert, K. H.*: Heterophase Polymerization: A Physical and Kinetic Comparision and Categorization. Vol. 112, pp. 115–134.
Hunkeler, D. see Macko, T.: Vol. 163, pp. 61–136.
Hunkeler, D. see Prokop, A.: Vol. 136, pp. 1–52; 53–74.
Hunkeler, D. see Wandrey, C.: Vol. 145, pp. 123–182.
Huval, C. C. see Dhal, P. K.: Vol. 192, pp. 9–58.

Iatrou, H. see Hadjichristidis, N.: Vol. 142, pp. 71–128.
Iatrou, H. see Hadjichristidis, N.: Vol. 189, pp. 1–124.

Ichikawa, T. see Yoshida, H.: Vol. 105, pp. 3–36.
Ihara, E. see Yasuda, H.: Vol. 133, pp. 53–102.
Ikada, Y. see Uyama, Y.: Vol. 137, pp. 1–40.
Ikehara, T. see Jinnuai, H.: Vol. 170, pp. 115–167.
Ilavsky, M.: Effect on Phase Transition on Swelling and Mechanical Behavior of Synthetic Hydrogels. Vol. 109, pp. 173–206.
Imai, M. see Kaji, K.: Vol. 191, pp. 187–240.
Imai, Y.: Rapid Synthesis of Polyimides from Nylon-Salt Monomers. Vol. 140, pp. 1–23.
Inomata, H. see Saito, S.: Vol. 106, pp. 207–232.
Inoue, S. see Sugimoto, H.: Vol. 146, pp. 39–120.
Irie, M.: Stimuli-Responsive Poly(N-isopropylacrylamide), Photo- and Chemical-Induced Phase Transitions. Vol. 110, pp. 49–66.
Ise, N. see Matsuoka, H.: Vol. 114, pp. 187–232.
Ishikawa, T.: Advances in Inorganic Fibers. Vol. 178, pp. 109–144.
Ito, H.: Chemical Amplification Resists for Microlithography. Vol. 172, pp. 37–245.
Ito, K. and *Kawaguchi, S.*: Poly(macronomers), Homo- and Copolymerization. Vol. 142, pp. 129–178.
Ito, K. see Kawaguchi, S.: Vol. 175, pp. 299–328.
Ito, S. and *Aoki, H.*: Nano-Imaging of Polymers by Optical Microscopy. Vol. 182, pp. 131–170.
Ito, Y. see Suginome, M.: Vol. 171, pp. 77–136.
Ivanov, A. E. see Zubov, V. P.: Vol. 104, pp. 135–176.

Jacob, S. and *Kennedy, J.*: Synthesis, Characterization and Properties of OCTA-ARM Polyisobutylene-Based Star Polymers. Vol. 146, pp. 1–38.
Jacobson, K., Eriksson, P., Reitberger, T. and *Stenberg, B.*: Chemiluminescence as a Tool for Polyolefin. Vol. 169, pp. 151–176.
Jaeger, W. see Bohrisch, J.: Vol. 165, pp. 1–41.
Jaffe, M., Chen, P., Choe, E.-W., Chung, T.-S. and *Makhija, S.*: High Performance Polymer Blends. Vol. 117, pp. 297–328.
Jancar, J.: Structure-Property Relationships in Thermoplastic Matrices. Vol. 139, pp. 1–66.
Jang, J.: Conducting Polymer Nanomaterials and Their Applications. Vol. 199, pp. 189–260.
Jen, A. K.-Y. see Kajzar, F.: Vol. 161, pp. 1–85.
Jerome, R. see Mecerreyes, D.: Vol. 147, pp. 1–60.
de Jeu, W. H. see Li, L.: Vol. 181, pp. 75–120.
de Jeu, W. H., Séréro, Y. and *Al-Hussein, M.*: Liquid Crystallinity in Block Copolymer Films for Controlling Polymeric Nanopatterns. Vol. 200, pp. 71–90.
Jiang, M., Li, M., Xiang, M. and *Zhou, H.*: Interpolymer Complexation and Miscibility and Enhancement by Hydrogen Bonding. Vol. 146, pp. 121–194.
Jin, J. see Shim, H.-K.: Vol. 158, pp. 191–241.
Jinnai, H., Nishikawa, Y., Ikehara, T. and *Nishi, T.*: Emerging Technologies for the 3D Analysis of Polymer Structures. Vol. 170, pp. 115–167.
Jo, W. H. and *Yang, J. S.*: Molecular Simulation Approaches for Multiphase Polymer Systems. Vol. 156, pp. 1–52.
Joanny, J.-F. see Holm, C.: Vol. 166, pp. 67–111.
Joanny, J.-F. see Thünemann, A. F.: Vol. 166, pp. 113–171.
Johannsmann, D. see Rühe, J.: Vol. 165, pp. 79–150.
Johansson, M. see Hult, A.: Vol. 143, pp. 1–34.
Joos-Müller, B. see Funke, W.: Vol. 136, pp. 137–232.
Jou, D., Casas-Vazquez, J. and *Criado-Sancho, M.*: Thermodynamics of Polymer Solutions under Flow: Phase Separation and Polymer Degradation. Vol. 120, pp. 207–266.
Jozefiak, T. H. see Dhal, P. K.: Vol. 192, pp. 9–58.

Kabanov, A. V., Batrakova, E. V., Sherman, S. and *Alakhov, V. Y.*: Polymer Genomics. Vol. 193, pp. 173–198.

Kaetsu, I.: Radiation Synthesis of Polymeric Materials for Biomedical and Biochemical Applications. Vol. 105, pp. 81–98.

Kaji, K., Nishida, K., Kanaya, T., Matsuba, G., Konishi, T. and *Imai, M.*: Spinodal Crystallization of Polymers: Crystallization from the Unstable Melt. Vol. 191, pp. 187–240.

Kaji, K. see Kanaya, T.: Vol. 154, pp. 87–141.

Kajzar, F., Lee, K.-S. and *Jen, A. K.-Y.*: Polymeric Materials and their Orientation Techniques for Second-Order Nonlinear Optics. Vol. 161, pp. 1–85.

Kakimoto, M. see Gaw, K. O.: Vol. 140, pp. 107–136.

Kaminski, W. and *Arndt, M.*: Metallocenes for Polymer Catalysis. Vol. 127, pp. 143–187.

Kammer, H. W., Kressler, H. and *Kummerloewe, C.*: Phase Behavior of Polymer Blends – Effects of Thermodynamics and Rheology. Vol. 106, pp. 31–86.

Kanaya, T. and *Kaji, K.*: Dynamcis in the Glassy State and Near the Glass Transition of Amorphous Polymers as Studied by Neutron Scattering. Vol. 154, pp. 87–141.

Kanaya, T. see Kaji, K.: Vol. 191, pp. 187–240.

Kandyrin, L. B. and *Kuleznev, V. N.*: The Dependence of Viscosity on the Composition of Concentrated Dispersions and the Free Volume Concept of Disperse Systems. Vol. 103, pp. 103–148.

Kaneko, M. see Ramaraj, R.: Vol. 123, pp. 215–242.

Kaneko, M. see Yagi, M.: Vol. 199, pp. 143–188.

Kang, E. T., Neoh, K. G. and *Tan, K. L.*: X-Ray Photoelectron Spectroscopic Studies of Electroactive Polymers. Vol. 106, pp. 135–190.

Kaplan, D. L. see Singh, A.: Vol. 194, pp. 211–224.

Kaplan, D. L. see Xu, P.: Vol. 194, pp. 69–94.

Karlsson, S. see Söderqvist Lindblad, M.: Vol. 157, pp. 139–161.

Karlsson, S.: Recycled Polyolefins. Material Properties and Means for Quality Determination. Vol. 169, pp. 201–229.

Kataoka, K. see Nishiyama, N.: Vol. 193, pp. 67–101.

Kato, K. see Uyama, Y.: Vol. 137, pp. 1–40.

Kato, M. see Usuki, A.: Vol. 179, pp. 135–195.

Kausch, H.-H. and *Michler, G. H.*: The Effect of Time on Crazing and Fracture. Vol. 187, pp. 1–33.

Kausch, H.-H. see Monnerie, L. Vol. 187, pp. 215–364.

Kautek, W. see Krüger, J.: Vol. 168, pp. 247–290.

Kawaguchi, S. see Ito, K.: Vol. 142, pp. 129–178.

Kawaguchi, S. and *Ito, K.*: Dispersion Polymerization. Vol. 175, pp. 299–328.

Kawata, S. see Sun, H.-B.: Vol. 170, pp. 169–273.

Kazanskii, K. S. and *Dubrovskii, S. A.*: Chemistry and Physics of Agricultural Hydrogels. Vol. 104, pp. 97–134.

Kennedy, J. P. see Jacob, S.: Vol. 146, pp. 1–38.

Kennedy, J. P. see Majoros, I.: Vol. 112, pp. 1–113.

Kennedy, K. A., Roberts, G. W. and *DeSimone, J. M.*: Heterogeneous Polymerization of Fluoroolefins in Supercritical Carbon Dioxide. Vol. 175, pp. 329–346.

Khalatur, P. G. and *Khokhlov, A. R.*: Computer-Aided Conformation-Dependent Design of Copolymer Sequences. Vol. 195, pp. 1–100.

Khokhlov, A., Starodybtzev, S. and *Vasilevskaya, V.*: Conformational Transitions of Polymer Gels: Theory and Experiment. Vol. 109, pp. 121–172.

Khokhlov, A. R. see Grosberg, A. Y.: Vol. 196, pp. 189–210.

Khokhlov, A. R. see Khalatur, P. G.: Vol. 195, pp. 1–100.

Khokhlov, A. R. see Kuchanov, S. I.: Vol. 196, pp. 129–188.
Khokhlov, A. R. see Okhapkin, I. M.: Vol. 195, pp. 177–210.
Kiefer, J., Hedrick, J. L. and *Hiborn, J. G.*: Macroporous Thermosets by Chemically Induced Phase Separation. Vol. 147, pp. 161–247.
Kihara, N. see Takata, T.: Vol. 171, pp. 1–75.
Kilian, H. G. and *Pieper, T.*: Packing of Chain Segments. A Method for Describing X-Ray Patterns of Crystalline, Liquid Crystalline and Non-Crystalline Polymers. Vol. 108, pp. 49–90.
Kim, J. see Quirk, R. P.: Vol. 153, pp. 67–162.
Kim, K.-S. see Lin, T.-C.: Vol. 161, pp. 157–193.
Kimmich, R. and *Fatkullin, N.*: Polymer Chain Dynamics and NMR. Vol. 170, pp. 1–113.
Kippelen, B. and *Peyghambarian, N.*: Photorefractive Polymers and their Applications. Vol. 161, pp. 87–156.
Kippenberger, A. M. see Bergbreiter, D. E.: Vol. 198, pp. 1–49.
Kirchhoff, R. A. and *Bruza, K. J.*: Polymers from Benzocyclobutenes. Vol. 117, pp. 1–66.
Kishore, K. and *Ganesh, K.*: Polymers Containing Disulfide, Tetrasulfide, Diselenide and Ditelluride Linkages in the Main Chain. Vol. 121, pp. 81–122.
Kitamaru, R.: Phase Structure of Polyethylene and Other Crystalline Polymers by Solid-State 13C/MNR. Vol. 137, pp. 41–102.
Klapper, M. see Rusanov, A. L.: Vol. 179, pp. 83–134.
Klee, D. and *Höcker, H.*: Polymers for Biomedical Applications: Improvement of the Interface Compatibility. Vol. 149, pp. 1–57.
Klemm, E., Pautzsch, T. and *Blankenburg, L.*: Organometallic PAEs. Vol. 177, pp. 53–90.
Klier, J. see Scranton, A. B.: Vol. 122, pp. 1–54.
v. Klitzing, R. and *Tieke, B.*: Polyelectrolyte Membranes. Vol. 165, pp. 177–210.
Kloeckner, J. see Wagner, E.: Vol. 192, pp. 135–173.
Klüppel, M.: The Role of Disorder in Filler Reinforcement of Elastomers on Various Length Scales. Vol. 164, pp. 1–86.
Klüppel, M. see Heinrich, G.: Vol. 160, pp. 1–44.
Knuuttila, H., Lehtinen, A. and *Nummila-Pakarinen, A.*: Advanced Polyethylene Technologies – Controlled Material Properties. Vol. 169, pp. 13–27.
Kobayashi, S. and *Ohmae, M.*: Enzymatic Polymerization to Polysaccharides. Vol. 194, pp. 159–210.
Kobayashi, S. see Uyama, H.: Vol. 194, pp. 51–67.
Kobayashi, S. see Uyama, H.: Vol. 194, pp. 133–158.
Kobayashi, S., Shoda, S. and *Uyama, H.*: Enzymatic Polymerization and Oligomerization. Vol. 121, pp. 1–30.
Kobayashi, T. see Abe, A.: Vol. 181, pp. 121–152.
Köhler, W. and *Schäfer, R.*: Polymer Analysis by Thermal-Diffusion Forced Rayleigh Scattering. Vol. 151, pp. 1–59.
Koenig, J. L. see Bhargava, R.: Vol. 163, pp. 137–191.
Koenig, J. L. see Andreis, M.: Vol. 124, pp. 191–238.
Koike, T.: Viscoelastic Behavior of Epoxy Resins Before Crosslinking. Vol. 148, pp. 139–188.
Kokko, E. see Löfgren, B.: Vol. 169, pp. 1–12.
Kokufuta, E.: Novel Applications for Stimulus-Sensitive Polymer Gels in the Preparation of Functional Immobilized Biocatalysts. Vol. 110, pp. 157–178.
Konishi, T. see Kaji, K.: Vol. 191, pp. 187–240.
Konno, M. see Saito, S.: Vol. 109, pp. 207–232.
Konradi, R. see Rühe, J.: Vol. 165, pp. 79–150.
Kopecek, J. see Putnam, D.: Vol. 122, pp. 55–124.

Korczagin, I., Lammertink, R. G. H., Hempenius, M. A., Golze, S. and *Vancso, G. J.*: Surface Nano- and Microstructuring with Organometallic Polymers. Vol. 200, pp. 91–118.
Koßmehl, G. see Schopf, G.: Vol. 129, pp. 1–145.
Kostoglodov, P. V. see Rusanov, A. L.: Vol. 179, pp. 83–134.
Kozlov, E. see Prokop, A.: Vol. 160, pp. 119–174.
Kramer, E. J. see Creton, C.: Vol. 156, pp. 53–135.
Kremer, K. see Baschnagel, J.: Vol. 152, pp. 41–156.
Kremer, K. see Holm, C.: Vol. 166, pp. 67–111.
Kressler, J. see Kammer, H. W.: Vol. 106, pp. 31–86.
Kricheldorf, H. R.: Liquid-Cristalline Polyimides. Vol. 141, pp. 83–188.
Krishnamoorti, R. see Giannelis, E. P.: Vol. 138, pp. 107–148.
Krüger, J. and *Kautek, W.*: Ultrashort Pulse Laser Interaction with Dielectrics and Polymers, Vol. 168, pp. 247–290.
Kuchanov, S. I.: Modern Aspects of Quantitative Theory of Free-Radical Copolymerization. Vol. 103, pp. 1–102.
Kuchanov, S. I. and *Khokhlov, A. R.*: Role of Physical Factors in the Process of Obtaining Copolymers. Vol. 196, pp. 129–188.
Kuchanov, S. I.: Principles of Quantitive Description of Chemical Structure of Synthetic Polymers. Vol. 152, pp. 157–202.
Kudaibergennow, S. E.: Recent Advances in Studying of Synthetic Polyampholytes in Solutions. Vol. 144, pp. 115–198.
Kuleznev, V. N. see Kandyrin, L. B.: Vol. 103, pp. 103–148.
Kulichkhin, S. G. see Malkin, A. Y.: Vol. 101, pp. 217–258.
Kulicke, W.-M. see Grigorescu, G.: Vol. 152, pp. 1–40.
Kumar, M. N. V. R., Kumar, N., Domb, A. J. and *Arora, M.*: Pharmaceutical Polymeric Controlled Drug Delivery Systems. Vol. 160, pp. 45–118.
Kumar, N. see Kumar, M. N. V. R.: Vol. 160, pp. 45–118.
Kummerloewe, C. see Kammer, H. W.: Vol. 106, pp. 31–86.
Kuznetsova, N. P. see Samsonov, G. V.: Vol. 104, pp. 1–50.
Kwon, Y. and *Faust, R.*: Synthesis of Polyisobutylene-Based Block Copolymers with Precisely Controlled Architecture by Living Cationic Polymerization. Vol. 167, pp. 107–135.

Labadie, J. W. see Hergenrother, P. M.: Vol. 117, pp. 67–110.
Labadie, J. W. see Hedrick, J. L.: Vol. 141, pp. 1–44.
Labadie, J. W. see Hedrick, J. L.: Vol. 147, pp. 61–112.
Lammertink, R. G. H. see Korczagin, I.: Vol. 200, pp. 91–118.
Lamparski, H. G. see O'Brien, D. F.: Vol. 126, pp. 53–84.
Laschewsky, A.: Molecular Concepts, Self-Organisation and Properties of Polysoaps. Vol. 124, pp. 1–86.
Laso, M. see Leontidis, E.: Vol. 116, pp. 283–318.
Lauprêtre, F. see Monnerie, L.: Vol. 187, pp. 35–213.
Lazár, M. and *Rychl, R.*: Oxidation of Hydrocarbon Polymers. Vol. 102, pp. 189–222.
Lechowicz, J. see Galina, H.: Vol. 137, pp. 135–172.
Léger, L., Raphaël, E. and *Hervet, H.*: Surface-Anchored Polymer Chains: Their Role in Adhesion and Friction. Vol. 138, pp. 185–226.
Lenz, R. W.: Biodegradable Polymers. Vol. 107, pp. 1–40.
Leontidis, E., de Pablo, J. J., Laso, M. and *Suter, U. W.*: A Critical Evaluation of Novel Algorithms for the Off-Lattice Monte Carlo Simulation of Condensed Polymer Phases. Vol. 116, pp. 283–318.
Lee, B. see Quirk, R. P.: Vol. 153, pp. 67–162.

Lee, K.-S. see Kajzar, F.: Vol. 161, pp. 1-85.
Lee, Y. see Quirk, R. P.: Vol. 153, pp. 67-162.
Lehtinen, A. see Knuuttila, H.: Vol. 169, pp. 13-27.
Leónard, D. see Mathieu, H. J.: Vol. 162, pp. 1-35.
Lesec, J. see Viovy, J.-L.: Vol. 114, pp. 1-42.
Levesque, D. see Weis, J.-J.: Vol. 185, pp. 163-225.
Li, L. and *de Jeu, W. H.*: Flow-induced mesophases in crystallizable polymers. Vol. 181, pp. 75-120.
Li, L. see Chan, C.-M.: Vol. 188, pp. 1-41.
Li, M., Coenjarts, C. and *Ober, C. K.*: Patternable Block Copolymers. Vol. 190, pp. 183-226.
Li, M. see Jiang, M.: Vol. 146, pp. 121-194.
Liang, G. L. see Sumpter, B. G.: Vol. 116, pp. 27-72.
Lienert, K.-W.: Poly(ester-imide)s for Industrial Use. Vol. 141, pp. 45-82.
Likhatchev, D. see Rusanov, A. L.: Vol. 179, pp. 83-134.
Lin, J. and *Sherrington, D. C.*: Recent Developments in the Synthesis, Thermostability and Liquid Crystal Properties of Aromatic Polyamides. Vol. 111, pp. 177-220.
Lin, T.-C., Chung, S.-J., Kim, K.-S., Wang, X., He, G. S., Swiatkiewicz, J., Pudavar, H. E. and *Prasad, P. N.*: Organics and Polymers with High Two-Photon Activities and their Applications. Vol. 161, pp. 157-193.
Linse, P.: Simulation of Charged Colloids in Solution. Vol. 185, pp. 111-162.
Lippert, T.: Laser Application of Polymers. Vol. 168, pp. 51-246.
Liu, Y. see Söderqvist Lindblad, M.: Vol. 157, pp. 139-161.
Long, T.-C. see Geil, P. H.: Vol. 180, pp. 89-159.
López Cabarcos, E. see Baltá-Calleja, F. J.: Vol. 108, pp. 1-48.
Lotz, B.: Analysis and Observation of Polymer Crystal Structures at the Individual Stem Level. Vol. 180, pp. 17-44.
Löfgren, B., Kokko, E. and *Seppälä, J.*: Specific Structures Enabled by Metallocene Catalysis in Polyethenes. Vol. 169, pp. 1-12.
Löwen, H. see Thünemann, A. F.: Vol. 166, pp. 113-171.
Lozinsky V. I.: Approaches to Chemical Synthesis of Protein-Like Copolymers. Vol. 196, pp. 87-128.
Luo, Y. see Schork, F. J.: Vol. 175, pp. 129-255.

Macko, T. and *Hunkeler, D.*: Liquid Chromatography under Critical and Limiting Conditions: A Survey of Experimental Systems for Synthetic Polymers. Vol. 163, pp. 61-136.
Maeda, H., Greish, K. and *Fang, J.*: The EPR Effect and Polymeric Drugs: A Paradigm Shift for Cancer Chemotherapy in the 21st Century. Vol. 193, pp. 103-121.
Majoros, I., Nagy, A. and *Kennedy, J. P.*: Conventional and Living Carbocationic Polymerizations United. I. A Comprehensive Model and New Diagnostic Method to Probe the Mechanism of Homopolymerizations. Vol. 112, pp. 1-113.
Makhaeva, E. E. see Okhapkin, I. M.: Vol. 195, pp. 177-210.
Makhija, S. see Jaffe, M.: Vol. 117, pp. 297-328.
Malmström, E. see Hult, A.: Vol. 143, pp. 1-34.
Malkin, A. Y. and *Kulichkhin, S. G.*: Rheokinetics of Curing. Vol. 101, pp. 217-258.
Maniar, M. see Domb, A. J.: Vol. 107, pp. 93-142.
Manias, E. see Giannelis, E. P.: Vol. 138, pp. 107-148.
Martin, H. see Engelhardt, H.: Vol. 165, pp. 211-247.
Marty, J. D. and *Mauzac, M.*: Molecular Imprinting: State of the Art and Perspectives. Vol. 172, pp. 1-35.

Mashima, K., Nakayama, Y. and *Nakamura, A.*: Recent Trends in Polymerization of a-Olefins Catalyzed by Organometallic Complexes of Early Transition Metals. Vol. 133, pp. 1–52.
Mathew, D. see Reghunadhan Nair, C. P.: Vol. 155, pp. 1–99.
Mathieu, H. J., Chevolot, Y., Ruiz-Taylor, L. and *Leónard, D.*: Engineering and Characterization of Polymer Surfaces for Biomedical Applications. Vol. 162, pp. 1–35.
Matsuba, G. see Kaji, K.: Vol. 191, pp. 187–240.
Matsuda T.: Photoiniferter-Driven Precision Surface Graft Microarchitectures for Biomedical Applications. Vol. 197, pp. 67–106.
Matsumura S.: Enzymatic Synthesis of Polyesters via Ring-Opening Polymerization. Vol. 194, pp. 95–132.
Matsumoto, A.: Free-Radical Crosslinking Polymerization and Copolymerization of Multivinyl Compounds. Vol. 123, pp. 41–80.
Matsumoto, A. see Otsu, T.: Vol. 136, pp. 75–138.
Matsuoka, H. and *Ise, N.*: Small-Angle and Ultra-Small Angle Scattering Study of the Ordered Structure in Polyelectrolyte Solutions and Colloidal Dispersions. Vol. 114, pp. 187–232.
Matsushige, K., Hiramatsu, N. and *Okabe, H.*: Ultrasonic Spectroscopy for Polymeric Materials. Vol. 125, pp. 147–186.
Mattice, W. L. see Rehahn, M.: Vol. 131/132, pp. 1–475.
Mattice, W. L. see Baschnagel, J.: Vol. 152, pp. 41–156.
Mattozzi, A. see Gedde, U. W.: Vol. 169, pp. 29–73.
Mauzac, M. see Marty, J. D.: Vol. 172, pp. 1–35.
Mays, W. see Xu, Z.: Vol. 120, pp. 1–50.
Mays, J. W. see Pitsikalis, M.: Vol. 135, pp. 1–138.
McGrath, J. E. see Hedrick, J. L.: Vol. 141, pp. 1–44.
McGrath, J. E., Dunson, D. L. and *Hedrick, J. L.*: Synthesis and Characterization of Segmented Polyimide-Polyorganosiloxane Copolymers. Vol. 140, pp. 61–106.
McLeish, T. C. B. and *Milner, S. T.*: Entangled Dynamics and Melt Flow of Branched Polymers. Vol. 143, pp. 195–256.
Mecerreyes, D., Dubois, P. and *Jerome, R.*: Novel Macromolecular Architectures Based on Aliphatic Polyesters: Relevance of the Coordination-Insertion Ring-Opening Polymerization. Vol. 147, pp. 1–60.
Mecham, S. J. see McGrath, J. E.: Vol. 140, pp. 61–106.
Meille, S. V. see Allegra, G.: Vol. 191, pp. 87–135.
Menzel, H. see Möhwald, H.: Vol. 165, pp. 151–175.
Meyer, T. see Spange, S.: Vol. 165, pp. 43–78.
Michler, G. H. see Kausch, H.-H.: Vol. 187, pp. 1–33.
Mikos, A. G. see Thomson, R. C.: Vol. 122, pp. 245–274.
Milner, S. T. see McLeish, T. C. B.: Vol. 143, pp. 195–256.
Mirous, B. K. see Brittain, W. J.: Vol. 198, pp. 125–147.
Mison, P. and *Sillion, B.*: Thermosetting Oligomers Containing Maleimides and Nadiimides End-Groups. Vol. 140, pp. 137–180.
Miyasaka, K.: PVA-Iodine Complexes: Formation, Structure and Properties. Vol. 108, pp. 91–130.
Miller, R. D. see Hedrick, J. L.: Vol. 141, pp. 1–44.
Minko, S. see Rühe, J.: Vol. 165, pp. 79–150.
Moeller, M. see Albrecht, K.: Vol. 200, pp. 57–70.
Möhwald, H., Menzel, H., Helm, C. A. and *Stamm, M.*: Lipid and Polyampholyte Monolayers to Study Polyelectrolyte Interactions and Structure at Interfaces. Vol. 165, pp. 151–175.
Monkenbusch, M. see Richter, D.: Vol. 174, pp. 1–221.

Monnerie, L., Halary, J. L. and *Kausch, H.-H.*: Deformation, Yield and Fracture of Amorphous Polymers: Relation to the Secondary Transitions. Vol. 187, pp. 215-364.
Monnerie, L., Lauprêtre, F. and *Halary, J. L.*: Investigation of Solid-State Transitions in Linear and Crosslinked Amorphous Polymers. Vol. 187, pp. 35-213.
Monnerie, L. see Bahar, I.: Vol. 116, pp. 145-206.
Moore, J. S. see Ray, C. R.: Vol. 177, pp. 99-149.
Mori, H. see Bohrisch, J.: Vol. 165, pp. 1-41.
Morishima, Y.: Photoinduced Electron Transfer in Amphiphilic Polyelectrolyte Systems. Vol. 104, pp. 51-96.
Morton, M. see Quirk, R. P.: Vol. 153, pp. 67-162.
Motornov, M. see Rühe, J.: Vol. 165, pp. 79-150.
Mourran, A. see Albrecht, K.: Vol. 200, pp. 57-70.
Mours, M. see Winter, H. H.: Vol. 134, pp. 165-234.
Müllen, K. see Grimsdale, A. C.: Vol. 199, pp. 1-82.
Müllen, K. see Scherf, U.: Vol. 123, pp. 1-40.
Müller, A. H. E. see Bohrisch, J.: Vol. 165, pp. 1-41.
Müller, A. H. E. see Förster, S.: Vol. 166, pp. 173-210.
Müller, A. J., Balsamo, V. and *Arnal, M. L.*: Nucleation and Crystallization in Diblock and Triblock Copolymers. Vol. 190, pp. 1-63.
Müller, M. and *Schmid, F.*: Incorporating Fluctuations and Dynamics in Self-Consistent Field Theories for Polymer Blends. Vol. 185, pp. 1-58.
Müller, M. see Thünemann, A. F.: Vol. 166, pp. 113-171.
Müller-Plathe, F. see Gusev, A. A.: Vol. 116, pp. 207-248.
Müller-Plathe, F. see Baschnagel, J.: Vol. 152, p. 41-156.
Mukerherjee, A. see Biswas, M.: Vol. 115, pp. 89-124.
Munz, M., Cappella, B., Sturm, H., Geuss, M. and *Schulz, E.*: Materials Contrasts and Nanolithography Techniques in Scanning Force Microscopy (SFM) and their Application to Polymers and Polymer Composites. Vol. 164, pp. 87-210.
Murat, M. see Baschnagel, J.: Vol. 152, p. 41-156.
Muthukumar, M.: Modeling Polymer Crystallization. Vol. 191, pp. 241-274.
Muzzarelli, C. see Muzzarelli, R. A. A.: Vol. 186, pp. 151-209.
Muzzarelli, R. A. A. and *Muzzarelli, C.*: Chitosan Chemistry: Relevance to the Biomedical Sciences. Vol. 186, pp. 151-209.
Mylnikov, V.: Photoconducting Polymers. Vol. 115, pp. 1-88.

Nagy, A. see Majoros, I.: Vol. 112, pp. 1-11.
Naji, A., Seidel, C. and *Netz, R. R.*: Theoretical Approaches to Neutral and Charged Polymer Brushes. Vol. 198, pp. 149-183.
Naka, K. see Uemura, T.: Vol. 167, pp. 81-106.
Nakamura, A. see Mashima, K.: Vol. 133, pp. 1-52.
Nakayama, Y. see Mashima, K.: Vol. 133, pp. 1-52.
Narasinham, B. and *Peppas, N. A.*: The Physics of Polymer Dissolution: Modeling Approaches and Experimental Behavior. Vol. 128, pp. 157-208.
Navarro, M., Aparicio, C., Charles-Harris, M., Ginebra, M. P., Engel, E. and *Planell, J. A.*: Development of a Biodegradable Composite Scaffold for Bone Tissue Engineering: Physicochemical, Topographical, Mechanical, Degradation, and Biological Properties. Vol. 200, pp. 209-232.
Nechaev, S. see Grosberg, A.: Vol. 106, pp. 1-30.
Neoh, K. G. see Kang, E. T.: Vol. 106, pp. 135-190.
Netz, R. R. see Holm, C.: Vol. 166, pp. 67-111.

Netz, R. R. see Naji, A.: Vol. 198, pp. 149–183.
Netz, R. R. see Rühe, J.: Vol. 165, pp. 79–150.
Newman, S. M. see Anseth, K. S.: Vol. 122, pp. 177–218.
Nijenhuis, K. te: Thermoreversible Networks. Vol. 130, pp. 1–252.
Ninan, K. N. see Reghunadhan Nair, C. P.: Vol. 155, pp. 1–99.
Nishi, T. see Jinnai, H.: Vol. 170, pp. 115–167.
Nishida, K. see Kaji, K.: Vol. 191, pp. 187–240.
Nishikawa, Y. see Jinnai, H.: Vol. 170, pp. 115–167.
Nishiyama, N. and *Kataoka, K.*: Nanostructured Devices Based on Block Copolymer Assemblies for Drug Delivery: Designing Structures for Enhanced Drug Function. Vol. 193, pp. 67–101.
Noid, D. W. see Otaigbe, J. U.: Vol. 154, pp. 1–86.
Noid, D. W. see Sumpter, B. G.: Vol. 116, pp. 27–72.
Nomura, M., Tobita, H. and *Suzuki, K.*: Emulsion Polymerization: Kinetic and Mechanistic Aspects. Vol. 175, pp. 1–128.
Northolt, M. G., Picken, S. J., Den Decker, M. G., Baltussen, J. J. M. and *Schlatmann, R.*: The Tensile Strength of Polymer Fibres. Vol. 178, pp. 1–108.
Novac, B. see Grubbs, R.: Vol. 102, pp. 47–72.
Novikov, V. V. see Privalko, V. P.: Vol. 119, pp. 31–78.
Nummila-Pakarinen, A. see Knuuttila, H.: Vol. 169, pp. 13–27.

Ober, C. K. see Li, M.: Vol. 190, pp. 183–226.
O'Brien, D. F., Armitage, B. A., Bennett, D. E. and *Lamparski, H. G.*: Polymerization and Domain Formation in Lipid Assemblies. Vol. 126, pp. 53–84.
Ogasawara, M.: Application of Pulse Radiolysis to the Study of Polymers and Polymerizations. Vol. 105, pp. 37–80.
Ohmae, M. see Kobayashi, S.: Vol. 194, pp. 159–210.
Ohno, K. see Tsujii, Y.: Vol. 197, pp. 1–47.
Okabe, H. see Matsushige, K.: Vol. 125, pp. 147–186.
Okada, M.: Ring-Opening Polymerization of Bicyclic and Spiro Compounds. Reactivities and Polymerization Mechanisms. Vol. 102, pp. 1–46.
Okada, K. see Hikosaka, M.: Vol. 191, pp. 137–186.
Okano, T.: Molecular Design of Temperature-Responsive Polymers as Intelligent Materials. Vol. 110, pp. 179–198.
Okay, O. see Funke, W.: Vol. 136, pp. 137–232.
Okhapkin, I. M., Makhaeva, E. E. and *Khokhlov, A. R.*: Water Solutions of Amphiphilic Polymers: Nanostructure Formation and Possibilities for Catalysis. Vol. 195, pp. 177–210.
Onuki, A.: Theory of Phase Transition in Polymer Gels. Vol. 109, pp. 63–120.
Oppermann, W. see Holm, C.: Vol. 166, pp. 1–27.
Oppermann, W. see Volk, N.: Vol. 166, pp. 29–65.
Osad'ko, I. S.: Selective Spectroscopy of Chromophore Doped Polymers and Glasses. Vol. 114, pp. 123–186.
Osakada, K. and *Takeuchi, D.*: Coordination Polymerization of Dienes, Allenes, and Methylenecycloalkanes. Vol. 171, pp. 137–194.
Otaigbe, J. U., Barnes, M. D., Fukui, K., Sumpter, B. G. and *Noid, D. W.*: Generation, Characterization, and Modeling of Polymer Micro- and Nano-Particles. Vol. 154, pp. 1–86.
Otsu, T. and *Matsumoto, A.*: Controlled Synthesis of Polymers Using the Iniferter Technique: Developments in Living Radical Polymerization. Vol. 136, pp. 75–138.

de Pablo, J. J. see Leontidis, E.: Vol. 116, pp. 283–318.
Padias, A. B. see Penelle, J.: Vol. 102, pp. 73–104.

Pascault, J.-P. see Williams, R. J. J.: Vol. 128, pp. 95–156.
Pasch, H.: Analysis of Complex Polymers by Interaction Chromatography. Vol. 128, pp. 1–46.
Pasch, H.: Hyphenated Techniques in Liquid Chromatography of Polymers. Vol. 150, pp. 1–66.
Pasut, G. and *Veronese, F. M.*: PEGylation of Proteins as Tailored Chemistry for Optimized Bioconjugates. Vol. 192, pp. 95–134.
Paul, W. see Baschnagel, J.: Vol. 152, pp. 41–156.
Paulsen, S. B. and *Barsett, H.*: Bioactive Pectic Polysaccharides. Vol. 186, pp. 69–101.
Pautzsch, T. see Klemm, E.: Vol. 177, pp. 53–90.
Penczek, P., Czub, P. and *Pielichowski, J.*: Unsaturated Polyester Resins: Chemistry and Technology. Vol. 184, pp. 1–95.
Penczek, P. see Batog, A. E.: Vol. 144, pp. 49–114.
Penczek, P. see Bogdal, D.: Vol. 163, pp. 193–263.
Penelle, J., Hall, H. K., Padias, A. B. and *Tanaka, H.*: Captodative Olefins in Polymer Chemistry. Vol. 102, pp. 73–104.
Peppas, N. A. see Bell, C. L.: Vol. 122, pp. 125–176.
Peppas, N. A. see Hassan, C. M.: Vol. 153, pp. 37–65.
Peppas, N. A. see Narasimhan, B.: Vol. 128, pp. 157–208.
Petersen, K. L. see Geil, P. H.: Vol. 180, pp. 89–159.
Pet'ko, I. P. see Batog, A. E.: Vol. 144, pp. 49–114.
Pheyghambarian, N. see Kippelen, B.: Vol. 161, pp. 87–156.
Pichot, C. see Hunkeler, D.: Vol. 112, pp. 115–134.
Picken, S. J. see Northolt, M. G.: Vol. 178, pp. 1–108.
Pielichowski, J. see Bogdal, D.: Vol. 163, pp. 193–263.
Pielichowski, J. see Penczek, P.: Vol. 184, pp. 1–95.
Pieper, T. see Kilian, H. G.: Vol. 108, pp. 49–90.
Pispas, S. see Hadjichristidis, N.: Vol. 142, pp. 71–128.
Pispas, S. see Hadjichristidis, N.: Vol. 200, pp. 37–56.
Pispas, S. see Pitsikalis, M.: Vol. 135, pp. 1–138.
Pitsikalis, M., Pispas, S., Mays, J. W. and *Hadjichristidis, N.*: Nonlinear Block Copolymer Architectures. Vol. 135, pp. 1–138.
Pitsikalis, M. see Hadjichristidis, N.: Vol. 142, pp. 71–128.
Pitsikalis, M. see Hadjichristidis, N.: Vol. 189, pp. 1–124.
Planell, J. A. see Navarro, M.: Vol. 200, pp. 209–232.
Pleul, D. see Spange, S.: Vol. 165, pp. 43–78.
Plummer, C. J. G.: Microdeformation and Fracture in Bulk Polyolefins. Vol. 169, pp. 75–119.
Pötschke, D. see Dingenouts, N.: Vol. 144, pp. 1–48.
Pokrovskii, V. N.: The Mesoscopic Theory of the Slow Relaxation of Linear Macromolecules. Vol. 154, pp. 143–219.
Pospíšil, J.: Functionalized Oligomers and Polymers as Stabilizers for Conventional Polymers. Vol. 101, pp. 65–168.
Pospíšil, J.: Aromatic and Heterocyclic Amines in Polymer Stabilization. Vol. 124, pp. 87–190.
Powers, A. C. see Prokop, A.: Vol. 136, pp. 53–74.
Prasad, P. N. see Lin, T.-C.: Vol. 161, pp. 157–193.
Priddy, D. B.: Recent Advances in Styrene Polymerization. Vol. 111, pp. 67–114.
Priddy, D. B.: Thermal Discoloration Chemistry of Styrene-co-Acrylonitrile. Vol. 121, pp. 123–154.
Privalko, V. P. and *Novikov, V. V.*: Model Treatments of the Heat Conductivity of Heterogeneous Polymers. Vol. 119, pp. 31–78.
Prociak, A. see Bogdal, D.: Vol. 163, pp. 193–263.

Prokop, A., Hunkeler, D., DiMari, S., Haralson, M. A. and *Wang, T. G.*: Water Soluble Polymers for Immunoisolation I: Complex Coacervation and Cytotoxicity. Vol. 136, pp. 1–52.
Prokop, A., Hunkeler, D., Powers, A. C., Whitesell, R. R. and *Wang, T. G.*: Water Soluble Polymers for Immunoisolation II: Evaluation of Multicomponent Microencapsulation Systems. Vol. 136, pp. 53–74.
Prokop, A., Kozlov, E., Carlesso, G. and *Davidsen, J. M.*: Hydrogel-Based Colloidal Polymeric System for Protein and Drug Delivery: Physical and Chemical Characterization, Permeability Control and Applications. Vol. 160, pp. 119–174.
Pruitt, L. A.: The Effects of Radiation on the Structural and Mechanical Properties of Medical Polymers. Vol. 162, pp. 65–95.
Pudavar, H. E. see Lin, T.-C.: Vol. 161, pp. 157–193.
Pudzich, R., Fuhrmann-Lieker, T. and *Salbeck, J.*: Spiro Compounds for Organic Electroluminescence and Related Applications. Vol. 199, pp. 83–142.
Pukánszky, B. and *Fekete, E.*: Adhesion and Surface Modification. Vol. 139, pp. 109–154.
Putnam, D. and *Kopecek, J.*: Polymer Conjugates with Anticancer Acitivity. Vol. 122, pp. 55–124.
Putra, E. G. R. see Ungar, G.: Vol. 180, pp. 45–87.

Quirk, R. P., Yoo, T., Lee, Y., M., Kim, J. and *Lee, B.*: Applications of 1,1-Diphenylethylene Chemistry in Anionic Synthesis of Polymers with Controlled Structures. Vol. 153, pp. 67–162.

Ramaraj, R. and *Kaneko, M.*: Metal Complex in Polymer Membrane as a Model for Photosynthetic Oxygen Evolving Center. Vol. 123, pp. 215–242.
Rangarajan, B. see Scranton, A. B.: Vol. 122, pp. 1–54.
Ranucci, E. see Söderqvist Lindblad, M.: Vol. 157, pp. 139–161.
Raphaël, E. see Léger, L.: Vol. 138, pp. 185–226.
Rastogi, S. and *Terry, A. E.*: Morphological implications of the interphase bridging crystalline and amorphous regions in semi-crystalline polymers. Vol. 180, pp. 161–194.
Ray, C. R. and *Moore, J. S.*: Supramolecular Organization of Foldable Phenylene Ethynylene Oligomers. Vol. 177, pp. 99–149.
Reddinger, J. L. and *Reynolds, J. R.*: Molecular Engineering of p-Conjugated Polymers. Vol. 145, pp. 57–122.
Reghunadhan Nair, C. P., Mathew, D. and *Ninan, K. N.*: Cyanate Ester Resins, Recent Developments. Vol. 155, pp. 1–99.
Reguera, J. see Rodríguez-Cabello, J. C.: Vol. 200, pp. 119–168.
Rehahn, M., Mattice, W. L. and *Suter, U. W.*: Rotational Isomeric State Models in Macromolecular Systems. Vol. 131/132, pp. 1–475.
Rehahn, M. see Bohrisch, J.: Vol. 165, pp. 1–41.
Rehahn, M. see Holm, C.: Vol. 166, pp. 1–27.
Reichert, K. H. see Hunkeler, D.: Vol. 112, pp. 115–134.
Reihmann, M. and *Ritter, H.*: Synthesis of Phenol Polymers Using Peroxidases. Vol. 194, pp. 1–49.
Reineker, P. see Holm, C.: Vol. 166, pp. 67–111.
Reitberger, T. see Jacobson, K.: Vol. 169, pp. 151–176.
Reiter, G. see Sommer, J.-U.: Vol. 200, pp. 1–36.
Ritter, H. see Reihmann, M.: Vol. 194, pp. 1–49.
Reynolds, J. R. see Reddinger, J. L.: Vol. 145, pp. 57–122.
Richter, D. see Ewen, B.: Vol. 134, pp. 1–130.

Richter, D., Monkenbusch, M. and *Colmenero, J.*: Neutron Spin Echo in Polymer Systems. Vol. 174, pp. 1–221.
Riegler, S. see *Trimmel, G.*: Vol. 176, pp. 43–87.
Ringsdorf, H. see *Duncan, R.*: Vol. 192, pp. 1–8.
Risse, W. see *Grubbs, R.*: Vol. 102, pp. 47–72.
Rivas, B. L. and *Geckeler, K. E.*: Synthesis and Metal Complexation of Poly(ethyleneimine) and Derivatives. Vol. 102, pp. 171–188.
Roberts, C. J. see *Ellis, J. S.*: Vol. 193, pp. 123–172.
Roberts, G. W. see *Kennedy, K. A.*: Vol. 175, pp. 329–346.
Robin, J. J.: The Use of Ozone in the Synthesis of New Polymers and the Modification of Polymers. Vol. 167, pp. 35–79.
Robin, J. J. see *Boutevin, B.*: Vol. 102, pp. 105–132.
Rodríguez-Cabello, J. C., Reguera, J., Girotti, A., Arias, F. J. and *Alonso, M.*: Genetic Engineering of Protein-Based Polymers: The Example of Elastinlike Polymers. Vol. 200, pp. 119–168.
Rodríguez-Pérez, M. A.: Crosslinked Polyolefin Foams: Production, Structure, Properties, and Applications. Vol. 184, pp. 97–126.
Roe, R.-J.: MD Simulation Study of Glass Transition and Short Time Dynamics in Polymer Liquids. Vol. 116, pp. 111–114.
Roovers, J. and *Comanita, B.*: Dendrimers and Dendrimer-Polymer Hybrids. Vol. 142, pp. 179–228.
Rothon, R. N.: Mineral Fillers in Thermoplastics: Filler Manufacture and Characterisation. Vol. 139, pp. 67–108.
de Rosa, C. see *Auriemma, F.*: Vol. 181, pp. 1–74.
Rozenberg, B. A. see *Williams, R. J. J.*: Vol. 128, pp. 95–156.
Rozkiewicz, D. I. see *Schönherr, H.*: Vol. 200, pp. 169–208.
Rühe, J., Ballauff, M., Biesalski, M., Dziezok, P., Gröhn, F., Johannsmann, D., Houbenov, N., Hugenberg, N., Konradi, R., Minko, S., Motornov, M., Netz, R. R., Schmidt, M., Seidel, C., Stamm, M., Stephan, T., Usov, D. and *Zhang, H.*: Polyelectrolyte Brushes. Vol. 165, pp. 79–150.
Ruckenstein, E.: Concentrated Emulsion Polymerization. Vol. 127, pp. 1–58.
Ruiz-Taylor, L. see *Mathieu, H. J.*: Vol. 162, pp. 1–35.
Rusanov, A. L.: Novel Bis (Naphtalic Anhydrides) and Their Polyheteroarylenes with Improved Processability. Vol. 111, pp. 115–176.
Rusanov, A. L., Likhatchev, D., Kostoglodov, P. V., Müllen, K. and *Klapper, M.*: Proton-Exchanging Electrolyte Membranes Based on Aromatic Condensation Polymers. Vol. 179, pp. 83–134.
Russel, T. P. see *Hedrick, J. L.*: Vol. 141, pp. 1–44.
Russum, J. P. see *Schork, F. J.*: Vol. 175, pp. 129–255.
Rychly, J. see *Lazár, M.*: Vol. 102, pp. 189–222.
Ryner, M. see *Stridsberg, K. M.*: Vol. 157, pp. 27–51.
Ryzhov, V. A. see *Bershtein, V. A.*: Vol. 114, pp. 43–122.

Sabsai, O. Y. see *Barshtein, G. R.*: Vol. 101, pp. 1–28.
Saburov, V. V. see *Zubov, V. P.*: Vol. 104, pp. 135–176.
Saito, S., Konno, M. and *Inomata, H.*: Volume Phase Transition of N-Alkylacrylamide Gels. Vol. 109, pp. 207–232.
Salbeck, J. see *Pudzich, R.*: Vol. 199, pp. 83–142.
Samsonov, G. V. and *Kuznetsova, N. P.*: Crosslinked Polyelectrolytes in Biology. Vol. 104, pp. 1–50.

Santa Cruz, C. see Baltá-Calleja, F. J.: Vol. 108, pp. 1–48.
Santos, S. see Baschnagel, J.: Vol. 152, p. 41–156.
Satchi-Fainaro, R., Duncan, R. and *Barnes, C. M.*: Polymer Therapeutics for Cancer: Current Status and Future Challenges. Vol. 193, pp. 1–65.
Satchi-Fainaro, R. see Duncan, R.: Vol. 192, pp. 1–8.
Sato, T. and *Teramoto, A.*: Concentrated Solutions of Liquid-Christalline Polymers. Vol. 126, pp. 85–162.
Schaller, C. see Bohrisch, J.: Vol. 165, pp. 1–41.
Schäfer, R. see Köhler, W.: Vol. 151, pp. 1–59.
Scherf, U. and *Müllen, K.*: The Synthesis of Ladder Polymers. Vol. 123, pp. 1–40.
Sherman, S. see Kabanov, A. V.: Vol. 193, pp. 173–198.
Schlatmann, R. see Northolt, M. G.: Vol. 178, pp. 1–108.
Schmid, F. see Müller, M.: Vol. 185, pp. 1–58.
Schmidt, M. see Förster, S.: Vol. 120, pp. 51–134.
Schmidt, M. see Rühe, J.: Vol. 165, pp. 79–150.
Schmidt, M. see Volk, N.: Vol. 166, pp. 29–65.
Scholz, M.: Effects of Ion Radiation on Cells and Tissues. Vol. 162, pp. 97–158.
Schönherr, H., Degenhart, G. H., Dordi, B., Feng, C. L., Rozkiewicz, D. I., Shovsky, A. and *Vancso, G. J.*: Organic and Macromolecular Films and Assemblies as (Bio)reactive Platforms: From Model Studies on Structure–Reactivity Relationships to Submicrometer Patterning. Vol. 200, pp. 169–208.
Schönherr, H. see Vancso, G. J.: Vol. 182, pp. 55–129.
Schopf, G. and *Koßmehl, G.*: Polythiophenes – Electrically Conductive Polymers. Vol. 129, pp. 1–145.
Schork, F. J., Luo, Y., Smulders, W., Russum, J. P., Butté, A. and *Fontenot, K.*: Miniemulsion Polymerization. Vol. 175, pp. 127–255.
Schulz, E. see Munz, M.: Vol. 164, pp. 97–210.
Schwahn, D.: Critical to Mean Field Crossover in Polymer Blends. Vol. 183, pp. 1–61.
Seppälä, J. see Löfgren, B.: Vol. 169, pp. 1–12.
Séréro, Y. see de Jeu, W. H.: Vol. 200, pp. 71–90.
Sturm, H. see Munz, M.: Vol. 164, pp. 87–210.
Schweizer, K. S.: Prism Theory of the Structure, Thermodynamics, and Phase Transitions of Polymer Liquids and Alloys. Vol. 116, pp. 319–378.
Scranton, A. B., Rangarajan, B. and *Klier, J.*: Biomedical Applications of Polyelectrolytes. Vol. 122, pp. 1–54.
Sefton, M. V. and *Stevenson, W. T. K.*: Microencapsulation of Live Animal Cells Using Polycrylates. Vol. 107, pp. 143–198.
Seidel, C. see Holm, C.: Vol. 166, pp. 67–111.
Seidel, C. see Naji, A.: Vol. 198, pp. 149–183.
Seidel, C. see Rühe, J.: Vol. 165, pp. 79–150.
El Seoud, O. A. and *Heinze, T.*: Organic Esters of Cellulose: New Perspectives for Old Polymers. Vol. 186, pp. 103–149.
Shabat, D. see Amir, R. J.: Vol. 192, pp. 59–94.
Shamanin, V. V.: Bases of the Axiomatic Theory of Addition Polymerization. Vol. 112, pp. 135–180.
Shcherbina, M. A. see Ungar, G.: Vol. 180, pp. 45–87.
Sheiko, S. S.: Imaging of Polymers Using Scanning Force Microscopy: From Superstructures to Individual Molecules. Vol. 151, pp. 61–174.
Sherrington, D. C. see Cameron, N. R.: Vol. 126, pp. 163–214.
Sherrington, D. C. see Lin, J.: Vol. 111, pp. 177–220.

Sherrington, D. C. see Steinke, J.: Vol. 123, pp. 81–126.
Shibayama, M. see Tanaka, T.: Vol. 109, pp. 1–62.
Shiga, T.: Deformation and Viscoelastic Behavior of Polymer Gels in Electric Fields. Vol. 134, pp. 131–164.
Shim, H.-K. and *Jin, J.:* Light-Emitting Characteristics of Conjugated Polymers. Vol. 158, pp. 191–241.
Shoda, S. see Kobayashi, S.: Vol. 121, pp. 1–30.
Shovsky, A. see Schönherr, H.: Vol. 200, pp. 169–208.
Siegel, R. A.: Hydrophobic Weak Polyelectrolyte Gels: Studies of Swelling Equilibria and Kinetics. Vol. 109, pp. 233–268.
de Silva, D. S. M. see Ungar, G.: Vol. 180, pp. 45–87.
Silvestre, F. see Calmon-Decriaud, A.: Vol. 207, pp. 207–226.
Sillion, B. see Mison, P.: Vol. 140, pp. 137–180.
Simon, F. see Spange, S.: Vol. 165, pp. 43–78.
Simon, G. P. see Becker, O.: Vol. 179, pp. 29–82.
Simon, P. F. W. see Abetz, V.: Vol. 189, pp. 125–212.
Simonutti, R. see Sozzani, P.: Vol. 181, pp. 153–177.
Singh, A. and *Kaplan, D. L.:* In Vitro Enzyme-Induced Vinyl Polymerization. Vol. 194, pp. 211–224.
Singh, A. see Xu, P.: Vol. 194, pp. 69–94.
Singh, R. P. see Sivaram, S.: Vol. 101, pp. 169–216.
Singh, R. P. see Desai, S. M.: Vol. 169, pp. 231–293.
Sinha Ray, S. see Biswas, M.: Vol. 155, pp. 167–221.
Sivaram, S. and *Singh, R. P.:* Degradation and Stabilization of Ethylene-Propylene Copolymers and Their Blends: A Critical Review. Vol. 101, pp. 169–216.
Slugovc, C. see Trimmel, G.: Vol. 176, pp. 43–87.
Smulders, W. see Schork, F. J.: Vol. 175, pp. 129–255.
Soares, J. B. P. see Anantawaraskul, S.: Vol. 182, pp. 1–54.
Sommer, J.-U. and *Reiter, G.:* The Formation of Ordered Polymer Structures at Interfaces: A Few Intriguing Aspects. Vol. 200, pp. 1–36.
Sozzani, P., Bracco, S., Comotti, A. and *Simonutti, R.:* Motional Phase Disorder of Polymer Chains as Crystallized to Hexagonal Lattices. Vol. 181, pp. 153–177.
Söderqvist Lindblad, M., Liu, Y., Albertsson, A.-C., Ranucci, E. and *Karlsson, S.:* Polymer from Renewable Resources. Vol. 157, pp. 139–161.
Spange, S., Meyer, T., Voigt, I., Eschner, M., Estel, K., Pleul, D. and *Simon, F.:* Poly(Vinylformamide-co-Vinylamine)/Inorganic Oxid Hybrid Materials. Vol. 165, pp. 43–78.
Stamm, M. see Möhwald, H.: Vol. 165, pp. 151–175.
Stamm, M. see Rühe, J.: Vol. 165, pp. 79–150.
Starodybtzev, S. see Khokhlov, A.: Vol. 109, pp. 121–172.
Stegeman, G. I. see Canva, M.: Vol. 158, pp. 87–121.
Steinke, J., Sherrington, D. C. and *Dunkin, I. R.:* Imprinting of Synthetic Polymers Using Molecular Templates. Vol. 123, pp. 81–126.
Stelzer, F. see Trimmel, G.: Vol. 176, pp. 43–87.
Stenberg, B. see Jacobson, K.: Vol. 169, pp. 151–176.
Stenzenberger, H. D.: Addition Polyimides. Vol. 117, pp. 165–220.
Stephan, T. see Rühe, J.: Vol. 165, pp. 79–150.
Stevenson, W. T. K. see Sefton, M. V.: Vol. 107, pp. 143–198.
Stridsberg, K. M., Ryner, M. and *Albertsson, A.-C.:* Controlled Ring-Opening Polymerization: Polymers with Designed Macromoleculars Architecture. Vol. 157, pp. 27–51.
Sturm, H. see Munz, M.: Vol. 164, pp. 87–210.

Suematsu, K.: Recent Progress of Gel Theory: Ring, Excluded Volume, and Dimension. Vol. 156, pp. 136–214.
Sugimoto, H. and *Inoue, S.*: Polymerization by Metalloporphyrin and Related Complexes. Vol. 146, pp. 39–120.
Suginome, M. and *Ito, Y.*: Transition Metal-Mediated Polymerization of Isocyanides. Vol. 171, pp. 77–136.
Sumpter, B. G., Noid, D. W., Liang, G. L. and *Wunderlich, B.*: Atomistic Dynamics of Macromolecular Crystals. Vol. 116, pp. 27–72.
Sumpter, B. G. see Otaigbe, J. U.: Vol. 154, pp. 1–86.
Sun, H.-B. and *Kawata, S.*: Two-Photon Photopolymerization and 3D Lithographic Microfabrication. Vol. 170, pp. 169–273.
Suter, U. W. see Gusev, A. A.: Vol. 116, pp. 207–248.
Suter, U. W. see Leontidis, E.: Vol. 116, pp. 283–318.
Suter, U. W. see Rehahn, M.: Vol. 131/132, pp. 1–475.
Suter, U. W. see Baschnagel, J.: Vol. 152, pp. 41–156.
Suzuki, A.: Phase Transition in Gels of Sub-Millimeter Size Induced by Interaction with Stimuli. Vol. 110, pp. 199–240.
Suzuki, A. and *Hirasa, O.*: An Approach to Artifical Muscle by Polymer Gels due to Micro-Phase Separation. Vol. 110, pp. 241–262.
Suzuki, K. see Nomura, M.: Vol. 175, pp. 1–128.
Swiatkiewicz, J. see Lin, T.-C.: Vol. 161, pp. 157–193.

Tagawa, S.: Radiation Effects on Ion Beams on Polymers. Vol. 105, pp. 99–116.
Taguet, A., Ameduri, B. and *Boutevin, B.*: Crosslinking of Vinylidene Fluoride-Containing Fluoropolymers. Vol. 184, pp. 127–211.
Takata, T., Kihara, N. and *Furusho, Y.*: Polyrotaxanes and Polycatenanes: Recent Advances in Syntheses and Applications of Polymers Comprising of Interlocked Structures. Vol. 171, pp. 1–75.
Takeuchi, D. see Osakada, K.: Vol. 171, pp. 137–194.
Tan, K. L. see Kang, E. T.: Vol. 106, pp. 135–190.
Tanaka, H. and *Shibayama, M.*: Phase Transition and Related Phenomena of Polymer Gels. Vol. 109, pp. 1–62.
Tanaka, T. see Penelle, J.: Vol. 102, pp. 73–104.
Tauer, K. see Guyot, A.: Vol. 111, pp. 43–66.
Tendler, S. J. B. see Ellis, J. S.: Vol. 193, pp. 123–172.
Tenhu, H. see Aseyev, V. O.: Vol. 196, pp. 1–86.
Teramoto, A. see Sato, T.: Vol. 126, pp. 85–162.
Terent'eva, J. P. and *Fridman, M. L.*: Compositions Based on Aminoresins. Vol. 101, pp. 29–64.
Terry, A. E. see Rastogi, S.: Vol. 180, pp. 161–194.
Theodorou, D. N. see Dodd, L. R.: Vol. 116, pp. 249–282.
Thomson, R. C., Wake, M. C., Yaszemski, M. J. and *Mikos, A. G.*: Biodegradable Polymer Scaffolds to Regenerate Organs. Vol. 122, pp. 245–274.
Thünemann, A. F., Müller, M., Dautzenberg, H., Joanny, J.-F. and *Löwen, H.*: Polyelectrolyte complexes. Vol. 166, pp. 113–171.
Tieke, B. see v. Klitzing, R.: Vol. 165, pp. 177–210.
Tobita, H. see Nomura, M.: Vol. 175, pp. 1–128.
Tokita, M.: Friction Between Polymer Networks of Gels and Solvent. Vol. 110, pp. 27–48.
Tomlinson, M. R. see Bhat, R. R.: Vol. 198, pp. 51–124.
Traser, S. see Bohrisch, J.: Vol. 165, pp. 1–41.
Tries, V. see Baschnagel, J.: Vol. 152, p. 41–156.

Trimmel, G., Riegler, S., Fuchs, G., Slugovc, C. and *Stelzer, F.*: Liquid Crystalline Polymers by Metathesis Polymerization. Vol. 176, pp. 43–87.

Tsujii, Y., Ohno, K., Yamamoto, S., Goto, A. and *Fukuda, T.*: Structure and Properties of High-Density Polymer Brushes Prepared by Surface-Initiated Living Radical Polymerization. Vol. 197, pp. 1–47.

Tsuruta, T.: Contemporary Topics in Polymeric Materials for Biomedical Applications. Vol. 126, pp. 1–52.

Uemura, T., Naka, K. and *Chujo, Y.*: Functional Macromolecules with Electron-Donating Dithiafulvene Unit. Vol. 167, pp. 81–106.

Ungar, G., Putra, E. G. R., de Silva, D. S. M., Shcherbina, M. A. and *Waddon, A. J.*: The Effect of Self-Poisoning on Crystal Morphology and Growth Rates. Vol. 180, pp. 45–87.

Usov, D. see Rühe, J.: Vol. 165, pp. 79–150.

Usuki, A., Hasegawa, N. and *Kato, M.*: Polymer-Clay Nanocomposites. Vol. 179, pp. 135–195.

Uyama, H. and *Kobayashi, S.*: Enzymatic Synthesis and Properties of Polymers from Polyphenols. Vol. 194, pp. 51–67.

Uyama, H. and *Kobayashi, S.*: Enzymatic Synthesis of Polyesters via Polycondensation. Vol. 194, pp. 133–158.

Uyama, H. see Kobayashi, S.: Vol. 121, pp. 1–30.

Uyama, Y.: Surface Modification of Polymers by Grafting. Vol. 137, pp. 1–40.

Vancso, G. J., Hillborg, H. and *Schönherr, H.*: Chemical Composition of Polymer Surfaces Imaged by Atomic Force Microscopy and Complementary Approaches. Vol. 182, pp. 55–129.

Vancso, G. J. see Korczagin, I.: Vol. 200, pp. 91–118.

Vancso, G. J. see Schönherr, H.: Vol. 200, pp. 169–208.

Varma, I. K. see Albertsson, A.-C.: Vol. 157, pp. 99–138.

Vasilevskaya, V. see Khokhlov, A.: Vol. 109, pp. 121–172.

Vaskova, V. see Hunkeler, D.: Vol. 112, pp. 115–134.

Verdugo, P.: Polymer Gel Phase Transition in Condensation-Decondensation of Secretory Products. Vol. 110, pp. 145–156.

Veronese, F. M. see Pasut, G.: Vol. 192, pp. 95–134.

Vettegren, V. I. see Bronnikov, S. V.: Vol. 125, pp. 103–146.

Vilgis, T. A. see Holm, C.: Vol. 166, pp. 67–111.

Viovy, J.-L. and *Lesec, J.*: Separation of Macromolecules in Gels: Permeation Chromatography and Electrophoresis. Vol. 114, pp. 1–42.

Vlahos, C. see Hadjichristidis, N.: Vol. 142, pp. 71–128.

Voigt, I. see Spange, S.: Vol. 165, pp. 43–78.

Volk, N., Vollmer, D., Schmidt, M., Oppermann, W. and *Huber, K.*: Conformation and Phase Diagrams of Flexible Polyelectrolytes. Vol. 166, pp. 29–65.

Volksen, W.: Condensation Polyimides: Synthesis, Solution Behavior, and Imidization Characteristics. Vol. 117, pp. 111–164.

Volksen, W. see Hedrick, J. L.: Vol. 141, pp. 1–44.

Volksen, W. see Hedrick, J. L.: Vol. 147, pp. 61–112.

Vollmer, D. see Volk, N.: Vol. 166, pp. 29–65.

Voskerician, G. and *Weder, C.*: Electronic Properties of PAEs. Vol. 177, pp. 209–248.

Waddon, A. J. see Ungar, G.: Vol. 180, pp. 45–87.

Wagener, K. B. see Baughman, T. W.: Vol. 176, pp. 1–42.

Wagner, E. and *Kloeckner, J.*: Gene Delivery Using Polymer Therapeutics. Vol. 192, pp. 135–173.

Wake, M. C. see Thomson, R. C.: Vol. 122, pp. 245–274.
Wandrey, C., Hernández-Barajas, J. and *Hunkeler, D.*: Diallyldimethylammonium Chloride and its Polymers. Vol. 145, pp. 123–182.
Wang, K. L. see Cussler, E. L.: Vol. 110, pp. 67–80.
Wang, S.-Q.: Molecular Transitions and Dynamics at Polymer/Wall Interfaces: Origins of Flow Instabilities and Wall Slip. Vol. 138, pp. 227–276.
Wang, S.-Q. see Bhargava, R.: Vol. 163, pp. 137–191.
Wang, T. G. see Prokop, A.: Vol. 136, pp. 1–52; 53–74.
Wang, X. see Lin, T.-C.: Vol. 161, pp. 157–193.
Watanabe, K. see Hikosaka, M.: Vol. 191, pp. 137–186.
Webster, O. W.: Group Transfer Polymerization: Mechanism and Comparison with Other Methods of Controlled Polymerization of Acrylic Monomers. Vol. 167, pp. 1–34.
Weder, C. see Voskerician, G.: Vol. 177, pp. 209–248.
Weis, J.-J. and *Levesque, D.*: Simple Dipolar Fluids as Generic Models for Soft Matter. Vol. 185, pp. 163–225.
Whitesell, R. R. see Prokop, A.: Vol. 136, pp. 53–74.
Williams, R. A. see Geil, P. H.: Vol. 180, pp. 89–159.
Williams, R. J. J., Rozenberg, B. A. and *Pascault, J.-P.*: Reaction Induced Phase Separation in Modified Thermosetting Polymers. Vol. 128, pp. 95–156.
Winkler, R. G. see Holm, C.: Vol. 166, pp. 67–111.
Winnik, F. M. see Aseyev, V. O.: Vol. 196, pp. 1–86.
Winter, H. H. and *Mours, M.*: Rheology of Polymers Near Liquid-Solid Transitions. Vol. 134, pp. 165–234.
Wittmeyer, P. see Bohrisch, J.: Vol. 165, pp. 1–41.
Wood-Adams, P. M. see Anantawaraskul, S.: Vol. 182, pp. 1–54.
Wu, C.: Laser Light Scattering Characterization of Special Intractable Macromolecules in Solution. Vol. 137, pp. 103–134.
Wu, C. see Zhang, G.: Vol. 195, pp. 101–176.
Wu, T. see Bhat, R. R.: Vol. 198, pp. 51–124.
Wunderlich, B. see Sumpter, B. G.: Vol. 116, pp. 27–72.

Xiang, M. see Jiang, M.: Vol. 146, pp. 121–194.
Xie, T. Y. see Hunkeler, D.: Vol. 112, pp. 115–134.
Xu, P., Singh, A. and *Kaplan, D. L.*: Enzymatic Catalysis in the Synthesis of Polyanilines and Derivatives of Polyanilines. Vol. 194, pp. 69–94.
Xu, P. see Geil, P. H.: Vol. 180, pp. 89–159.
Xu, Z., Hadjichristidis, N., Fetters, L. J. and *Mays, J. W.*: Structure/Chain-Flexibility Relationships of Polymers. Vol. 120, pp. 1–50.

Yagci, Y. and *Endo, T.*: N-Benzyl and N-Alkoxy Pyridium Salts as Thermal and Photochemical Initiators for Cationic Polymerization. Vol. 127, pp. 59–86.
Yagi, M. and *Kaneko, M.*: Charge Transport and Catalysis by Molecules Confined in Polymeric Materials and Application to Future Nanodevices for Energy Conversion. Vol. 199, pp. 143–188.
Yamaguchi, I. see Yamamoto, T.: Vol. 177, pp. 181–208.
Yamamoto, T.: Molecular Dynamics Modeling of the Crystal-Melt Interfaces and the Growth of Chain Folded Lamellae. Vol. 191, pp. 37–85.
Yamamoto, T., Yamaguchi, I. and *Yasuda, T.*: PAEs with Heteroaromatic Rings. Vol. 177, pp. 181–208.
Yamamoto, S. see Tsujii, Y.: Vol. 197, pp. 1–47.

Yamaoka, H.: Polymer Materials for Fusion Reactors. Vol. 105, pp. 117–144.
Yamazaki, S. see Hikosaka, M.: Vol. 191, pp. 137–186.
Yannas, I. V.: Tissue Regeneration Templates Based on Collagen-Glycosaminoglycan Copolymers. Vol. 122, pp. 219–244.
Yang, J. see Geil, P. H.: Vol. 180, pp. 89–159.
Yang, J. S. see Jo, W. H.: Vol. 156, pp. 1–52.
Yasuda, H. and *Ihara, E.*: Rare Earth Metal-Initiated Living Polymerizations of Polar and Nonpolar Monomers. Vol. 133, pp. 53–102.
Yasuda, T. see Yamamoto, T.: Vol. 177, pp. 181–208.
Yaszemski, M. J. see Thomson, R. C.: Vol. 122, pp. 245–274.
Yoo, T. see Quirk, R. P.: Vol. 153, pp. 67–162.
Yoon, D. Y. see Hedrick, J. L.: Vol. 141, pp. 1–44.
Yoshida, H. and *Ichikawa, T.*: Electron Spin Studies of Free Radicals in Irradiated Polymers. Vol. 105, pp. 3–36.

Zhang, G. and *Wu, C.*: Folding and Formation of Mesoglobules in Dilute Copolymer Solutions. Vol. 195, pp. 101–176.
Zhang, H. see Rühe, J.: Vol. 165, pp. 79–150.
Zhang, Y.: Synchrotron Radiation Direct Photo Etching of Polymers. Vol. 168, pp. 291–340.
Zhao, B. see Brittain, W. J.: Vol. 198, pp. 125–147.
Zheng, J. and *Swager, T. M.*: Poly(arylene ethynylene)s in Chemosensing and Biosensing. Vol. 177, pp. 151–177.
Zhou, H. see Jiang, M.: Vol. 146, pp. 121–194.
Zhou, Z. see Abe, A.: Vol. 181, pp. 121–152.
Zubov, V. P., Ivanov, A. E. and *Saburov, V. V.*: Polymer-Coated Adsorbents for the Separation of Biopolymers and Particles. Vol. 104, pp. 135–176.

Subject Index

ABC terpolymers   48
Adsorption   57
AFM   1, 176
Alkaline phosphatase   214
Anionic polymerization   37
Araneus diadematus   129

Bioabsorbable composite scaffold   209, 223
Biocompatible bilayer-vesicle architectures   196
Biointerfacing   169
Biscyclopentadienide   93
Block copolymers   1, 37, 41, 57, 71
– –, amphiphilic   50
– –, self-assembly   102
Bombyx mori   128
Bond-Fluctuation-Model (BFM)   4
Bone tissue engineering   209

Calcium phosphate soluble glasses   212
Capillary force lithography   100
Chemical force microscopy (CFM)   183
Collagen   221
Composite scaffolds   223
Crystallization   1, 13

Dendrimers, nanopatterning   195
Dewetting   99
Diblock copolymers   73
– –, end-functionalized   41
– –, junction-point-functionalized   43
3-(Dimethylamino)propyllithium   41
1-(4-Dimethylaminophenyl)-1-phenyl-ethylene   42
Dimethylsila[1]ferrocenophane   93
1,2-Dimyristoyl-*sn*-glycero-3-phosphatidylcholine   173

Dip-pen nanolithography (DPN)   170, 191
n,*n*-Dithiobis(*N*-hydroxysuccinimidyl-*n*-alkanoate)   170
Dodecyltrimethylammonium bromide (DTMAB)   52
Dragline spider silks   126

Elastinlike polymers   119
Etch resistance   91
Extracellular matrix (ECM) proteins   220

Fibronectin   221
Fluorinated polymers   84
Fresnel reflectivity   85

Genetic engineering   119
GISAXS   86
Gutenberg method, polymer design/production   123

Integrins   221
Inverse temperature transition   135
Inverted chemical force microscopy (iCFM)   170

Keratins   127

Langmuir–Blodget   60
LC block copolymers, amorphous-side-chain   52
Lipid bilayers   196
Liposomes   173

Macroporosity   224
6-[4-(4-Methoxyphenyl)phenoxy]hexylmethacrylate (MPPHM)   52
Micelles   57
Micropatterning   169

Microphase separation 37
Miktoarm stars 46

Nanolithography 91
–, dip-pen (DPN) 170, 191
Nanopatterning 57, 169
NanoPATterns, surface-induced 57
Nanostructure 71
– ordering 3
Nephila clavipes dragline 126
NHS ester SAMs, aminolysis 182
– –, hydrolysis 178

Octadecanethiol (ODT) 184
Order-disorder transition 45
Organometallic polymers 91
Osteoblast cell culture 209
Osteocalcin 214

P2VP-PS-P2VP 39
PAMAM dendrimers 191
PB_h-PEO 31
PB-PS-PMMA 39
PChEMA 72, 77
PEO 31, 50, 93
PEO-PBO-PEO 39
Phosphatase 214
PIBVE-*b*-PLC 72
PMMA-*b*-PF8H2A 72
Poly(aminoalkyl methacryate) 93
Poly(dimethylsiloxane) 93, 98
Poly(ethylene oxide) (PEO) 31, 50, 93
Poly(ferrocenyldimethylsilane), reactive ion etch barrier 94
Poly(ferrocenylmethylphenylsilane) 96
Poly(ferrocenylphenylphosphine) 93
Poly(ferrocenylsilane), homopolymers, lithography 97
–, lithography 102
–, polyions 108
Poly(α-hydroxy acids) 210
Poly(*N*-hydroxysuccinimidyl methacrylate) (PNHSMA) 173
Poly(isobutylvinylether) (PIBVE) 80
Poly(*N*-isopropylacrylamides) (PNIPAM) 136
Poly(methyl methacrylate) 93
Poly(VPGVG) 136
Polyamidoamine (PAMAM) dendrimers 170

Polybutadiene (PB) 51
Polyisoprene 93
Polylactic acid (PLA) 210
Polymer crystallization 1
Polymer thin films 169
Polypropylene imine dendrimers 192
Polystyrene 93
Postpolymerization functionalization 43
Printing, organometallic polymers, soft lithography 97
Protein adsorption 209
Protein-based polymers 119
PS-*b*-P2VP 57
PS-*b*-P4VP 57
PS-*b*-PChEMA 72
PS-P2VP-P*t*BMA 39
PS-PB-PCL 40
PS-PB-PMMA 39
PS-PB-P*t*BMA 39
PS-PEO 31
PSLi 52
PVP 57

Resists, self-assembling 106

SAMs 169
Scaffolds, bioabsorbable composite 209, 223
–, composite 223
Scaffold porosity 224
Scanning probe lithography (SPL) 170
Self-assembled monolayers (SAMs) 169
Self-assembling resists 106
Self-assembly 37, 57, 119
Self-organization 41
SFM 57
Silks 126, 128
SINPATs 57, 61
Smart polymers 119
– –, surfaces 45
Smectic liquid crystals 71
Spider silks 126
Striped substrates 1
Styrene 52
Styrene-isoprene-methyl methacrylate (SIM) 48
Sulfobetaine zwitterion 43
Surfaces 37
–, patterned 174
–, reactivity 169

Thin films   71, 91, 169
Thiols, SAM   197
Ti/SINPAT   57

Vesicles   173
–, biocompatible bilayer architectures   196

–, unrolling   199
Vitronectin   221

Wettability   209

X-ray reflectivity (XR)   77

Printing: Krips bv, Meppel
Binding: Stürtz, Würzburg

DATE DUE

GAYLORD — PRINTED IN U.S.A.

SCI QD 281 .P6 F6 no.200

Ordered polymeric
nanostructures at surfaces